UNDERSTANDI BOTANY

FOR

HIGH SCHOOLS AND COLLEGES

Prof. Carl William Matthias, Ph.d Plant Biology

Copyright © 2023 by Professor Carl William Matthias

All rights reserved. No part of this publication may be reproduced, stored in a retrieval system, or transmitted, in any form or by any means, electronic, mechanical, photocopying, recording, or otherwise, without the prior written permission of the publisher.

FOREWORD

I am delighted to write this foreword for Professor Carl William Matthias's outstanding botany textbook. With decades of experience as a respected botanist, Professor Matthias has crafted a comprehensive and insightful work that is deserving of high praise. Botany, the study of plants and their biological processes, holds great significance in addressing global challenges like food security, biodiversity conservation, and environmental sustainability. This textbook expertly explores these topics, offering a valuable resource for high school and college students venturing into the captivating world of botany.

In this extraordinary textbook, Professor Carl William Matthias demonstrates a profound mastery of the subject matter and a remarkable ability to distill complex concepts into accessible and engaging prose. The meticulous attention to detail combined with a passion for plant science, shines through every chapter, making this book an invaluable resource for high school and college students venturing into the captivating realm of botany.

From the foundations of plant anatomy and physiology to the exploration of plant ecology, genetics, and evolution, this textbook provides a comprehensive framework for understanding the intricate world of plants. Through vivid descriptions and illustrations, insightful diagrams, photographs and thought-provoking examples, Professor Carl William Matthias skillfully guides readers on plant science.

The book also incorporates the latest advancements in the field, integrating modern research findings and cutting-edge technologies. Students will gain access to up-to-date knowledge and insights that will enhance their understanding of the subject.

In closing, I commend Professor Matthias for their dedication and remarkable contribution to the field of botany. It is an honor to introduce this textbook, which will undoubtedly serve as a beacon of knowledge and inspiration for years to come.

PREFACE

Welcome to the world of plant! This textbook is a comprehensive resource designed to provide high school and college students with a solid foundation in the field of plant biology. Botany is a diverse discipline that explores the structure, function, classification, evolution, and ecological roles of plants. It encompasses the study of plant anatomy, physiology, genetics, ecology, and their interactions with the environment. Understanding plants is essential not only for appreciating the natural world but also for addressing critical global challenges such as food security, climate change, and conservation.

In writing this textbook, my aim was to create a comprehensive and accessible resource that would inspire curiosity, foster a deep understanding of plant life, and ignite a passion for botanical science. The content is organized in a logical progression, taking you from the basic principles of plant biology to more complex topics, while providing a balance between theoretical knowledge and practical applications. Each chapter offers a combination of scientific rigour and readability, with clear explanations, vivid illustrations, and real-world examples. Additionally, I have included interactive elements to enhance your learning experience and promote active engagement with the subject matter.

It is important to note that this textbook represents the collective knowledge and research of numerous scientists and scholars who have dedicated their lives to the study of botany. Their contributions, along with the wealth of scientific literature and research, form the foundation upon which this textbook is built. Whether you are a budding botanist, an aspiring biologist, or simply someone interested in the wonders of the natural world, I hope this textbook will serve as a valuable companion on your botanical journey.

Wishing you an exciting and enriching learning experience.

Professor Carl William Matthias.

ACKNOWLEDGEMENT

I would like to express my heartfelt appreciation to the individuals and organizations who have supported and inspired me throughout the process of creating this botany textbook. My sincere gratitude goes to my family and friends for their unwavering support, encouragement, and understanding during the long hours spent researching, writing, and editing this textbook.

I extend my deepest thanks to my mentors, advisers, and professors who have provided invaluable guidance, expertise, and encouragement throughout my academic and writing journey. I would also like to acknowledge the reviewers and experts in the field of botany who generously shared their insights, expertise, and constructive feedback, helping to refine the content and ensure its accuracy.

A special mention goes to the staff and librarians at various academic institutions who assisted me in accessing the necessary research materials, books, journals, and other resources essential for the completion of this textbook.

I am grateful to the publishers and the production team involved in the publication of this textbook for their professionalism, support, and dedication to producing a high-quality educational resource.

Finally, I would like to thank all the students, teachers, and educators who will utilize this textbook in their classrooms and contribute to the advancement of botanical knowledge.

This textbook represents the culmination of countless hours of research, writing, and collaboration.

CONTENTS

CHAPTER 1
THE CELL STRUCTURE AND FUNCTIONS... 1

CHAPTER 2
MICROSCOPES..33

CHAPTER 3
BASIC PRINCIPLE OF GENETICS AND HEREDITARY {REPRODUCTION}................... 59

CHAPTER 4
CELL DIVISION.. 80

CHAPTER 5
MENDELIAN GENETICS...99

CHAPTER 6
MOLECULAR GENETICS... 108

CHAPTER 7
MOVEMENT OF MATERIALS IN AND OUT OF CELLS......................................118

CHAPTER 8
PHYSIOLOGY.. 125

CHAPTER 9
PLANT HORMONES AND GROWTH REGULATIONS....................................... 143

CHAPTER 10
ELEMENTS OF ECOLOGY, TYPES OF HABITAT 161

CHAPTER 11
ENVIRONMENTAL POLLUTION... 197

CHAPTER 12
PLANT FORMS AND EVOLUTION..217

CHAPTER 13
CRYPTOGAMS..232

CHAPTER 14
GYMNOSPERMS.. 244

CHAPTER 15
ANGIOSPERMS-MORPHOLOGY...258

CHAPTER 16
PLANTS ANATOMY...307

CHAPTER 17
INTRODUCTORY TOPICS IN ETHNOBOTANY..345

CHAPTER 18
BIODIVERSITY...358

DETAILED TABLE OF CONTENT

CHAPTER 1 .... - 1 -

THE CELL STRUCTURE AND FUNCTIONS .... - 1 -

    1.1 HISTORY: THE CELL THEORY .... - 1 -

        1.1.1 MODERN CELL THEORY .... - 3 -

        1.1.2 DEVELOPMENT OF CELL BIOLOGY RESEARCH .... - 4 -

        1.1.3 TIME LINE IN CELL BIOLOGY RESEARCH .... - 5 -

    1.2 PROKARYOTES AND EUKARYOTES .... - 6 -

        1.2.1 CHARACTERISTICS OF PROKARYOTES .... - 8 -

        1.2.2 CHARACTERISTICS OF EUKARYOTES .... - 10 -

        1.2.3 DIFFERENCES BETWEEN PROKARYOTES AND EUKARYOTES .... - 11 -

    1.3 THE CELL – View From Microscope .... - 13 -

        1.3.1 CHARACTERISTICS OF CELLS .... - 15 -

        1.4 DIFFERENCE BETWEEN PLANT CELL AND ANIMAL CELL .... - 32 -

CHAPTER 2 .... - 33 -

MICROSCOPES .... - 33 -

    2.1 DEVELOPMENT OF MICROSCOPES .... - 34 -

    2.2 TYPES OF MICROSCOPES .... - 42 -

        2.2.1 DISSECTION or STEREO MICROSCOPE .... - 42 -

        2.2.2 COMPOUND MICROSCOPE .... - 44 -

        2.2.5 SCANNING ELECTRON MICROSCOPE (SEM) .... - 49 -

        2.2.6 TRANSMISSION ELECTRON MICROSCOPE (TEM) .... - 51 -

    2.3 CONCEPTS IN MICROSCOPY .... - 53 -

    2.3.1 PARTS OF MICROSCOPES .... - 54 -

2.3.2 DIFFERENCES BETWEEN LIGHT AND TRANSMISSION ELECTRON MICROSCOPE: .... - 56 -

2.3.3 MICROSCOPIC TECHNIQUES .... - 56 -

CHAPTER 3 .... - 59 -

BASIC PRINCIPLE OF GENETICS AND HEREDITARY {REPRODUCTION} - 59 -

3.1   REPRODUCTION .... - 59 -

3.2 ASEXUAL REPRODUCTION .... - 60 -

3.3 ARTIFICIAL METHODS OF VEGETATIVE PROPAGATION .... - 72 -

3.4 SEXUAL REPRODUCTION .... - 78 -

ALTERNATION OF GENERATION .... - 79 -

CHAPTER 4 .... - 80 -

CELL DIVISION .... - 80 -

4.1 CHROMOSOME DESCRIPTION .... - 80 -

4.2 MITOTIC CELL DIVISION (MITOSIS) .... - 82 -

4.2.1 CHARACTERISTICS OF MITOSIS .... - 83 -

4.3 MEIOTIC CELL DIVISION (MEIOSIS) .... - 90 -

4.3.1 CHARACTERISTICS OF MEIOSIS .... - 90 -

CHAPTER 5 .... - 99 -

MENDELIAN GENETICS .... - 99 -

5.1 SHORT HISTORY ON GREGORY MENDEL .... - 99 -

5.2 MENDEL'S WORK ON GARDEN PEA (PISUM SATIVUM) .... - 100 -

5.3 SOME IMPORTANT GENETIC TERMS .... - 101 -

5.4 SIMPLE MONOHYBRID EXPERIMENT .... - 103 -

5.5 DIHYBRID INHERITANCE .... - 105 -

The First Law (Law Of Segregation): .... - 107 -

The Second Law (Law Of Independent Assortment): .... - 107 -

CHAPTER 6 .... - 108 -

MOLECULAR GENETICS .... - 108 -

    6.1 TRANSFORMATION .... - 109 -

    6.2 SUMMARY OF GRIFFITH'S EXPERIMENT .... - 110 -

    6.3 SEMI-CONSERVATIVE DNA REPLICATION .... - 111 -

    6.4 THE NATURE OF DNA .... - 113 -

    6.5 RIBONUCLEIC ACID (RNA) .... - 114 -

    6.6 THE GENETIC CODE .... - 115 -

    6.7 CHARACTERISTICS OF THE GENETIC CODES .... - 117 -

CHAPTER 7 .... - 118 -

MOVEMENT OF MATERIALS IN AND OUT OF CELLS .... - 118 -

    7.1 CELLULAR TRANSPORT .... - 118 -

    7.2 CELL MEMBRANE/PLASMA MEMBRANE .... - 119 -

    7.3 FLUID MOSAIC MODEL .... - 121 -

    7.4 TYPES OF TRANSPORT PROTEIN .... - 123 -

CHAPTER 8 .... - 125 -

PHYSIOLOGY .... - 125 -

    8.1 SIMPLE DIFFUSION .... - 126 -

    8.2 DIFFUSION .... - 128 -

    8.3 PASSIVE TRANSPORT: SIMPLE DIFFUSION AND FACILITATED DIFFUSION. .... - 129 -

    8.4 ACTIVE TRANSPORT .... - 132 -

    8.5 OSMOSIS .... - 134 -

    8.6 CELLS IN SOLUTIONS .... - 137 -

    8.6.1 CELL IN ISOTONIC SOLUTION .... - 138 -

    8.6.2 CELL IN HYPOTONIC SOLUTION .... - 139 -

    8.6.3 CELL IN HYPERTONIC SOLUTION .... - 139 -

  8.7 ENDOCYTOSIS AND EXOCYTOSIS .... - 140 -

  CELLULAR TRANSPORT MECHANISMS .... - 141 -

CHAPTER 9 .... - 143 -

PLANT HORMONES AND GROWTH REGULATION .... - 143 -

  9.1 INTRODUCTION TO PLANT HORMONES .... - 143 -

  9.2 AUXINS .... - 144 -

  9.3 GIBBERELLINS .... - 147 -

  9.4 CYTOKININS .... - 150 -

  9.5 ABSCISIC ACID (ABA) .... - 151 -

  9.6 ETHYLENE .... - 153 -

  9.7 PLANT HORMONE INTERACTIONS AND SIGNAL TRANSDUCTION .... - 155 -

  9.8 APPLICATIONS OF PLANT HORMONES IN AGRICULTURE AND HORTICULTURE .... - 157 -

  9.9 FUTURE PERSPECTIVES AND RESEARCH FRONTIERS IN PLANT HORMONES .... - 158 -

CHAPTER 10 .... - 161 -

ELEMENTS OF ECOLOGY, TYPES OF HABITAT .... - 161 -

  10.1 ECOLOGY .... - 161 -

  10.2 UNITS FOR THE STUDY OF ECOLOGY. .... - 162 -

  10.3 ENERGY FLOW AND MATERIAL OR NUTRIENT CYCLING .... - 163 -

  10.4 FOOD CHAINS, FOOD WEBS AND TROPHIC LEVELS .... - 164 -

    TROPHIC LEVELS .... - 165 -

  10.5 BIOGEOCHEMICAL CYCLE .... - 169 -

GASEOUS CYCLES. ....- 170 -

10.6 INPUTS AND LOSSES OF ELEMENTS ....- 172 -

10.7 HUMAN IMPACT ON BIOGEOCHEMICAL CYCLES ....- 173 -

10.8 HABITAT, MICROHABITAT, ECOLOGICAL NICHE. ....- 175 -

10.9 TERRESTIAL/LAND HABITATS ....- 177 -

    DESERTS. ....- 178 -

    TUNDRA ....- 180 -

    FORESTS. ....- 181 -

    GRASSLANDS. ....- 184 -

10.10 AQUATIC HABITATS ....- 188 -

    FRESHWATERS ....- 188 -

    STREAMS AND RIVERS ....- 189 -

    LAKES AND PONDS ....- 190 -

    FRESHWATER MARSHES. ....- 192 -

    MARINE HABITATS ....- 193 -

    ESTUARIES AND SEASHORES. ....- 194 -

    DELTAS ....- 195 -

CHAPTER 11 ....- 197 -

ENVIRONMENTAL POLLUTION ....- 197 -

11.1 NATURAL SOURCES OF POLLUTION. ....- 197 -

11.2 ANTHROPOGENIC/ MADE-MADE SOURCES. ....- 199 -

11.3 POLLUTANTS ....- 200 -

11.4 TYPES OF POLLUTION ....- 202 -

11.4.1 AIR POLLUTION ....- 202 -

11.4.2 GLOBAL WARMING: 'GREENHOUSE EFFECT' .... - 205 -

11.4.3 OZONE LAYER DEPLETION .... - 206 -

11.4.4 WATER POLLUTION .... - 208 -

    11.4.5 EUTROPHICATION .... - 209 -

    11.4.6 EFFECTS OF EUTROPHICATION IN WATER BODY .... - 212 -

11.5 PESTICIDE CONTAMINATION .... - 213 -

    11.5.1 ALTERNATIVE METHODS TO USE OF PESTICIDES .... - 215 -

CHAPTER 12 .... - 217 -

PLANT FORMS AND EVOLUTION .... - 217 -

12.1 FORMS OF PLANTS .... - 217 -

12.2 VARIETY CLASSIFICATION IN THE PLANT KINGDOM .... - 221 -

12.3 NOMENCLATURE AND HIERARCHIC RELATIONSHIPS OF ORGANISMS .... - 223 -

12.4 BOTANICAL NAMES OF SOME PLANTS .... - 224 -

12.5 EVOLUTION OF PLANTS .... - 227 -

12.6 CLASSIFICATION INTO THE PLANT AND ANIMAL KINGDOM .... - 228 -

CHAPTER 13 .... - 232 -

CRYPTOGAMS .... - 232 -

13.1 ALGAE .... - 232 -

13.2 LICHENS .... - 239 -

13.3 BRYOPHYTES .... - 241 -

13.4 PTERIDOPHYTES .... - 242 -

CHAPTER 14 .... - 244 -

GYMNOSPERMS .... - 244 -

14.1 WHAT ARE PLANTS? .... - 244 -

XIII

14.2 USEFULNESS OF PLANTS TO MAN .... - 245 -

14.3 ALTERNATION OF GENERATIONS .... - 247 -

14.4 HIGHLIGHTS OF PLANT PHYLOGENY .... - 248 -

14.5 THE CHALLENGES OF LAND COLONIZATION: THE ORIGIN OF VASCULAR TISSUES. ...... - 249 -

14.6 THE SEED PLANTS AND THE SIGNIFICANCE OF THE SEED HABIT .... - 251 -

14.7 THE GYMNOSPERMS .... - 252 -

    14.7.1 LIFE CYCLE OF GYMNOSPERMS .... - 252 -

14.8 THE ANGIOSPERMS .... - 254 -

    14.8.1 THE ANGIOSPERM LIFE CYCLE .... - 255 -

CHAPTER 15 .... - 258 -

ANGIOSPERMS-MORPHOLOGY .... - 258 -

15.1 PLANT MORPHOLOGY .... - 260 -

15.2 ROOT MORPHOLOGY .... - 262 -

15.3 STEM MORPHOLOGY .... - 267 -

15.4 LEAF MORPHOLOGY .... - 277 -

15.5 INFLORESCENCE .... - 288 -

15.6 FRUITS .... - 296 -

CHAPTER 16 .... - 307 -

PLANTS ANATOMY .... - 307 -

16.1 WHAT IS PLANT ANATOMY? .... - 307 -

16.2 MERISTEMS OR MERISTEMATIC TISSUE .... - 308 -

16.3 PERMANENT TISSUES .... - 315 -

    1. PARENCHYMA .... - 317 -

    2. COLLENCHYMA .... - 319 -

    3. SCLERENCHYMA .... - 321 -

  16.4 SECRETORY TISSUE .... - 336 -

  16.5 DERMAL TISSUE .... - 339 -

CHAPTER 17 .... - 345 -

INTRODUCTORY TOPICS IN ETHNOBOTANY .... - 345 -

  17.1 PLANTS AND PEOPLE .... - 345 -

  17.2 DEFINITIONS AND CONCEPTS: ETHNOBOTANY .... - 346 -

  17.3 DIFFERENT ASPECTS OF ETHNOBOTANY .... - 348 -

  17.4 IMPORTANCE OF ETHNOBOTANY .... - 350 -

  17.5 AIMS AND OBJECTIVES OF ETHNOBOTANY .... - 351 -

  17.6 SCOPE OF ETHNOBOTANY .... - 353 -

  17.7 DESCRIPTION OF SOME USEFUL PLANTS. .... - 355 -

CHAPTER 18 .... - 358 -

BIODIVERSITY .... - 358 -

  18.1 DEFINITION OF BIODIVERSITY .... - 358 -

  18.2 IMPORTANCE OF BIODIVERSITY .... - 359 -

  18.3 HUMAN IMPACT ON BIODIVERSITY .... - 360 -

  18.4 THREAT TO BIODIVERSITY .... - 361 -

  18.5 SOCIETY'S ROLE IN SUPPORTING BIODIVERSITY .... - 363 -

  18.6 BIODIVERSITY CONSERVATION METHODS .... - 364 -

  18.7 CONSERVATION AT THE NATIONAL LEVEL .... - 366 -

GLOSSARY .... - 367 -

| Professor Carl William Matthias

CHAPTER 1

THE CELL STRUCTURE AND FUNCTIONS

1.1 HISTORY: THE CELL THEORY

Cell theory is the historic scientific theory, now universally accepted, that living organisms are made up of cells, that they are the basic structural/organizational unit of all organisms, and that all cells come from pre-existing cells. The initial development of the theory, during the mid-17th century, was made possible by advances in microscopy; the study of cells is called cell biology.

As the invention of the telescope made the Cosmos (bigger "worlds") accessible to human, the microscope opened up "smaller worlds" by showing composition of living forms, previously not seen by naked eyesThe cell was first discovered and named by Robert Hooke in 1665. He remarked that it looked strangely similar to **cellula or small rooms of monks.**

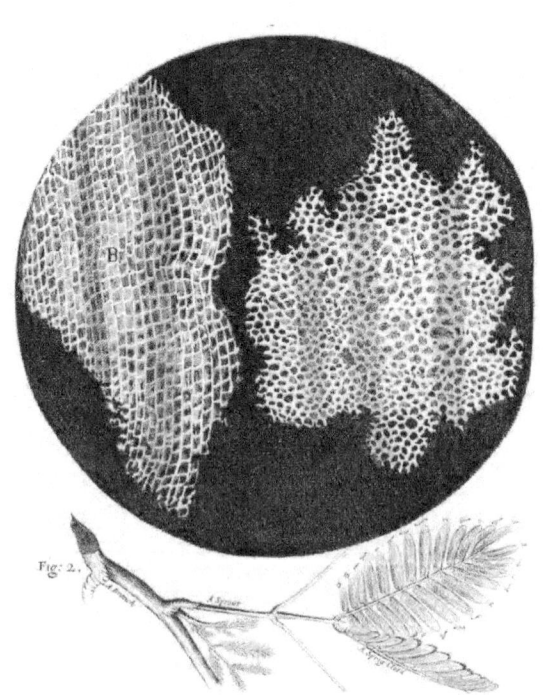

However, what Hooke actually saw was the dead cell walls of plant cells (cork), and he published these description in Micrographia. The first scientist to observe a live cell with a microscope was Anton van Leeuwenhoek, in 1674 while working on the algae Spirogyra. After the first observations of life under the microscope, it took two centuries of research before the 'cell theory', the idea that all living things are composed of cells or their products, was formulated. Cells are the basic unit of structure in all organisms and also the basic unit of reproduction. Hints at the idea that the cell is the basic component of living organisms emerged well before 1838–39, which was when the cell theory was officially formulated.

In 1838, the botanist Matthias Jakob Schleiden (1804–1881) suggested that every structural element of plants is composed of cells or their products. The following year, 1839 a similar conclusion was elaborated for animals by the zoologist Theodor Schwann (1810–1882). He stated that "the elementary parts of all tissues are formed of cells" and that "there is one universal principle of development for the elementary parts of organisms... and this principle is in the formation of cells". The conclusions of Schleiden and Schwann are considered to represent the official formulation of 'cell theory' and their names are almost as closely linked to cell theory as are those of Watson and Crick with the structure of DNA. Cell theory stimulated a reductionistic approach to biological problems and became the most general structural paradigm in biology. It emphasized the concept of the unity of life and brought about the concept of organisms as "republics of living elementary units"

The cell theory states that; cell is the basic unit of structure and function in living organisms.

Schwann in 1839 summarized his observations into three conclusions about cells:

- The cell is the unit of structure, physiology, and organization in living things.
- The cell retains a dual existence as a distinct entity and a building block in the construction of organisms.

- Cells form by free-cell formation, similar to the formation of crystals (spontaneous generation).

We know today that the first two tenets are correct, but the third is clearly wrong. The correct interpretation of cell formation by division was finally promoted by others and formally enunciated in Rudolph Virchow's powerful dictum, "Omnis cellula e cellula". - " which means "All cells only arise from pre-existing cells"

BASIC COMPONENTS OF THE CELL THEORY

- All organisms are composed of one or more cells. (Schleiden & Schwann – 1838-39).
- The cell is the basic unit of life in all living things. (Schleiden & Schwann – 1838-39).
- All cells are produced by the division of preexisting cells. (Virchow - 1858).

1.1.1 MODERN CELL THEORY

1. All known living things are made up of cells.
2. The cell is the structural & functional unit of all living things.
3. All cells come from pre-existing cells by division. (spontaneous generation does not occur).
4. Cells contains hereditary information which is passed from cell to cell during cell division.
5. All cells are basically the same in chemical composition.
6. All energy flow (metabolism & biochemistry) of life occurs within cells.

1.1.2 DEVELOPMENT OF CELL BIOLOGY RESEARCH

There is a rapid growth of Molecular Biology in the mid-20th century which lead to the explosion of cell biology research in the 1950's

Now, it is possible to maintain, grow, and manipulate cells outside of living organisms.

The first of such cell to be cultured was derived from cervical cancer cells taken from one Henrietta Lacks, who died of cancer in 1951 and this was done by George Otto Gey and co-workers

The cell line, referred to as HeLa cells, have now become the watershed in the study of cell biology as the knowledge on the structure of DNA was the significant breakthrough of molecular biology.

A LOT OF PROGRESS IN THE STUDY OF CELLS IN THE RECENT DECADE WHICH INCLUDES:

- The characterization of the minimal media requirements for cells and development of sterile cell culture techniques.

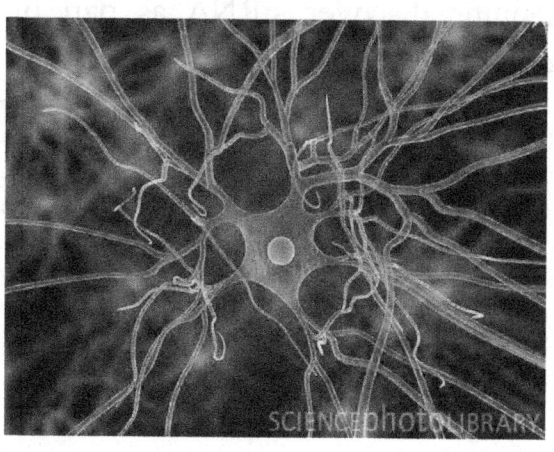

- Advances in the study and use of Electron Microscopy
- Advances in the development of Transfection Methods
- Discovery of Green Fluorescent Protein (GFP) in jellyfish
- Discovery of small interfering RNA (siRNA), among others

1.1.3 TIME LINE IN CELL BIOLOGY RESEARCH

- 1595 – Jansen credited with 1st compound microscope
- 1655 – Hooke described 'cells' in cork.
- 1674 – Leeuwenhoek discovered protozoa.
- 1838 – Schleiden and Schwann proposed cell theory.
- 1858 – Rudolf Virchow expounds his theory
- 1939 – Siemens produced the first commercial TEM
- 1965 – Cambridge Instruments produced the first SEM
- 1995 – Tsien identifies mutant of GFP
- 1999 – Hamilton and Baulcombe discover siRNA
- 1981 – Transgenic mice and fruit flies are produced. Mouse embryonic stem cell line established
- 1998 – Mice are cloned from somatic cells
- 1999 – Hamilton and Baulcombe discover siRNA as part of post-transcriptional gene silencing (PTGS) in plants
- 2000 – Human genome DNA sequence draft.

1.2 PROKARYOTES AND EUKARYOTES

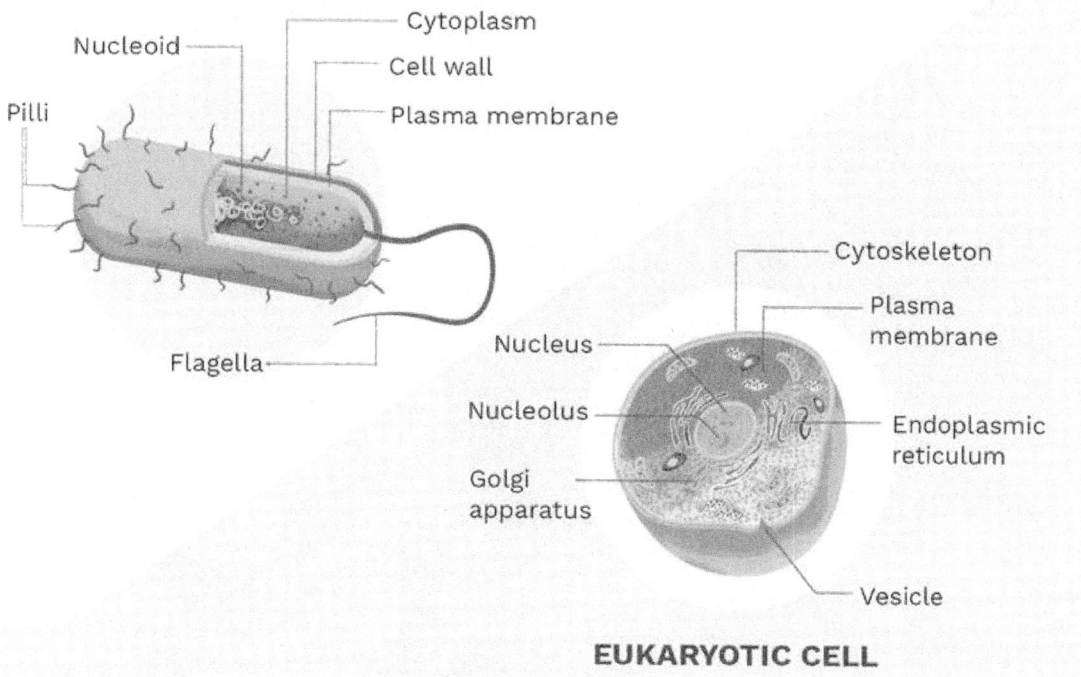

All living organisms can be sorted into one of two groups depending on the fundamental structure of their cells: the prokaryotes and the eukaryotes.

Prokaryotes are organisms made up of cells that lack a cell nucleus or any membrane-encased organelles.

Eukaryotes are organisms made up of cells that possess a membrane-bound nucleus that holds genetic material as well as membrane-bound organelles.

PROKARYOTE (Pro = "before", karyon = "nucleus"), This is a success story spanning at least 3.5 billion years, they were the earliest organisms and they have lived and evolved all alone on earth for 2 billion years. In terms of metabolic and numbers, they still dominate the biosphere, outnumbering all eukaryotes combined. They are probably the smallest living organisms, ranging in size from 0.15 μm (mycoplasmas) to 0.25 μm (chlamydiae) to 0.45 μm (rickettsiae) to about 2.0 μm (many of the bacteria).

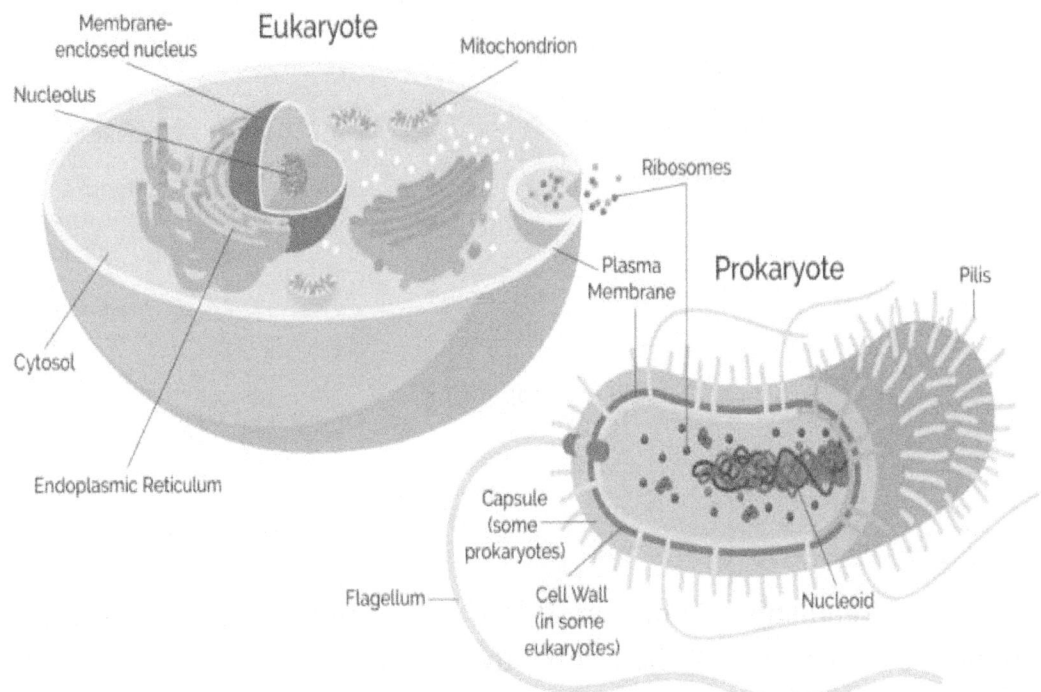

EUKARYOTE (Eu = "true", karyon = "nucleus") They originates by symbiosis among the prokarytes; they are organisms whose cells contain cellular bodies (organelles) enclosed within membranes. Their cells are generally larger and more complex than prokaryotic cells. The organelles function in the activities of the cell and are compartments for localizing metabolic function. Microscopic protozoa, unicellular algae, and fungi have eukaryotic cells.

1.2.1 CHARACTERISTICS OF PROKARYOTES

- They are organisms with a single cell.
- They have distinctive cell walls made of peptidoglycan, a large polymer composed of N-acetylglucosamine and N-acetylmuramic acid.
- They lack distinct nuclei & organelles bounded by membrane.
- They are the first forms of life and yet, they remain essential to life in very many ways.
- They are made up of Eubacteria (Monera) and the Archaebacteria (Archeans).
- They consist of autotrophs and heterotrophs.

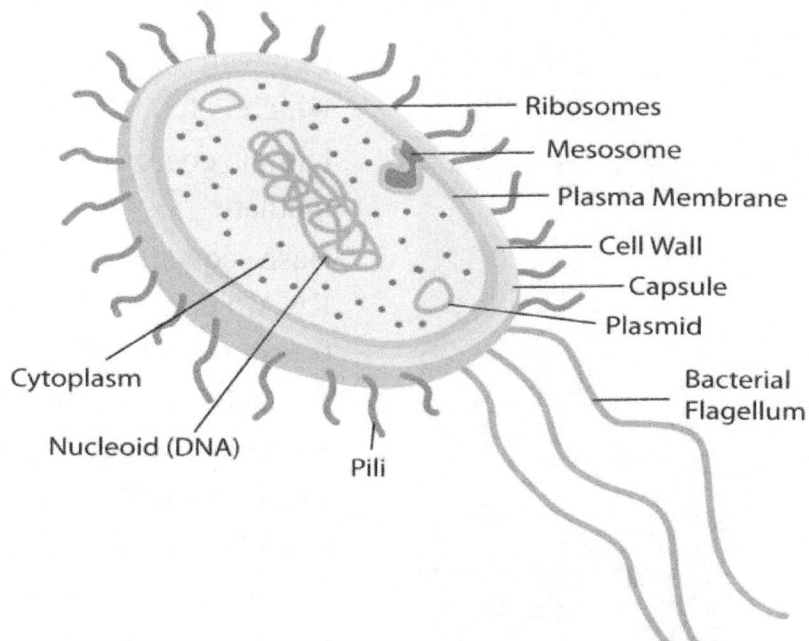

Some bacteria, including E. coli, Salmonella, and Listeria, are found in foods and can cause disease; others are actually helpful to human digestion and other functions.

Some bacteria, occur in spherical forms called cocci (singular, coccus) or in rod-like forms called bacilli (singular, bacillus). Some bacteria have a comma shape (vibrio), or a flexible, wavy shape (spirochete), or a corkscrew shape (spirillum).

Some prokaryotes have a variety of shapes and sizes and are said to be pleomorphic e.g. Rickettsiae and mycoplasmas.

ARCHAEANS were discovered to be a unique life form which is capable of living indefinitely in extreme environments such as, hot springs, salt ponds hydrothermal vents or arctic ice.

Archaeans are micro-organisms that converts and use CO_2 and H_2 from the environment to produce methane – METHANOGENS e.g. *Metanococcus jannaschii* and they have;

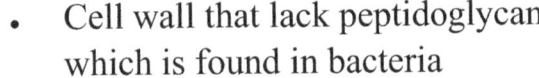

- Cell wall that lack peptidoglycan which is found in bacteria
- Unique coenzymes to produce methane
- Base sequence of two types of RNA, i.e. ribosomal (rRNA) and transfer RNA (tRNA) that is distinctively different from others
- Bacteria-like genes and operons (a functional unit of key nucleotide sequences of DNA)
- Eukaryotic-like information processing, secretion systems and eukaryotic protein synthesis.

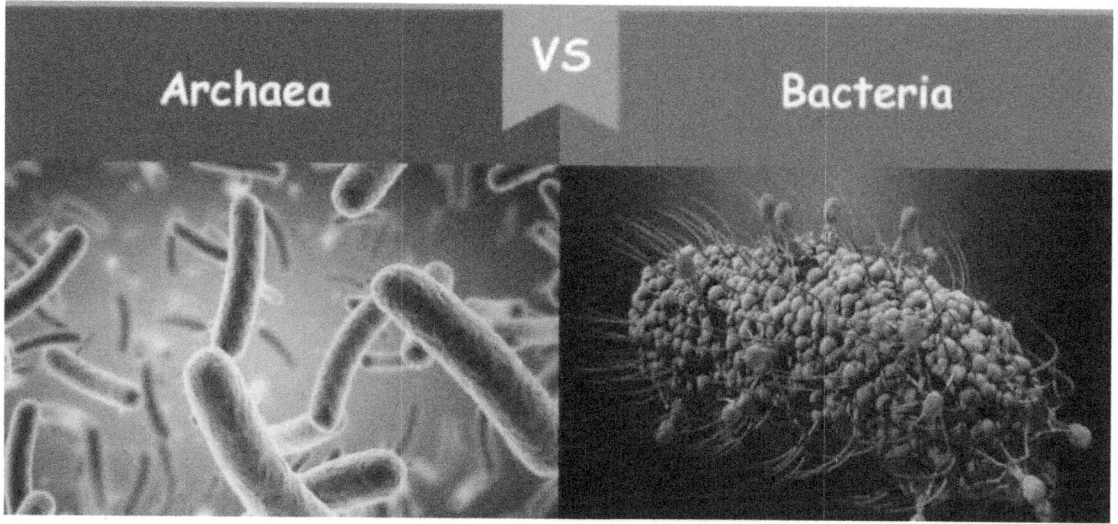

1.2.2 CHARACTERISTICS OF EUKARYOTES

- Cells are found in plants, animals, fungi and protista. They have Plasma membrane
- Glycocalyx (components external to the plasma membrane)
- Cytoplasm (semifluid)
- Cytoskeleton - microfilaments and microtubules that suspend organelles give shape, and allow motion
- Presence of characteristic membranes that encloses subcellular organelles.
- In these cells, organelles create compartments where specific biochemical reactions can occur.
- Some eukaryotic cells contain stored nutrients and pigment molecules

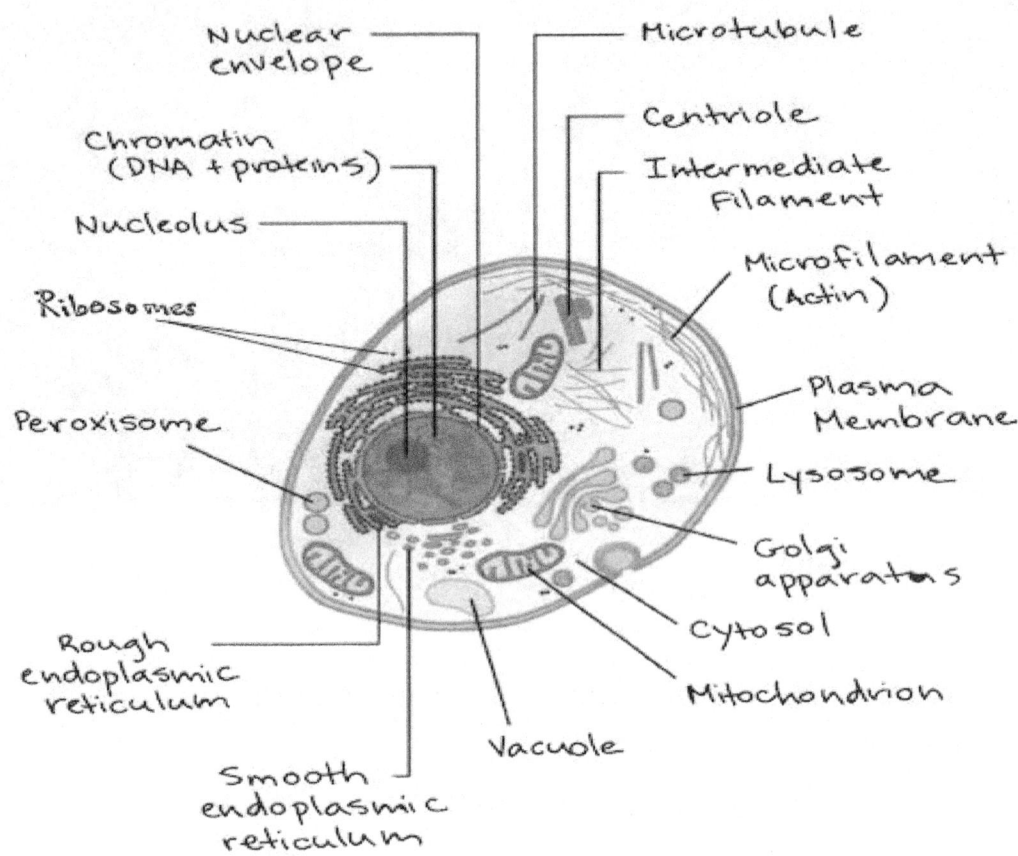

1.2.3 DIFFERENCES BETWEEN PROKARYOTES AND EUKARYOTES

Feature	Prokaryote	Eukaryote
Organisms	Bacteria, Achaeans	Protoctists, fungi, plants, animals
Cell size	Diameter 0.2 – 10 µm	10 – 100 µm most common
Form	Mainly unicellular	Multicellular except Protoctista
Origin	3.5 billion years ago	1.5 billion years ago
Organelles	Few, no envelope	Many, envelope-bound

Cell walls	Present most times, but chemically complex	When present, usually chemically simple
Genetic materials	DNA is single, circular and naked (no proteins)	DNA is multiple, linear and associated with proteins
Cell division	Mostly binary fission	Mitosis, meiosis or both
Photosynthesis	No chloroplast, takes place on membranes without stacking	Chloroplasts containing membranes present, they are stacked into lamellae or grana

1.3 THE CELL – View From Microscope

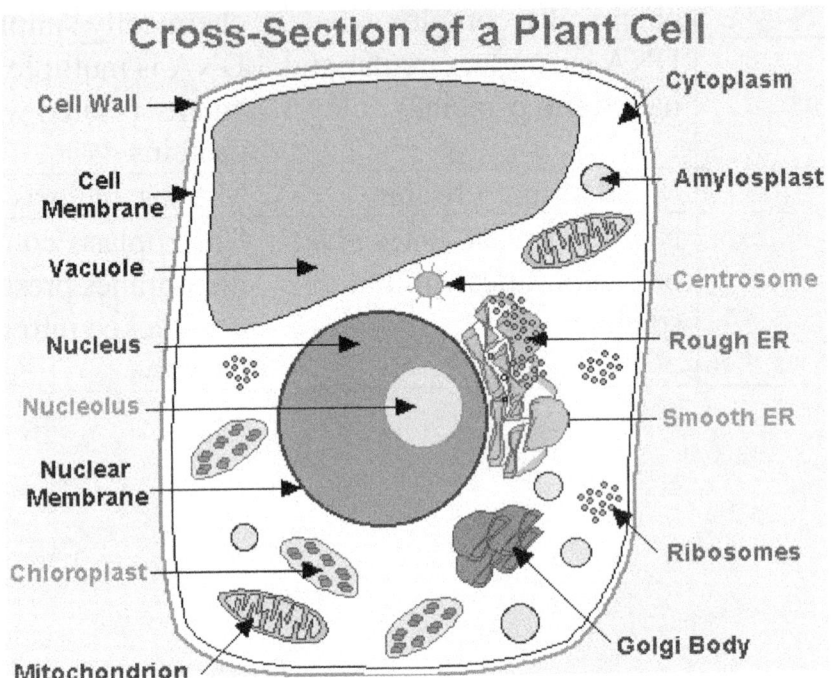

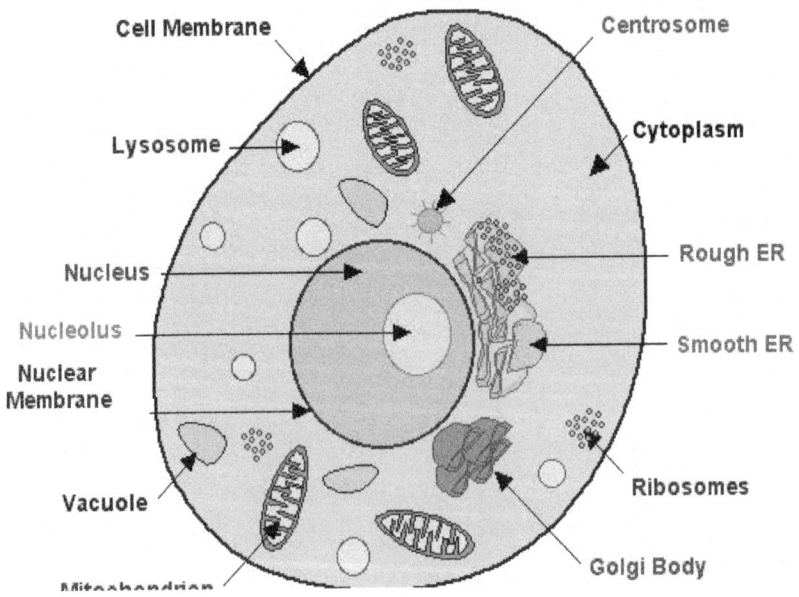

A cell is defined as the smallest, basic unit of life that is responsible for all of life's processes. Cells are the structural, functional, and biological units of all living beings. A cell can replicate itself independently. Hence, they are known as the building blocks of life. Each cell contains a fluid called the cytoplasm, which is enclosed by a membrane. Also present in the cytoplasm are several biomolecules like proteins, nucleic acids and lipids. Moreover, cellular structures called cell organelles are suspended in the cytoplasm.

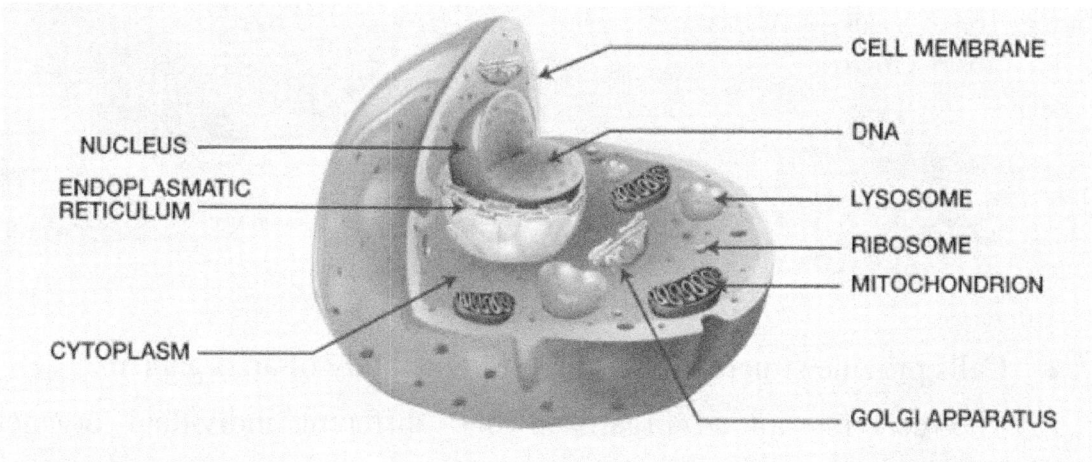

All organisms are made up of cells. They may be made up of a single cell (unicellular), or many cells (multicellular). Mycoplasmas are the smallest known cells. Cells are the building blocks of all living beings. They provide structure to the body and convert the nutrients taken from the food into energy.

Cells are complex and their components perform various functions in an organism. They are of different shapes and sizes, pretty much like bricks of the buildings. Our body is made up of cells of different shapes and sizes.

Cells comprise several cell organelles that perform specialized functions to carry out life processes. Every organelle has a specific structure. The hereditary material of the organisms is also present in the cells

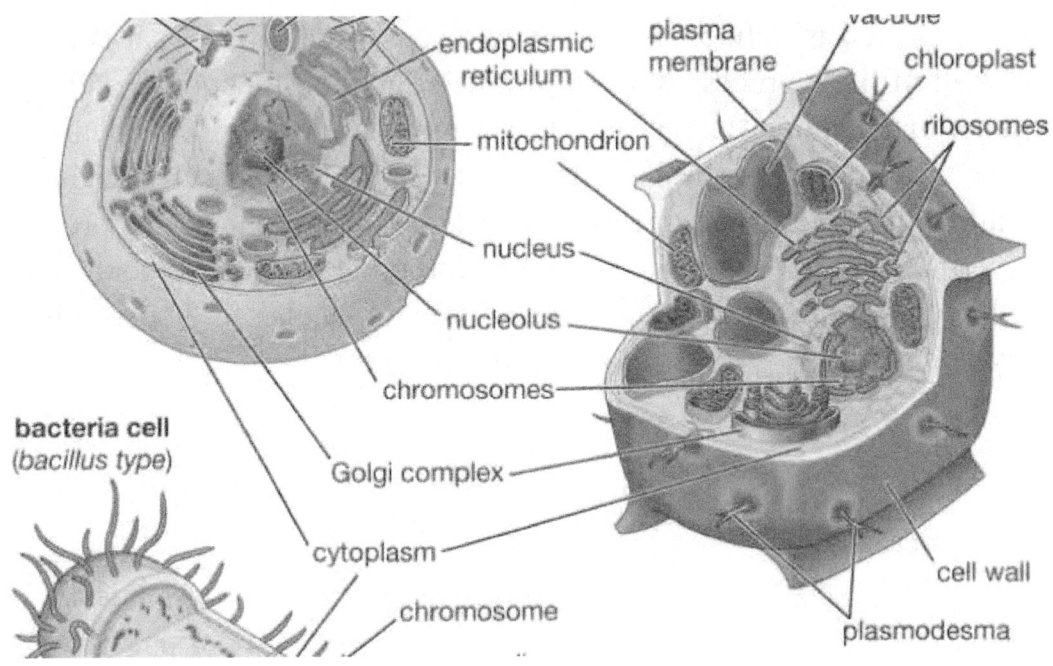

1.3.1 CHARACTERISTICS OF CELLS

- Cells provide structure and support to the body of an organism.
- The cell interior is organized into different individual organelles surrounded by a separate membrane.

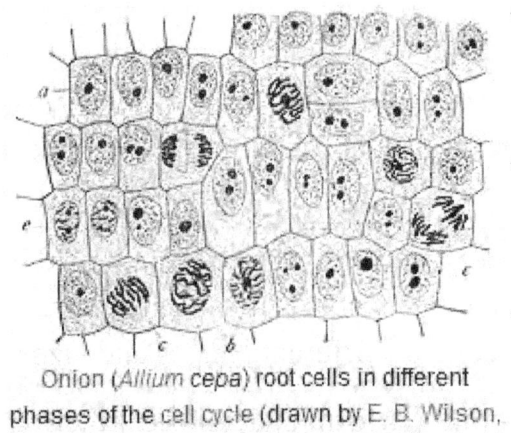

Onion (*Allium cepa*) root cells in different phases of the cell cycle (drawn by E. B. Wilson, 1900)

- The nucleus (major organelle) holds genetic information necessary for reproduction and cell growth.
- Every cell has one nucleus and membrane-bound organelles in the cytoplasm.
- Mitochondria, a double membrane-bound organelle is mainly responsible for the energy transactions vital for the survival of the cell.
- Lysosomes digest unwanted materials in the cell.

- Endoplasmic reticulum plays a significant role in the internal organization of the cell by synthesizing selective molecules and processing, directing and sorting them to their appropriate locations.

1.3.2 CELL WALL:

One of the most important distinguishing features of plant cells is the pr ence of a cell wall. It differs in chemical composition from those of prokaryotes and fungi due to its cellulose content. Cell walls are significantly thicker than plasma membranes and were visible even to early microscopists, including Robert Hooke, who originally identified the structures in a sample of cork, and then coined the term cells in the 1660s.

The thickness, as well as the composition and organization, of cell walls can vary significantly. Many plant cells have both a primary cell wall, which accommodates the cell as it grows, and a secondary cell wall they develop inside the primary wall after the cell has stopped growing. The primary cell wall is thinner and more pliant than the secondary cell wall, and is sometimes retained in an unchanged or slightly modified state without the addition of the secondary wall, even after the growth process has ended. The relative rigidity of the cell wall renders plants sedentary, unlike animals whose lack of this type of structure allows their cells more flexibility, which is necessary for locomotion.

The cell wall not only provides structural support and protection to plant cells but also plays an important role in communication between cells and in plant development. For instance, the secondary cell wall provides additional strength and rigidity to cells, allowing them to support the weight of the plant. Moreover, the cell wall is responsible for the maintenance of plant shape, which is important for photosynthesis and efficient gas exchange.

The cell wall also contains other compounds such as lignin, hemicellulose, and pectin. The composition of the cell wall can vary depending on the plant species, tissue type, and developmental stage. For example, the cell wall of wood is made up of a high proportion of lignin, which makes it tough and resistant to decay.

Recent research has also shown that the cell wall plays a crucial role in plant immunity. The plant cell wall acts as a barrier that prevents pathogens from invading plant tissues. The wall is also involved in the recognition of invading pathogens, triggering an immune response in plants. Understanding the molecular mechanisms involved in plant cell wall synthesis and the immune response can lead to the development of new methods to protect crops from diseases.

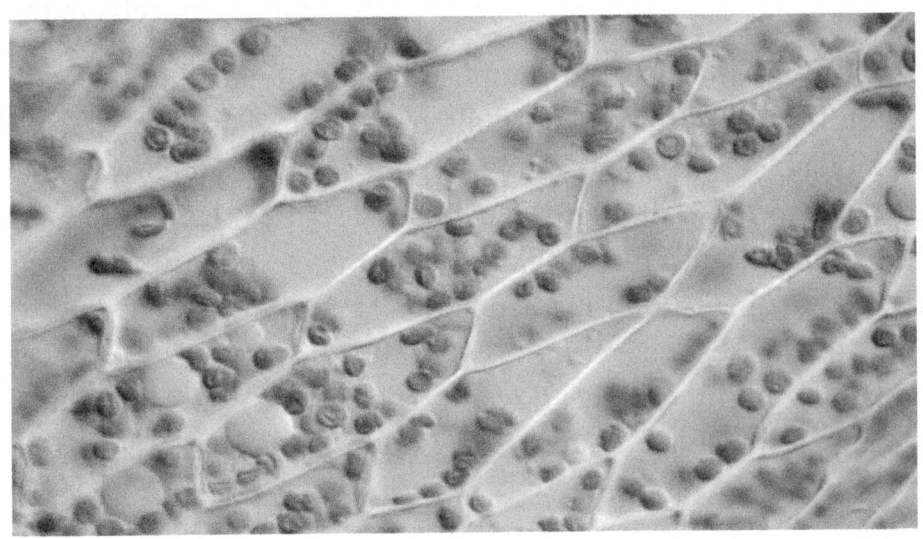

Image of moss leaf cells showing cell walls (between cells) and chloroplasts (green). Alan Phillips/E+/Getty Images

CELL WALL (CHEMICAL COMPONENT): The main chemical components of the primary plant cell wall include **cellulose,** a complex carbohydrate made up of several thousand glucose molecules. the cell wall contains two groups of branched polysaccharides, the **pectins** and **cross-linking glycans**. Organized into a network with the **cellulose micro-fibrils**, the cross-linking **glycans** increase the tensile strength of the cellulose, whereas the coextensive network of **pectins** provides the cell wall with the

ability to resist compression. In addition to these networks, a small amount of **protein** can be found in all plant primary
cell walls. Some of this protein is thought to increase that mechanical strength and part of it consists of enzymes, which initiate reactions form, remodel, or breakdown the structural networks of the wall

The chemical components of the secondary plant cell wall differ from those of the primary wall and are usually laid down after the primary wall has been formed. Secondary cell walls contain significant amounts of hemicelluloses, which are also complex carbohydrates, as well as lignin, a complex polymer that provides strength and rigidity to the cell wall. The exact composition and structure of the secondary cell wall vary depending on the plant species, tissue type, and environmental conditions. The secondary cell wall plays an important role in the mechanical support of plants and provides protection against physical and biological stresses.

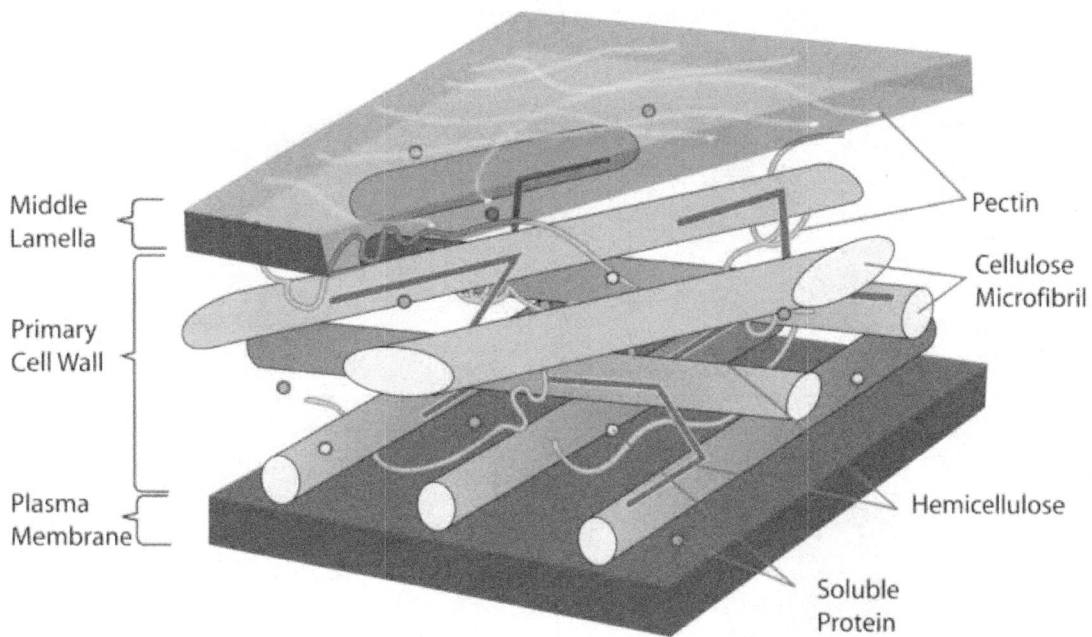

1.3.3 CELL MEMBRANES:

The cell membrane (also called the plasma membrane or plasmalemma) is one biological membrane separating the interior of a cell from the outside environment. The cell membrane surrounds all cells and it is selectively-permeable, controlling the movement of substances in and out of cells. It contains a wide variety of biological molecules, primarily proteins and lipids, which are involved in a variety of cellular processes such as cell adhesion, ion channel conductance and cell signalling. The plasma membrane also serves as the attachment point for the intracellular cytoskeleton and, if present, the extracellular cell wall.

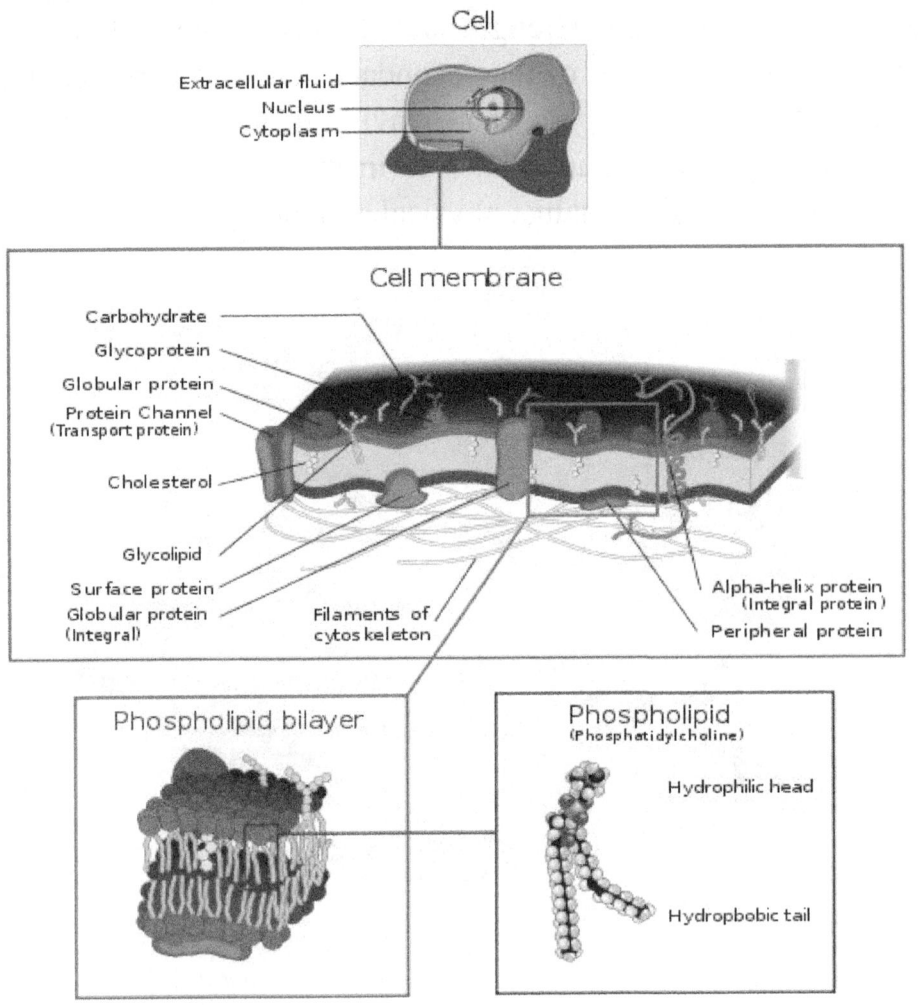

The structure of the cell membrane is often described as a fluid-mosaic model, where the phospholipid bilayer acts as a fluid foundation for the mosaic of embedded proteins, lipids, and carbohydrates. The phospholipids that make up the majority of the membrane have hydrophobic tails that face inward and hydrophilic heads that face outward, allowing them to form a stable barrier between the cell and its environment. The embedded proteins in the membrane perform diverse functions, such as transporting molecules across the membrane, acting as enzymes or receptors, and providing structural support. The carbohydrates on the outer surface of the membrane can act as signaling molecules or markers for cellular identification.

1.3.4 THE NUCLEUS:

This is found in all eukaryotic cells except in matures sieve tube elements of the phloem. Under normal circumstances, cells contain only one nucleus. Nuclei are conspicuous because they are the largest cell organelles and they were the first to be described by microscopist and they are typically about 10 μm in diameter. The nucleus is surrounded nuclear envelope and contains chromatin and one or more nucleoli. The nucleus is a highly specialized organelle that serves as the information processing and administrative centre of the cell. This organelle has two major functions: it stores the cell's hereditary material, or DNA, and it coordinates the cell's activities, which include growth, intermediary metabolism, protein synthesis, and reproduction (cell division).

The chromatin inside the nucleus is made up of DNA, which is tightly coiled around histone proteins. The DNA contains the instructions for making all the proteins and other molecules needed by the cell, and the histones help to package the DNA into a compact structure that can fit inside the nucleus. During cell division, the chromatin condenses into distinct chromosomes, which can be easily separated and distributed to the daughter cells. The nucleolus is a specialized region within the nucleus that is responsible for producing ribosomes, the cellular machinery that synthesizes proteins. The nucleolus is composed of RNA, proteins, and small nucleolar RNA molecules, which work together to transcribe and process the ribosomal RNA needed for ribosome assembly. The nucleus is an essential organelle that plays a vital role in the survival and function of all eukaryotic cells.

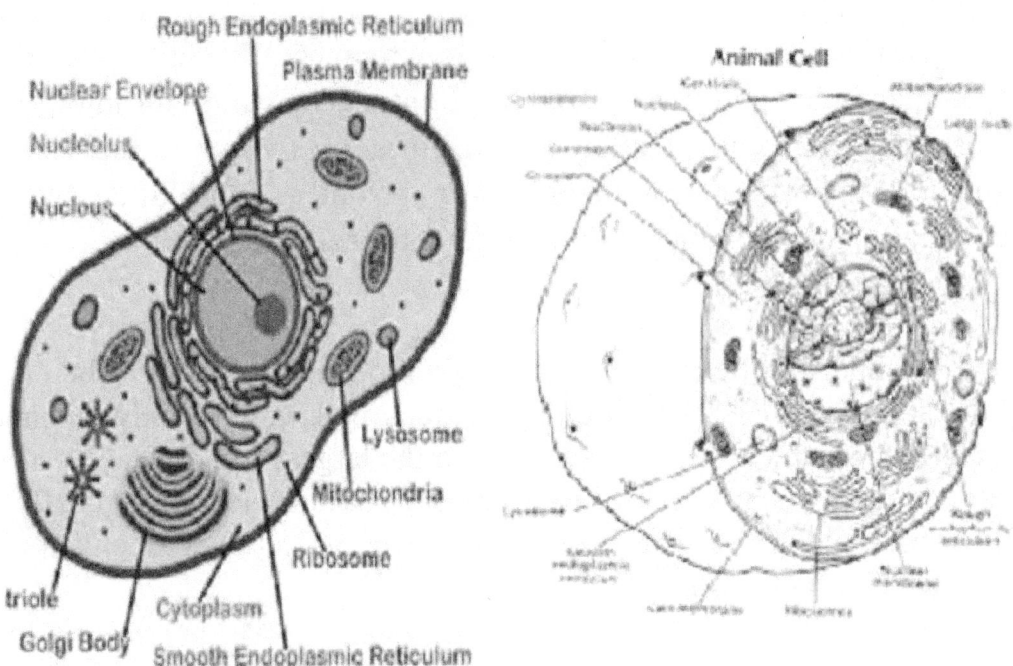

Nuclear envelope: The envelope composed of double membranes, the outer membrane is continuous with endoplasmic reticulum (ER) and it may be covered with ribosome which is pivotal to protein synthesis.

Nuclear pores: This is the passage through the nuclear envelope it allows the exchange of substances between the nucleus and the cytoplasm, the fusion of both membranes helps in the passage of the molecules through the pore.

Nucleus

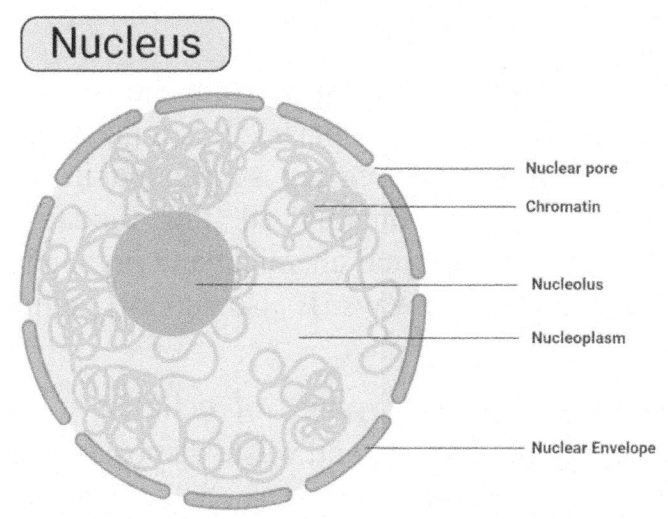

Chromatin materials: this term means **"coloured material"**, this was because it stained easily and becomes conspicuous during microscopic observation of cell. It consist of coils of **DNA** that is bound to basic proteins referred to as histones, this becomes condensed and coiled tightly into thread-like material called **CHROMOSOMES** during nuclear and cell division.

Nucleolus: A dark, rounded structure found within the nucleus. In cells, there may be one or more nucleoli, they are conspicuously stained within the nucleus due to the presence of large amount of **DNA** and **RNA** it contained. The major function is in Ribosome formation.

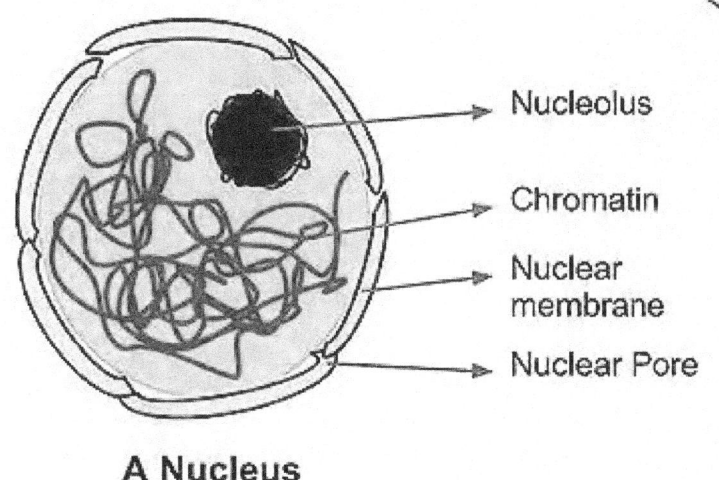

A Nucleus

1.3.5 CYTOPLASM:

This is the living content of an eukaryotic cell, apart from the Nucleus. In combination with the nucleus, it forms the protoplasm. It is aqueous (about 90% of water) other organelles and structures are suspended in it. The cytoplasm is made of solutions that contains all the fundamentals biochemical of life. It also acts as store for vital chemical and organelles, as well as been a site for most metabolic pathways. It also support cellular or cytoplasmic streaming.

The cytoplasm is also involved in a wide range of cellular activities, including protein synthesis, energy production, and waste disposal. Many of these activities take place within specialized organelles, such as the mitochondria, which produce ATP, the energy currency of the cell, or the lysosomes, which break down cellular waste and foreign materials. The endoplasmic reticulum and the Golgi apparatus, both located in the cytoplasm, are responsible for processing and packaging proteins and lipids for transport within the cell or secretion outside the cell. The cytoskeleton, a network of protein fibers that runs throughout the cytoplasm, provides structural support and helps to maintain the shape of the cell.

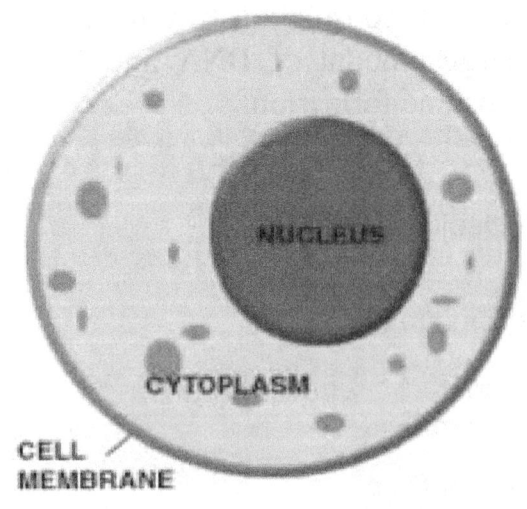

The cytoplasm is also involved in cell signaling and communication. Many signaling molecules, such as hormones and growth factors, are synthesized in the cytoplasm and are transported to their target cells by various mechanisms. The cytoplasm also contains numerous enzymes, which catalyze chemical reactions and play crucial roles in cellular regulation and homeostasis. The cytoplasm is involved in cell division, with the spindle fibers, which are responsible for separating the chromosomes during mitosis being assembled and disassembled in the cytoplasm.

1.3.6 ENDOPLASMIC RETICULUM (ER):

This is a system of membranes that runs through the cytoplasm in all eukaryotic cells. It is an extensive network that could be seen even under light microscope. They are made up of flattened, membrane-bound sac-like structure called cisternae. They may be covered with ribosome or not. When covered with ribosome (Rough ER), they function in the transport of proteins, whereas when not covered (Smooth ER) they function in the synthesis of lipids.

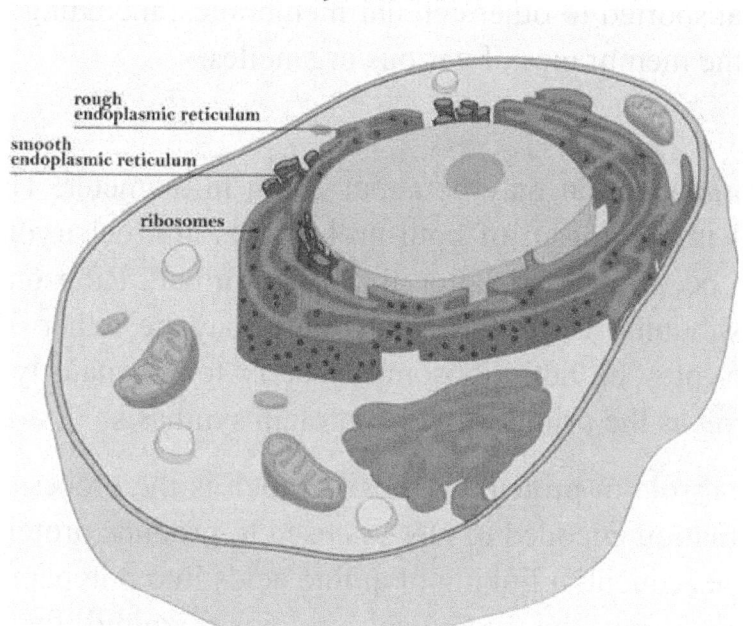

Endoplasmic Reticulum

Endoplasmic reticulum plays a crucial role in calcium homeostasis. Calcium ions are important signaling molecules that regulate many cellular processes, and the ER is a major site of calcium storage and release in the cell. The smooth ER is particularly important in this regard, as it contains specialized calcium pumps and channels that help to regulate the concentration of calcium ions in the cytoplasm.

Important function of the endoplasmic reticulum is its involvement in the process of protein folding and quality control. Proteins that are synthesized in the rough ER undergo a series of modifications, including folding, glycosylation, and disulfide bond formation, which are essential for their proper function. The ER also contains a quality control mechanism that ensures that misfolded or damaged proteins are correctly identified and either repaired or degraded.

The endoplasmic reticulum is also involved in the synthesis of membrane lipids, which are important components of all cellular membranes. The smooth ER contains a variety of enzymes that are involved in lipid synthesis, including phospholipids, cholesterol, and steroid hormones. These lipids are then transported to other cellular membranes, including the plasma membrane and the membranes of various organelles.

1.3.7 RIBOSOMES:

These are very small organelles, it may be about 20nm in diameter. They occur in large numbers in cytoplasm of both prokaryotes and eukaryotes, though much more in occurrence in latter than the former. Ribosomes consist of two major subunits, large and small, they may be either 70S found mostly in prokaryotes or 80S ribosomes that is found majorly in eukaryotes. They function as the primary sites of protein synthesis.

Ribosomes play a critical role in protein synthesis, which is the process by which the genetic information encoded in DNA is used to produce proteins. This process involves the sequential linking of amino acids into polypeptide chains, which ultimately form the functional proteins required for all cellular processes. Ribosomes achieve this by reading the genetic code carried by messenger RNA (mRNA) and catalyzing the formation of peptide bonds between adjacent amino acids.

Ribosomes are composed of two major subunits that come together during protein synthesis. The small subunit is responsible for recognizing and binding to the mRNA, while the large subunit contains the active site responsible for peptide bond formation. The exact structure and composition of ribosomes can vary depending on the organism and the type of cell in which they are found.

Ribosomes also play an important role in regulating gene expression and responding to various cellular stresses. For example, cells can adjust the number and activity of their ribosomes in response to changes in nutrient availability or other environmental factors, allowing them to fine-tune their protein production as needed.

Ribosomes have become important targets for many antibiotics and other drugs, as they are essential for bacterial and fungal growth. By targeting the ribosome, these drugs can selectively inhibit the growth of pathogenic microorganisms while leaving the host cells unharmed, making them valuable tools in the fight against infectious disease.

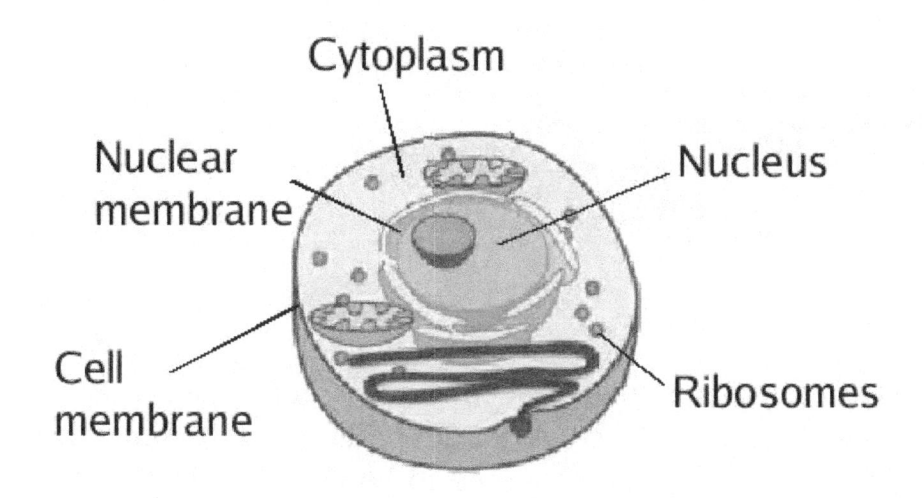

1.3.8 LYSOSOMES:

This occurs mostly in eukaryotes, they are simple sacs covered by a single membrane. They contains digestive enzymes such as proteases, nucleases and lipases which function during breakdown of proteins, nucleic acids and lipids respectively.

Lysosomes also play a crucial role in the recycling of cellular waste materials. They are responsible for breaking down and recycling cellular components that are no longer functional or necessary, such as damaged organelles, through a process called autophagy. Lysosomes also participate in the process of phagocytosis, where they fuse with vesicles containing ingested material, such as bacteria or other pathogens, to break them down and eliminate them from the cell. Lysosomal storage disorders, such as Tay-Sachs disease, result from defects in lysosomal enzymes that impair the breakdown of specific macromolecules and lead to the accumulation of undigested substances within the cell.

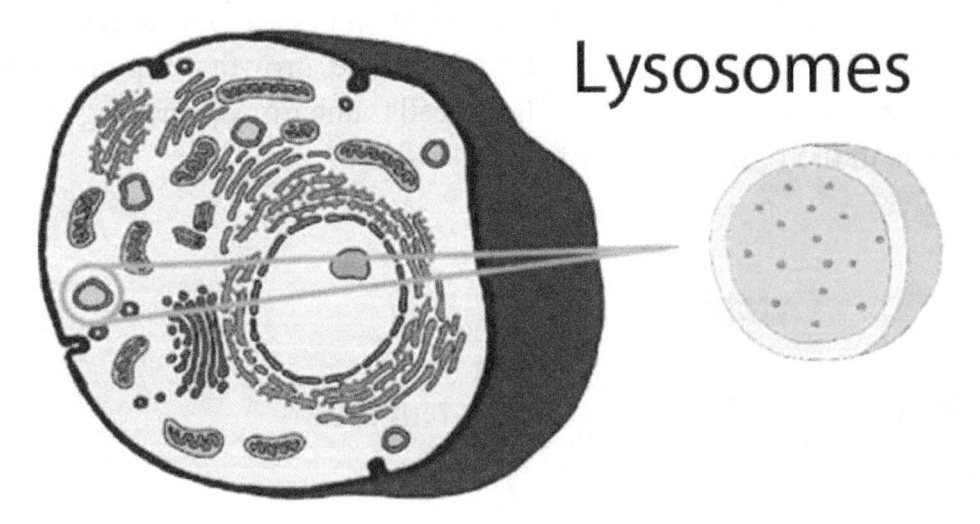

1.3.9 THE VACUOLES:

The vacuole is a membrane-bound sac that plays roles in intracellular digestion and the release of cellular waste products. In animal cells, vacuoles are generally small. Vacuoles tend to be large in plant cells and play a role in turgor pressure. When a plant is well-watered, water collects in cell vacuoles producing rigidity in the plant. Without sufficient water, pressure in the vacuole is reduced and the plant wilts. In unicellular organisms, there is contractile vacuoles that helps in water removal.

Vacuoles in plant cells can also store nutrients and pigments. For example, the vacuoles in plant cells can store pigments such as anthocyanins which give flowers and fruits their coloration. Vacuoles can also store amino acids, sugars, ions, and other compounds that are important for the plant's growth and survival.

Interestingly, some plant species have evolved unique vacuoles with specialized functions. For example, the latex in rubber trees is stored in large vacuoles. In some carnivorous plants, such as the Venus flytrap, vacuoles contain digestive enzymes that help to break down prey. These specialized vacuoles have adapted to meet the specific needs of the plants in which they occur.

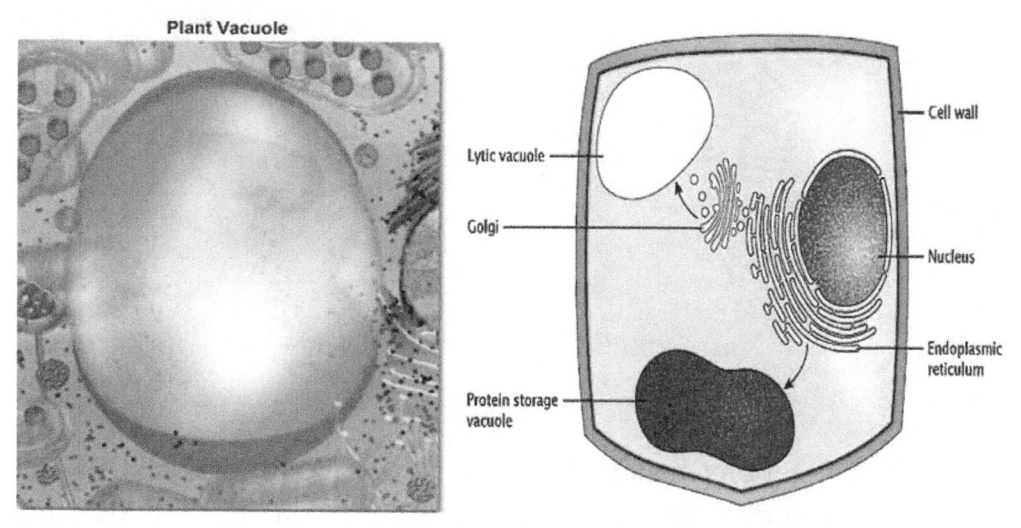

1.3.10 GOLGI APPARATUS:

This was discovered by Camilo Golgi in 1898. It is a membrane-bound structure with a single membrane. It is actually a stack of membrane-bound vesicles that are important in packaging macromolecules (protein) for transport elsewhere in the cell. The stack of larger vesicles is surrounded by numerous smaller vesicles containing those packaged macromolecules. The enzymatic or hormonal contents of lysosomes, peroxisomes and secretory vesicles are packaged in membrane-bound vesicles at the periphery of the Golgi apparatus.

This organelle receives newly synthesized proteins and lipids from the endoplasmic reticulum (ER) and then modifies and sorts them before sending them to their final destination within the cell or for secretion outside the cell. The Golgi apparatus has a distinct polarity, with the cis-Golgi facing the ER and the trans-Golgi facing the plasma membrane. This polarity allows for efficient and directional transport of molecules through the Golgi stack. Defects in Golgi function have been linked to a variety of diseases, including neurological disorders and cancer.

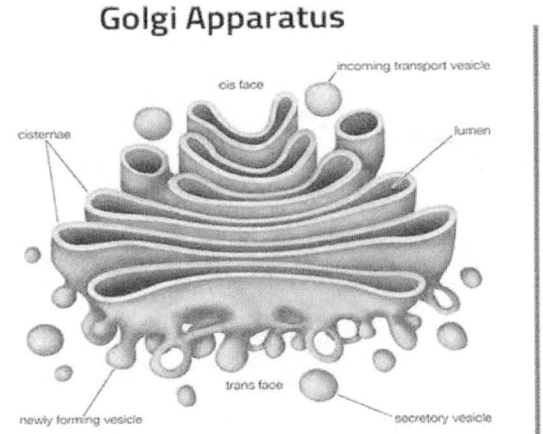

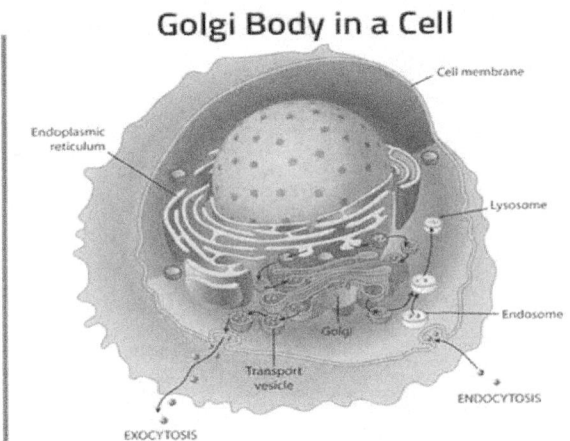

1.3.11 MICROTUBULES:

These are conveyer belts inside the cells. They move vesicles, granules, organelles like mitochondria, and chromosomes via special attachment proteins. They also serve a cytoskeletal role. Structurally, they are linear polymers of tubulin which is a globular protein. These linear polymers are called protofilaments. They may work alone, or join with other proteins to form more complex structures called cilia, flagella or centrioles.

Microtubules play a critical role in many cellular processes such as mitosis, cell division, and intracellular transport. They are highly dynamic structures that can rapidly assemble and disassemble, which allows them to quickly respond to changes in the cellular environment. In addition to their structural role, microtubules are also involved in signaling pathways, such as regulating the activity of enzymes and transcription factors. Mutations or defects in microtubule-associated proteins have been implicated in a range of diseases, including cancer, neurodegenerative disorders, and developmental abnormalities.

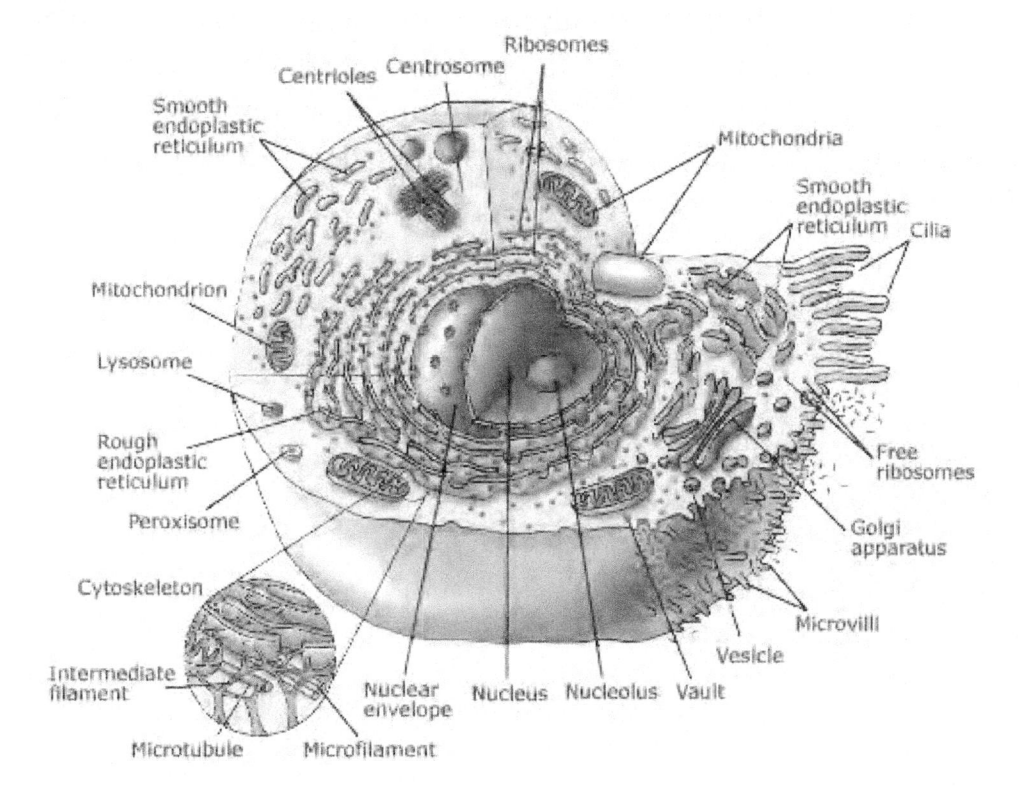

1.3.12   MITOCHONDRIA:

This is found nearly in all eukaryotic cells. In some cases, the cell has a single large mitochondrion, but more often, a typical cell has 100s or even 1000s of them. They are enclosed by double membranes of phospholipids layer embedded with proteins. The outer membrane is smooth but the inner layer is convoluted and referred to as CRISTAE. They provide the energy a cell needs to move, divide, produce secretory products, contract - in short, they are the power centers of the cell. They are about the size of bacteria but may have different shapes depending on the cell type.

Mitochondria are also known as the "cellular powerhouses" because they are responsible for producing most of the ATP (adenosine triphosphate) which is the energy currency of the cell. ATP is synthesized by a process called oxidative phosphorylation, which takes place in the inner membrane of the mitochondria. Mitochondria also play an important role in regulating cellular metabolism and apoptosis (programmed cell death). Additionally, mitochondria have their own DNA, called mitochondrial DNA, which is inherited only from the mother. Mutations in mitochondrial DNA have been linked to various diseases and aging.

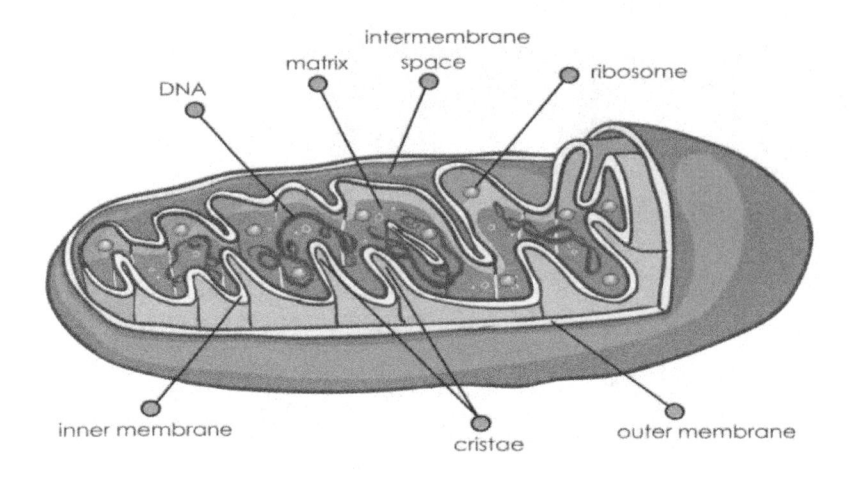

1.4 DIFFERENCE BETWEEN PLANT CELL AND ANIMAL CELL

S. No	Plant cell	Animal Cell
1	Usually they are larger than animal cells	Usually smaller than plant cells
2	Cell wall present in addition to plasma membrane and consists of middle lamellae, primary and secondary walls	Cell wall absent
3	Plasmodesmata present	Plasmodesmata absent
4	Chloroplast present	Chloroplast absent
5	Vacuole large and permanent	Vacuole small and temporary
6	Tonoplast present around vacuole	Tonoplast absent
7	Centrioles absent except motile cells of lower plants	Centrioles present
8	Nucleus present along the periphery of the cell	Nucleus at the centre of the cell
9	Lysosomes are rare	Lysosomes present
10	Storage material is starch grains	Storage material is a glycogen granules

CHAPTER 2

MICROSCOPES

The world of botany is an intricate tapestry of intricate structures and hidden wonders. As scientists and enthusiasts delve deeper into the realm of plants, one indispensable tool emerges to unveil the hidden secrets that lie beyond the reach of the naked eye—the microscope. Derived from the Greek words **"mikrós,"** meaning small, and **"skopeîn,"** meaning to look or see, the microscope serves as a gateway to a hidden universe teeming with minute organisms, intricate cellular structures, and fascinating details that might otherwise go unnoticed.

Microscopy, the scientific study of small objects and structures using a microscope, has revolutionized our understanding of the botanical world. It enables researchers and students alike to explore the intricate complexities that lie within the seemingly ordinary plants that surround us. Through the power of magnification, the microscope reveals a hidden world of beauty, diversity, and remarkable adaptations.

The term "microscopic" itself embodies the essence of the microscope's capabilities. It refers to objects or details that are invisible to the human eye unless aided by this remarkable instrument. With the microscope as our guide, we gain access to a world of wonders that would otherwise remain hidden from our view. From the delicate structures of plant cells to the intricate network of vascular tissues, the microscope allows us to observe and understand the intricate mechanisms that drive the life processes of plants.

- **Microscope** as a term was coined from two Greek words; **mikrós**, "small" and **skopeîn**, "to look" or "see" is a laboratory instrument used to examine objects that are too small to be seen by the naked eye.
- **Microscopy** is the science of investigating small objects and structures using a microscope.
- **Microscopic** means being invisible to the eye unless aided by a microscope.

2.1 DEVELOPMENT OF MICROSCOPES

About 1590, two Dutch spectacle makers, Zaccharias Janssen and his son Hans, discovered that nearby objects appeared greatly enlarged. That was the forerunner of the compound microscope and of the Telescope. The serendipitous discovery made by Zaccharias Janssen and his son Hans in 1590 laid the foundation for a remarkable scientific revolution. As spectacle makers in the Netherlands, they stumbled upon a peculiar phenomenon that would forever change the way we perceive the world. Through a simple arrangement of lenses, they found that nearby objects appeared greatly enlarged, offering a glimpse into a realm previously unseen.

This accidental revelation became the precursor to two extraordinary inventions—the compound microscope and the telescope. Building upon the Janssens' initial observation, scientists and inventors around the world began to explore and refine these optical devices, harnessing their power to explore the hidden wonders of the microscopic and celestial realms.

The compound microscope, as its name suggests, employs a combination of lenses to achieve magnification. It consists of an objective lens, placed close to the object being observed, and an eyepiece lens, through which the enlarged image is viewed. By carefully aligning and adjusting these lenses, researchers could peer into a realm where intricate structures and tiny organisms were brought into focus, unveiling a level of detail that was previously unimaginable.

Simultaneously, the telescope, inspired by the Janssens' discovery, revolutionized our understanding of the cosmos. By utilizing a similar arrangement of lenses, but with a different configuration optimized for long-distance observation, astronomers were able to gaze far beyond the confines of Earth. The telescope allowed humanity to explore distant celestial bodies, unravel the mysteries of our universe, and broaden our understanding of the cosmos.

The compound microscope and the telescope were more than just scientific instruments; they were gateways to new worlds. They opened up avenues for exploration and discovery, enabling scientists, naturalists, and astronomers to peer into realms that were previously inaccessible. These inventions became pivotal in propelling our understanding of the natural world and shaping the fields of botany, microbiology, astronomy, and countless others.

Since the breakthrough made by the Janssens, the compound microscope and telescope have undergone significant advancements. Modern microscopes now incorporate sophisticated techniques such as fluorescence microscopy, confocal microscopy, and electron microscopy, providing even greater resolution and clarity. Telescopes have grown in size and power, enabling us to peer further into the cosmos and unravel the mysteries of distant galaxies and celestial phenomena.

In 1609, the renowned Italian scientist Galileo Galilei, often referred to as the "father of modern physics and astronomy," became aware of the early experiments involving lenses and magnification. Intrigued by the possibilities, Galileo dedicated himself to understanding the principles behind these optical instruments and improving upon their design.

Drawing upon his deep knowledge of mathematics, physics, and the behavior of light, Galileo formulated the fundamental principles of lenses and optics. He realized that by carefully combining lenses of different shapes and focal lengths, it was possible to create a much more refined and powerful instrument for observation.

Galileo's keen intellect and meticulous craftsmanship led him to construct a vastly superior instrument—a telescope that not only magnified distant objects but also featured a focusing device. This innovation allowed for sharper and clearer images, enabling Galileo to make groundbreaking observations of celestial bodies and revolutionize our understanding of the universe.

With his new and improved telescope, Galileo turned his gaze towards the night sky. He observed the moon's craters and mountains, discovering that the lunar surface was not the smooth, perfect sphere believed at the time. He also observed the four largest moons of Jupiter, providing evidence that heavenly bodies could orbit something other than Earth—a significant blow to the prevailing geocentric view of the universe.

During the period from 1632 to 1723, a remarkable figure emerged in the field of microscopy: Anton van Leeuwenhoek of Holland;. Often referred to as the "father of microscopy," Leeuwenhoek made ground breaking contributions that forever transformed our understanding of the microscopic world. His insatiable curiosity and innovative methods propelled him to unprecedented discoveries.

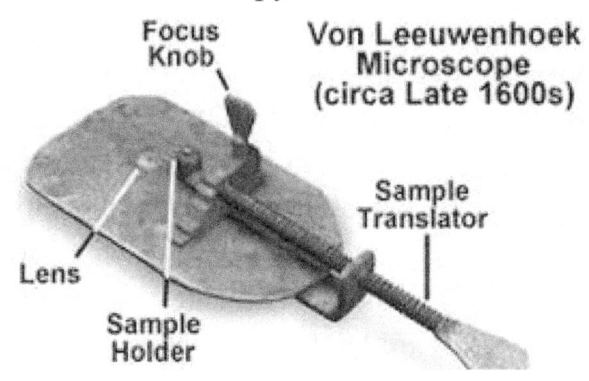

Leeuwenhoek's pursuit of knowledge led him to teach himself the art of grinding and polishing tiny lenses with extraordinary curvature. These lenses, when combined, achieved magnifications of up to **270 diameters,** surpassing anything that was known at the time. With these finely crafted lenses, Leeuwenhoek constructed his own microscopes, laying the groundwork for his remarkable observations and revelations.

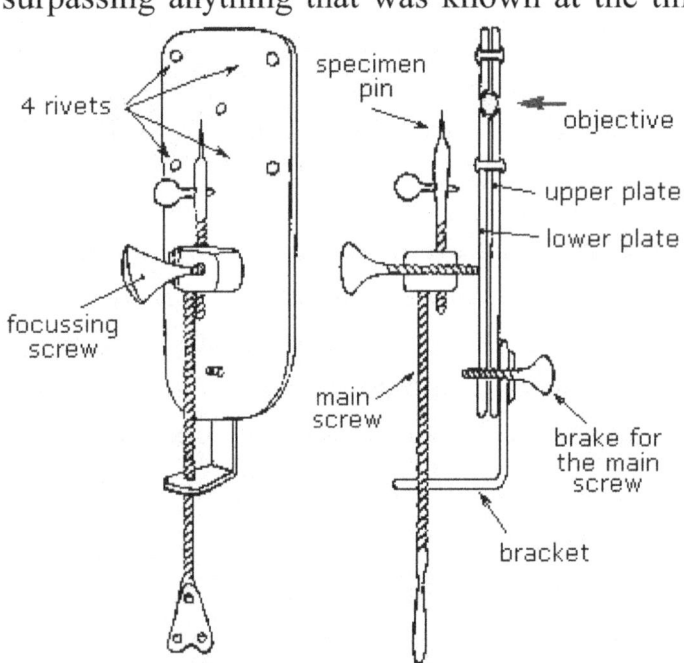

Through his meticulously crafted lenses, Leeuwenhoek became the first person to witness and describe the existence of bacteria, yeast cells, and the astonishing diversity of life thriving in a single drop of water. His observations extended to the intricate circulation of blood corpuscles within capillaries, a ground breaking discovery that revolutionized our understanding of human physiology.

Leeuwenhoek's insatiable curiosity knew no bounds, and he applied his lenses to a wide range of subjects, both living and non-living. His keen observations and meticulous record-keeping culminated in over a hundred letters sent to esteemed scientific institutions such as the Royal Society of England and the French Academy. In these letters, Leeuwenhoek shared his remarkable findings, documenting his observations on an extraordinary variety of topics, further enriching the scientific community's understanding of the microscopic realm.

His contributions were not only transformative but also inspirational, paving the way for future scientists and researchers to explore the hidden intricacies of the natural world. Leeuwenhoek's pioneering studies set a precedent for meticulous observation and the pursuit of knowledge, fostering a scientific spirit that continues to drive discoveries to this day.

- In 1632-1723 Anton van Leeuwenhoek of Holland; The father of microscopy. He taught himself new methods for grinding and polishing tiny lenses of great curvature which gave magnifications up to 270 diameters, the finest known at that time. These led to the building of his microscopes and discoveries for which he is famous.
- Anton van Leeuwenhoek was the first to see and describe bacteria, yeast cells, the teeming life in a drop of water, and the circulation of blood corpuscles in capillaries.
- He used his lenses to make pioneer studies on an extraordinary variety of things, both living and non-living, and reported his findings in over a hundred letters to the Royal Society of England and the French Academy.

During the 1660s, another prominent figure emerged in the field of microscopy: Robert Hooke, often referred to as the "**English father of microscopy.**" Hooke played a crucial role in the validation and advancement of Anton van Leeuwenhoek's discoveries concerning the existence of microscopic organisms in a drop of water.

Inspired by Leeuwenhoek's work, Hooke sought to replicate and build upon the Dutch scientist's achievements. He meticulously crafted his own version of Leeuwenhoek's light microscope, endeavoring to observe and study the hidden microscopic world. However, Hooke did not merely stop at replication; he sought to improve upon the original design, refining the microscope to enhance its capabilities and expand its potential applications.

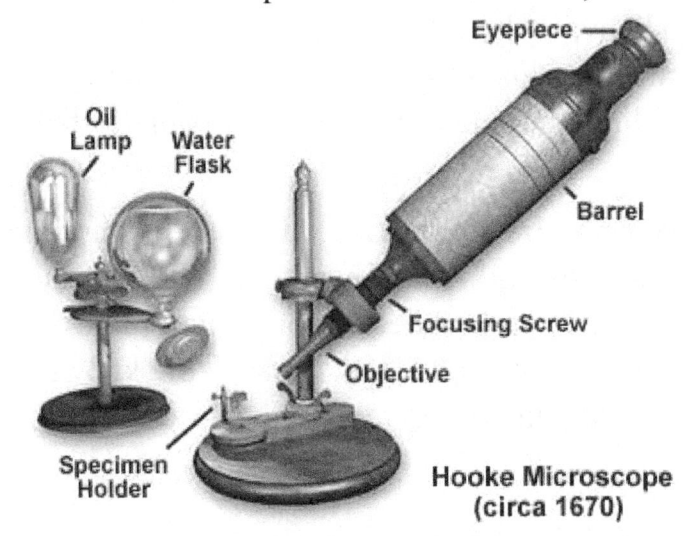

Through his innovative modifications, Hooke made significant contributions to the field of microscopy. He introduced improvements in lens quality, lighting techniques, and specimen preparation methods. These enhancements allowed for clearer, more detailed observations and opened up new avenues for scientific exploration.

Hooke's microscope not only enabled him to reconfirm Leeuwenhoek's discoveries of micro-organisms in a drop of water but also facilitated his own observations. He explored various biological samples, meticulously documenting his findings in his seminal work, "Micrographia," published in 1665. This influential publication contained detailed illustrations and descriptions of various microscopic structures, including the first known depiction of plant cells in cork. Hooke's observations and insights provided vital evidence for the existence of cells, laying the foundation for the field of cell biology.

Beyond his contributions to microscopy, Hooke was a multifaceted scientist who made significant advancements in various disciplines. He was a prominent figure in physics, astronomy, and architecture, among others.

However, his work in microscopy remains particularly noteworthy, as it advanced our understanding of the intricate world that lies beyond the limits of human vision.

The efforts of Robert Hooke, along with the pioneering work of Anton van Leeuwenhoek, propelled the field of microscopy forward, establishing a solid foundation for future scientists to build upon. Their contributions set the stage for further advancements in the design and capabilities of microscopes, paving the way for the extraordinary discoveries and breakthroughs that would follow in the centuries to come.

- In 1660s Robert Hooke the English father of microscopy, re-confirmed Anton van Leeuwenhoek's discoveries of the existence of tiny living organisms in a drop of water.
- Hooke made a copy of Leeuwenhoek's light microscope and then improved upon his design.

In the middle of the 19th century, Charles A. Spencer emerged as a key figure in the field of microscopy. Spencer's contributions to the development of microscopes further pushed the boundaries of magnification and resolution. He dedicated his efforts to constructing instruments of exceptional quality and performance, setting new standards in the field.

Spencer's microscopes were renowned for their ability to achieve remarkably high magnifications. With ordinary light, his instruments could provide magnifications of up to 1250 diameters, allowing scientists to delve even deeper into the microscopic world. However, Spencer's innovations did not stop there. He realized that using blue light, which has a shorter wavelength, could further enhance the resolution and clarity of the observed specimens. By harnessing blue light, his microscopes could achieve astonishing magnifications of up to 5000 diameters, enabling researchers to explore the intricacies of plant structures with unprecedented detail.

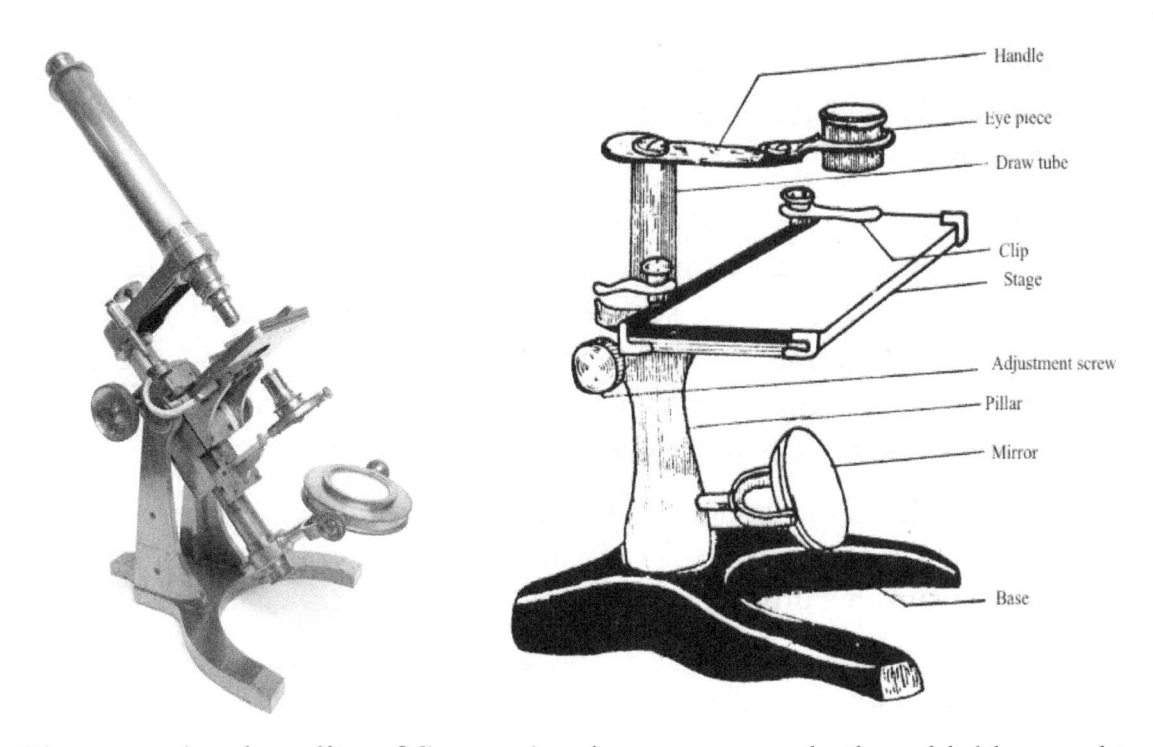

The exceptional quality of Spencer's microscopes made them highly sought after by scientists and researchers across various fields, including botany. These instruments were crucial in unraveling the complexities of plant anatomy, morphology, and reproductive structures. Spencer's microscopes allowed botanists to examine delicate plant tissues, study cellular structures, and investigate the minute details of floral organs, pollen grains, and plant seeds.

The advancements made by Charles A. Spencer marked a significant milestone in the history of microscopy, particularly in the context of botany. The increased magnification and resolution provided by his microscopes revolutionized the study of plants, enabling botanists to make precise observations, conduct detailed analyses, and contribute to our understanding of plant diversity and adaptations.

- Charles A. Spencer in the middle of 19th century built the finest instrument and give magnifications up to 1250 diameters with ordinary light and up to 5000 diameters with blue light.

A simple microscope only has one type of lens, usually objective lens, but a compound microscope contain both objective lens and ocular lens.

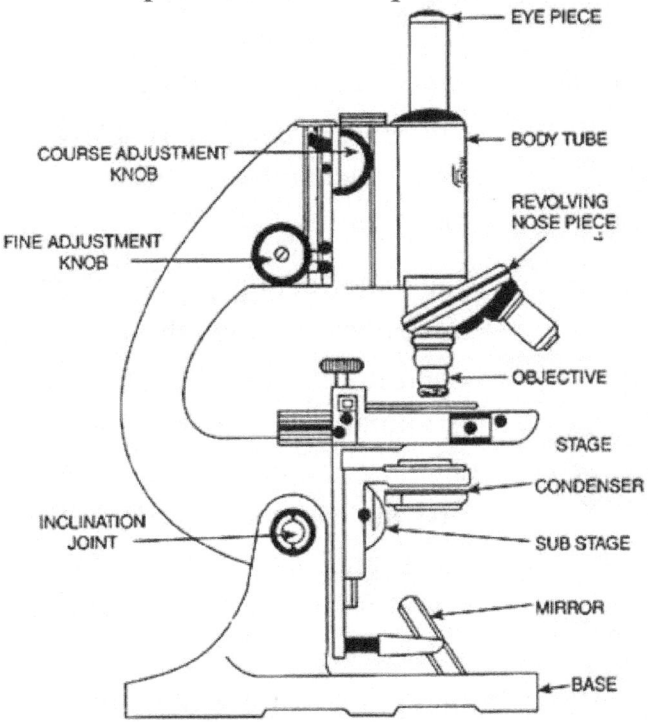

2.2 TYPES OF MICROSCOPES

There are four main types of microscopes that a biologist uses:
1. Optical Dissection Microscope
2. Optical Compound – Light, Fluorescence and Confocal microscopes
3. Scanning Electron Microscope (SEM)
4. Transmission Electron Microscope (TEM)

2.2.1 DISSECTION or STEREO MICROSCOPE

It is light illuminated. The image that appears is three dimensional. It is used for dissection to get a better look at the larger specimen You cannot see individual cells because it has a low magnification.

This type of microscope is specifically designed to facilitate the examination of larger specimens with higher clarity and depth perception. Unlike other microscopes, the dissecting microscope utilizes light illumination to create a three-dimensional image, providing researchers with a more realistic and comprehensive view of the specimen.

The dissecting microscope is particularly valuable in botany when it comes to the dissection and examination of plant structures. It enables botanists to study the intricate details of leaves, flowers, fruits, and other plant parts without the need for invasive procedures. By observing these larger structures under the dissecting microscope, botanists can gain a better understanding of the organization, arrangement, and unique characteristics of different plant species.

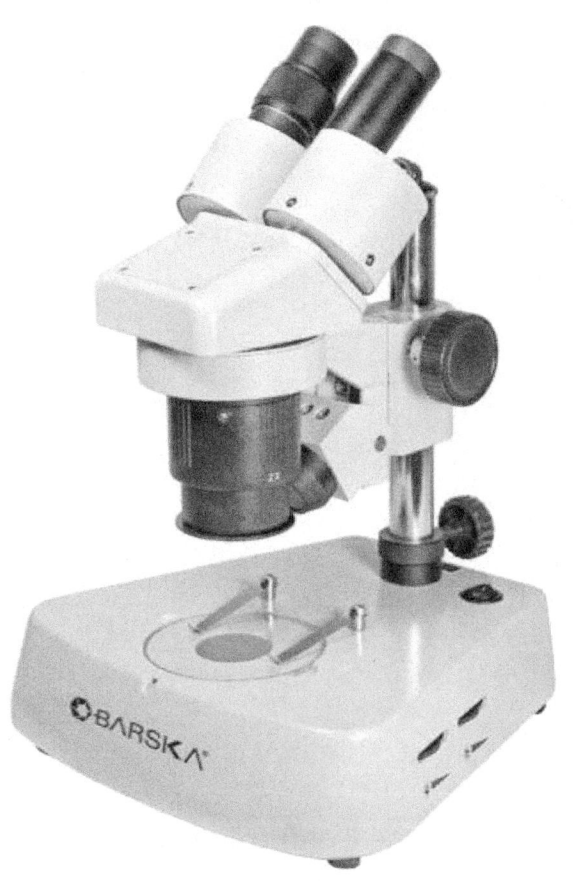

STEREO MICROSCOPE

Although the dissecting microscope offers excellent visualization and three-dimensional imaging, it does have limitations in terms of magnification. Due to its lower magnification power, the dissecting microscope is not suitable for observing individual cells or studying cellular-level structures. Its primary focus is on providing a detailed examination of larger plant parts rather than delving into the microscopic world of cells and organelles.

The dissecting microscope is an essential tool for teaching and educational purposes in botany. Its three-dimensional imaging capability provides students with a more interactive and immersive learning experience. It enables them to closely examine plant specimens, grasp their physical characteristics, and develop a deeper understanding of plant diversity.

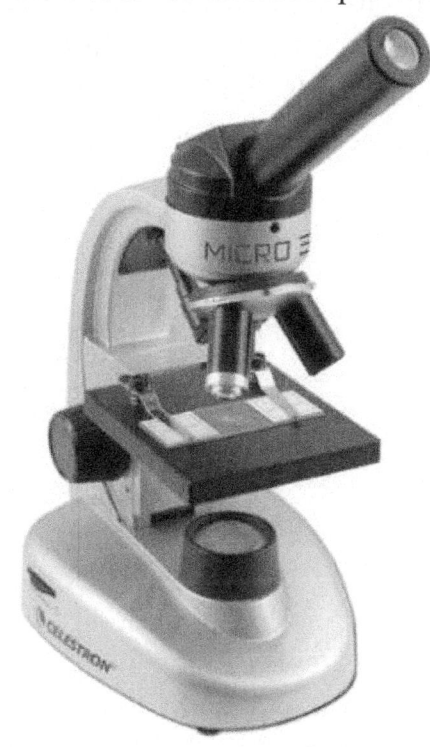

2.2.2 COMPOUND MICROSCOPE

It is also light illuminated. The image seen with this type of microscope is two-dimensional. This microscope is the most commonly used. You can view individual cells, even living ones. It has high magnification (from 4x - 100x). However, it has a low resolution.

The compound microscope holds a prominent place as the most commonly used tool for microscopic examination. Similar to the dissecting microscope the compound microscope also utilizes light illumination. However, there are distinct differences in the image produced and the applications for which it is best suited. It generates a two-dimensional image, allowing researchers to observe and study microscopic specimens in greater detail. This microscope is specifically designed to explore the intricate structures and processes occurring at the cellular level. With the compound microscope, botanists can delve into the fascinating world of individual plant cells, studying their morphology, internal structures, and cellular dynamics.

COMPOUND MICROSCOPE

The compound microscope's ability to view individual cells, including living ones, is of paramount importance in botany. It enables scientists to investigate various cellular aspects such as cell division, organelle structure, and the movement of cellular components. By observing living cells under the compound microscope, botanists can gain insights into essential biological processes, such as photosynthesis, respiration, and cell-to-cell communication.

One of the key advantages of the compound microscope is its high magnification capabilities, ranging from 4x to 100x or even higher with the use of oil immersion lenses. This magnification range allows for detailed examination of cellular structures and provides researchers with a closer look at the intricacies of plant tissues, such as the arrangement of cells in different plant organs.

However, it is worth noting that the compound microscope does have its limitations, particularly in terms of resolution. The resolving power of the compound microscope determines its ability to distinguish two closely spaced objects as separate entities. Due to the diffraction of light, the resolution of a compound microscope is inherently limited, leading to blurring or lack of clarity when observing extremely small structures or fine details.

2.2.3 FLUORESCENCE MICROSCOPY

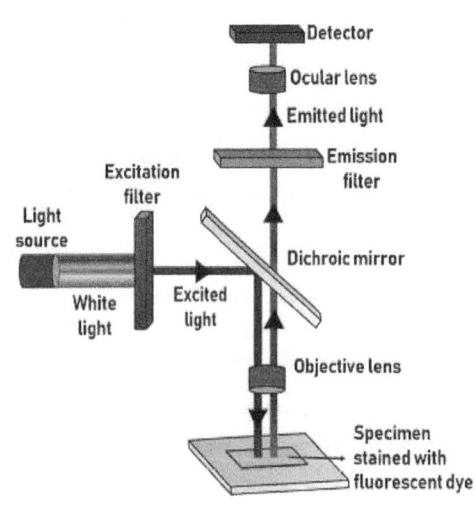

It is a light microscope used to study properties of organic or inorganic substances using the phenomena of fluorescence and phosphorescence instead of, or in addition to, reflection and absorption. Fluorescence illumination and observation is the most rapidly expanding microscopy technique employed today, both in the medical and biological sciences, a fact which has spurred the development of more sophisticated microscopes and numerous fluorescence accessories. Epi-fluorescence, or incident light fluorescence, has now become the method of choice in many applications.

Fluorescence microscopy involves the use of specialized microscopes and accessories designed to detect and visualize fluorescence emitted by certain substances. It is a rapidly expanding technique in both the medical and biological sciences, revolutionizing our understanding of plant structures and processes. This surge in popularity has driven the development of more sophisticated microscopes and a wide array of fluorescence-specific accessories to meet the growing demands of researchers in the field of botany.

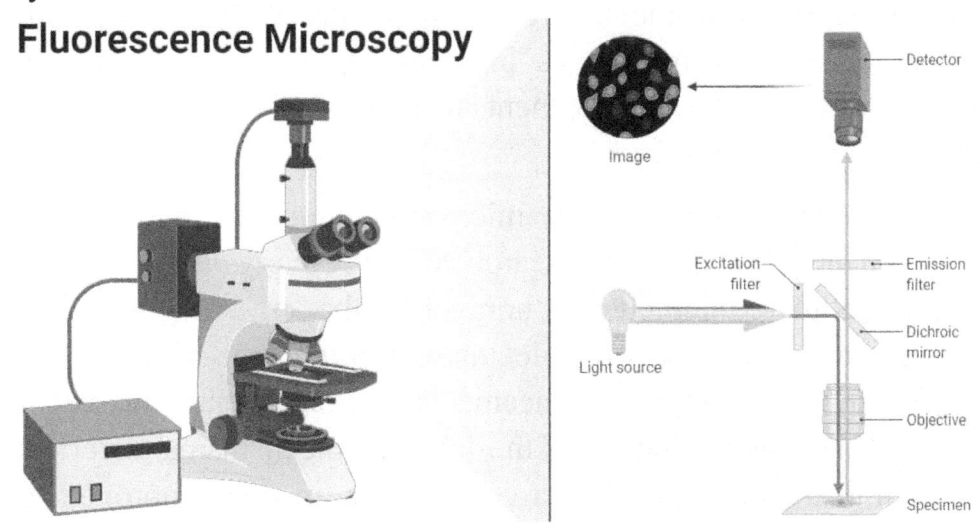

Fluorescence Microscopy

One of the key advantages of fluorescence microscopy is its ability to selectively label specific molecules or structures within plant cells. By using fluorescent dyes or probes that bind to particular targets of interest, researchers can visualize and track the localization, distribution, and interactions of various components within plants. This approach has proven invaluable in studying cellular organelles, such as chloroplasts and mitochondria, as well as specific proteins, nucleic acids, and metabolites involved in crucial plant processes.

Epi-fluorescence, also known as incident light fluorescence, has become the method of choice in many applications within botany. In this technique, the excitation light is directed onto the specimen from above, while the emitted fluorescence is captured by the microscope. This configuration enhances the specificity and sensitivity of fluorescence detection, enabling researchers to obtain high-quality images and precise information about the targeted molecules or structures.

The versatility of fluorescence microscopy extends beyond static imaging. With the aid of advanced techniques such as confocal microscopy and fluorescence lifetime imaging microscopy (FLIM), researchers can explore dynamic processes within living plant cells. These techniques allow for the visualization of real-time events, such as intracellular transport, signal transduction, and even changes in gene expression, providing a deeper understanding of plant development, responses to environmental cues, and cellular communication.

The integration of fluorescence microscopy into the field of botany has opened up new avenues of research and discovery. It has shed light on intricate cellular interactions, unraveled the mysteries of subcellular structures, and provided valuable insights into plant physiology, growth, and adaptation. The rapid advancements in fluorescence microscopy have broadened our understanding of the intricate mechanisms underlying plant life, paving the way for innovative solutions in crop improvement, plant breeding, and sustainable agriculture.

2.2.4 CONFOCAL MICROSCOPY

It is an optical imaging technique for increasing optical resolution and contrast of a micrograph by means of using a spatial pinhole to block out-of-focus light in image formation.

The key principle behind confocal microscopy is the use of a pinhole aperture that is placed at a specific plane within the microscope's optical path. This pinhole acts as a spatial filter, allowing only the focused light from the exact plane of interest to pass through while blocking the out-of-focus light that would otherwise contribute to blurring and reduced image quality. By eliminating the out-of-focus light, confocal microscopy enables researchers to obtain clearer, high-resolution images with improved contrast and a greater level of detail.

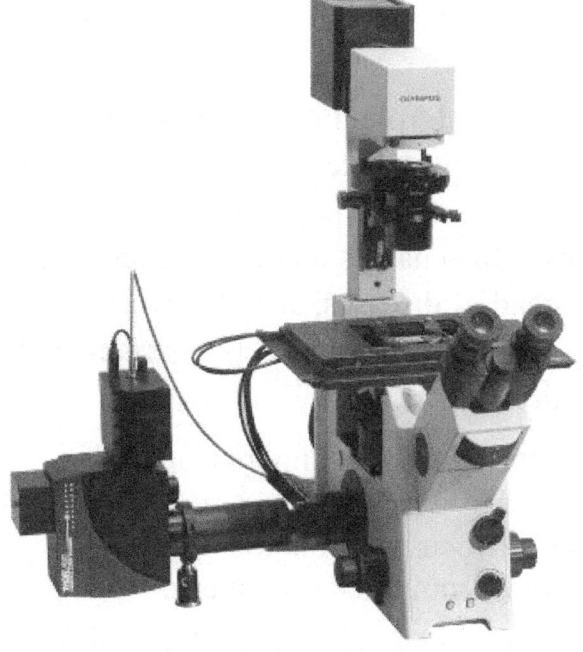

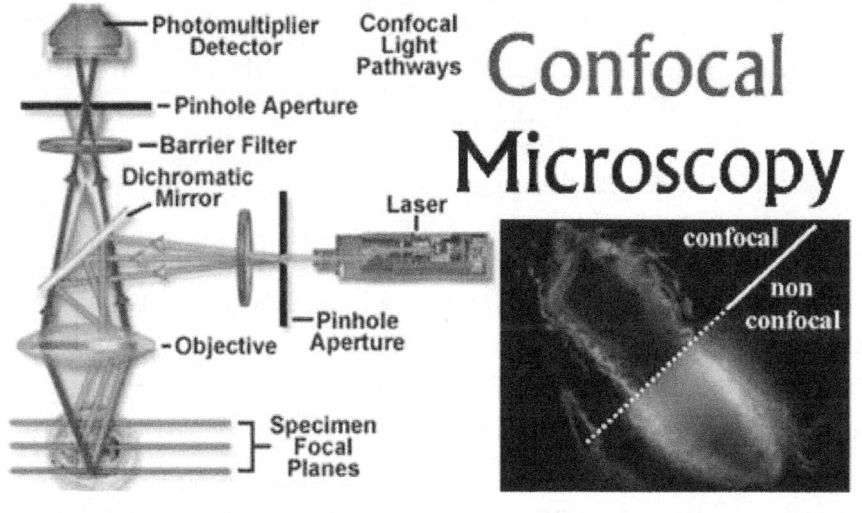

One of the major advantages of confocal microscopy in botany is its ability to generate optical sections of plant specimens at different depths within the sample. By acquiring a series of optical sections through the specimen, researchers can reconstruct three-dimensional images, providing a more comprehensive understanding of the internal structures and spatial relationships within plants. This capability has been instrumental in studying complex plant tissues, such as root systems, vascular networks, and intricate cellular architectures. Confocal microscopy is also widely utilized in the field of plant physiology to investigate dynamic processes within living cells and tissues. With the advent of fluorescent probes and dyes, researchers can selectively label specific components or molecules within plants, such as calcium ions, reactive oxygen species, or fluorescently tagged proteins. By combining confocal microscopy with these fluorescent markers, scientists can visualize and track dynamic changes in real-time, shedding light on physiological events like signal transduction, cell signaling, and developmental processes.

Confocal microscopy has played a crucial role in plant pathology and disease research. By employing fluorescently labeled antibodies or genetic markers, researchers can detect and localize specific pathogens or disease-related molecules within plant tissues. This enables a deeper understanding of plant-pathogen interactions, the spread of diseases, and the host's defense responses, ultimately leading to the development of effective strategies for disease management and crop protection

It offers several advantages over conventional optical microscopy:
1. Controllable depth of field,
2. Elimination of image degrading out-of-focus information,
3. Ability to collect serial optical sections from thick specimens.

    The key to the Confocal approach is the use of spatial filtering to eliminate out-of-focus light or flare in specimens that are thicker than the plane of focus. There has been a tremendous explosion in the popularity of Confocal microscopy in recent years, due in part to the relative ease with which extremely high-quality images can be obtained from specimens prepared for conventional optical microscopy, and in its great number of applications in many areas of current research interest.

2.2.5   SCANNING ELECTRON MICROSCOPE (SEM)

It is a type of electron microscope that produces images of a sample by scanning the surface with a focused beam of electrons. The electrons interact with atoms in the sample, producing various signals that contain information about the surface topography and composition of the sample.
They use electron illumination. The image is seen in three dimension. It has high magnification and high resolution. The specimen is coated in gold and the electrons bounce off to give an exterior view of the specimen. The pictures are in black and white.

Operating on the principles of electron illumination, the SEM scans the sample's surface with a fine electron beam. As the electrons interact with the atoms in the sample, various signals are emitted and collected, providing valuable information about the topography and composition of the specimen. These signals are then transformed into a visual representation of the sample, which can be observed in three dimensions.

Scanning Electron Microscope (SEM)

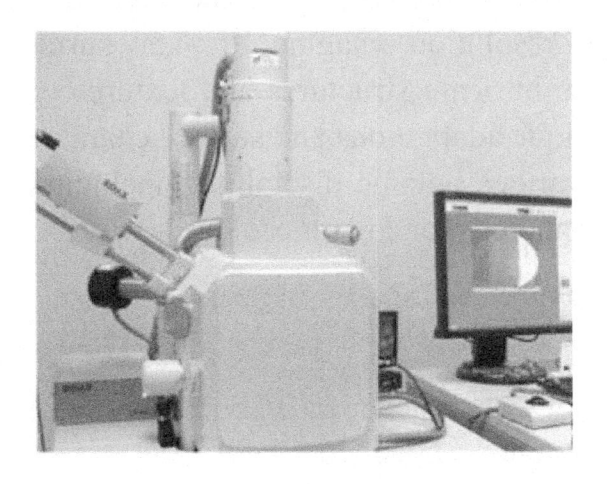

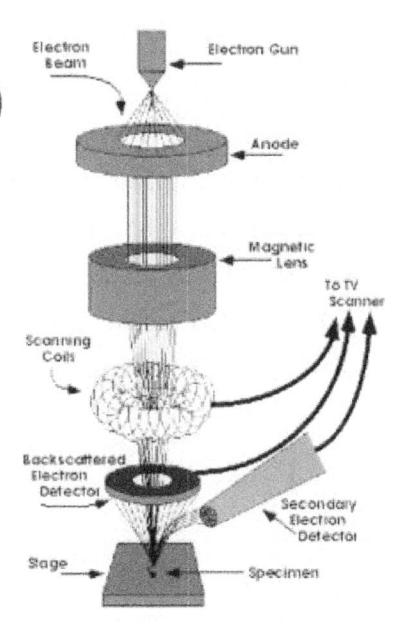

One advantage of SEM is its ability to achieve high magnification and resolution. The focused electron beam allows for incredibly detailed imaging, enabling researchers to observe minute surface structures and features of plant specimens with exceptional clarity. This level of magnification is particularly useful when studying fine structures such as the surface of leaves, pollen grains, or trichomes (hair-like structures) on plant surfaces.

To prepare the sample for SEM imaging, it is typically coated with a thin layer of conductive material, such as gold. This coating serves two purposes Firstly, it enhances the conductivity of the sample, allowing for better interaction with the electron beam. Secondly, it provides a reflective surface that enables the electrons to bounce off the specimen, generating a detailed exterior view.

SEM images are captured in black and white. The contrasting shades in the images result from the detection of secondary or backscattered electrons, which are sensitive to variations in the sample's composition and density. These grayscale images are highly informative and offer a wealth of data regarding the surface morphology and elemental composition of the plant specimen.

The Scanning Electron Microscope has revolutionized the study of botany by providing detailed and high-resolution imaging of plant surfaces. It allows researchers to explore the intricate structures and textures of plant samples, unveiling a world of unique adaptations and surface characteristics SEM has proven invaluable in various botanic disciplines, including plant taxonomy, palynology, plant morphology, and plant ecology.

2.2.6 TRANSMISSION ELECTRON MICROSCOPE (TEM)

It is a microscopy technique in which a beam of electrons is transmitted through a specimen to form an image. The specimen is most often an ultrathin section less than 100 nm thick or a suspension on a grid. They can magnify objects up to 2 million times. It has high magnification and high resolution. TEM gives a two-dimensional view of the image.

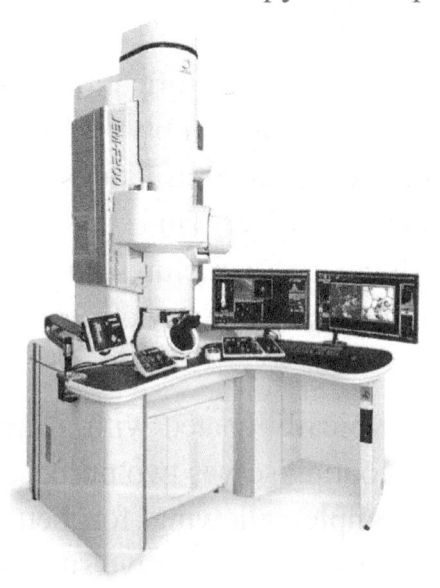

In TEM, the specimen is typically prepared as an ultrathin section, which is a slice of the specimen that is less than 100 nanometers thick. Alternatively, suspensions of particles or cells can be deposited on a grid for analysis The thinness of the specimen allows the electrons to pass through it with minimal scattering, resulting in clear and highly resolved images of the internal structures.

Advantage of TEM is its ability to achieve extremely high magnification. With the capability to magnify objects up to 2 million times, TEM enables researchers to explore the finest details of plant cells, organelles, and even individual macromolecules. This level of magnification is crucial for studying the subcellular structures, such as chloroplasts, mitochondria, endoplasmic reticulum, and cell walls, providing insights into their organization, function, and interactions within plant cells. TEM offers exceptional resolution, allowing for the visualization of minute structures within plant specimens. The high-resolution imaging capabilities of TEM enable researchers to study ultrastructural features such as membrane systems, vesicles, protein complexes, and even individual molecules. This level of resolution is essential for understanding the precise organization and arrangement of components within plant cells, shedding light on their physiological processes and cellular functions.

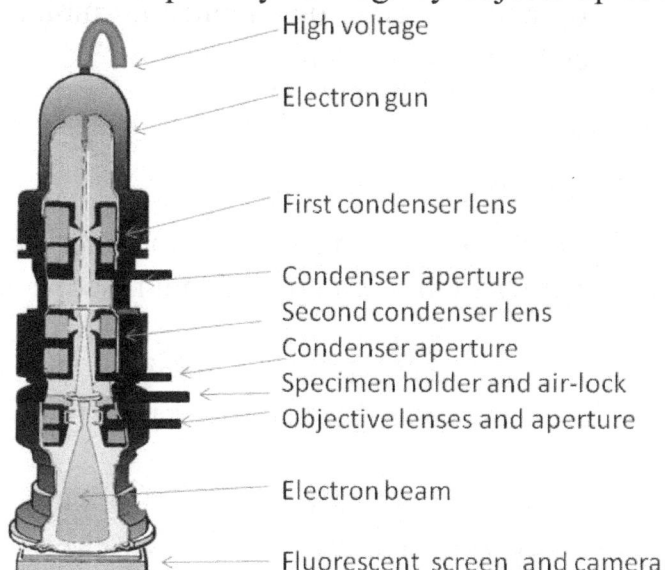

TEM provides a two-dimensional view of the image. The transmitted electrons are captured on a photographic film or an electronic detector, forming a projection image of the specimen's internal structures. This image represents a slice through the specimen, providing valuable information about the morphology, organization, and spatial relationships of the structures within the plane of the specimen.

The Transmission Electron Microscope has revolutionized our understanding of the internal structures and ultrastructure of plant cells. Its ability to visualize cellular components with high magnification and resolution has paved the way for research in plant biology, including investigations into organelle dynamics, cell development, molecular interactions, and cellular responses to environmental stimuli.

2.3 CONCEPTS IN MICROSCOPY

Some terminologies used in microscopy:

- Magnification is referring to the ratio of the size seen in the microscope to the actual size of the specimen. On a compound microscope it is usually between 4x and 100x
- Resolution is the clarity and detail seen. It is the minimal distance between two points in which they can be seen separately (i.e.: not blurred)
- Field of view refers to how much you actually see when looking in a microscope. As field of view increases, magnification decreases
- Depth of field is the number of layers you see. Total magnification is the product of the objective lens and the ocular (10x)
- Parfocal is a term used when describing compound microscopes. this means that the focus is maintained when changing the magnification. This way you don't have to re-focus when changing powers.
- Numerical Aperture (NA) is a measure of the lens' ability to gather and focus light. It determines the resolution of the microscope and ranges from 0 to 1. Higher NA results in higher resolution.
- Working distance is the distance between the objective lens and the specimen. It determines how thick the sample can be and how much space there is for manipulating the specimen.
- Refractive Index is a measure of how much a substance bends light. It affects the path of light through the lenses and specimen and can be adjusted with immersion oil.

2.3.1 PARTS OF MICROSCOPES

- Ocular lens (eyepiece): the lens at the top that you look through, this is point where the object is viewed, it contain systems of lens and gives the final magnification to the object. They are usually 10X or 15X power.

Microscope Parts

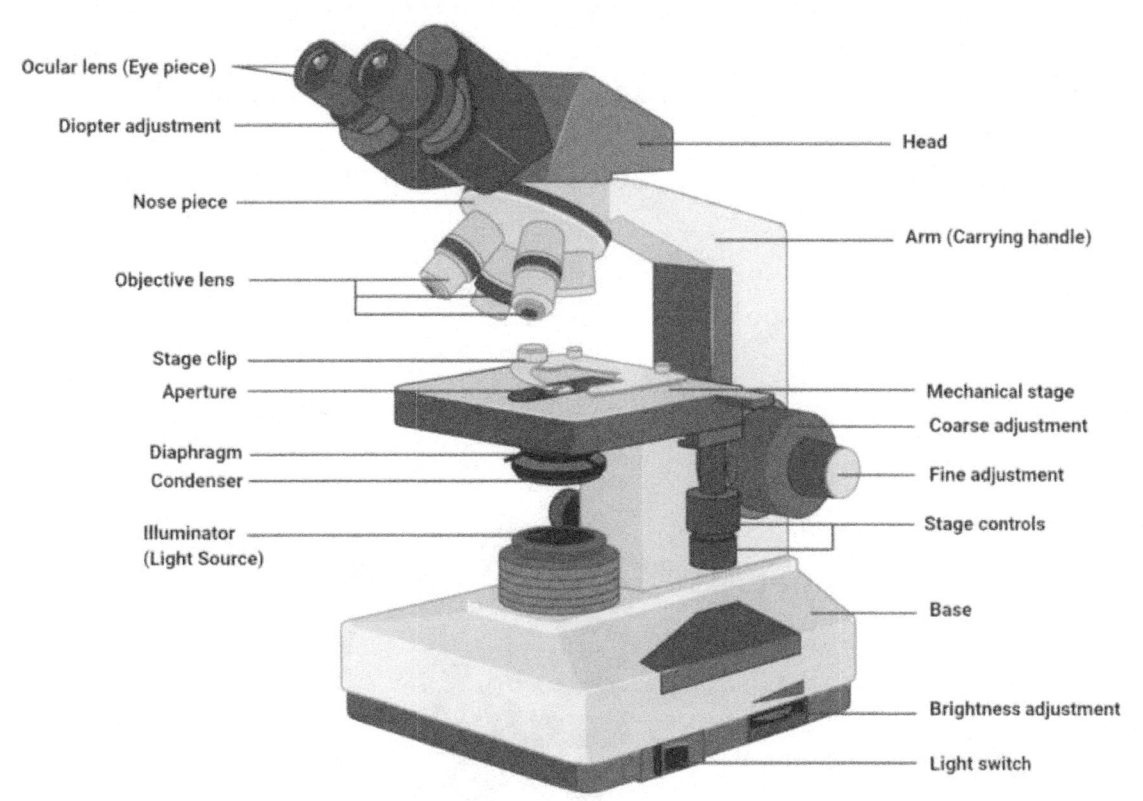

- Ocular lens (Eye piece)
- Diopter adjustment
- Nose piece
- Objective lens
- Stage clip
- Aperture
- Diaphragm
- Condenser
- Illuminator (Light Source)
- Head
- Arm (Carrying handle)
- Mechanical stage
- Coarse adjustment
- Fine adjustment
- Stage controls
- Base
- Brightness adjustment
- Light switch

- Body tube: this is a tube that connects the eyepiece to the objective lenses
- Revolving Nosepiece: houses the system of objective lenses, and it is used to change the lens to the required one
- Arm: This is used to carry the microscope its supports the tube and connects it to the base

- Objectives: This a system of lenses that first magnifies the object on the stage, usually you will find 3 or 4 objective lenses on a microscope. The shortest lens is the lowest power, the longest one is the lens with the greatest power.
- Stage: The flat platform where you place your slides
- Stage clips: This hold the slide containing the specimen on the stage
- Coarse adjustment knob: It is used to focus the specimen
- Fine adjustment knob: used to fine tune the focus on specimen
- Diaphragm: helps to adjust the circumference of light from source, it is part of condenser
- Light source: This is the source of light that illuminate the object, it could be mirror.
- Base: this the base of the microscope, it support the instrument.

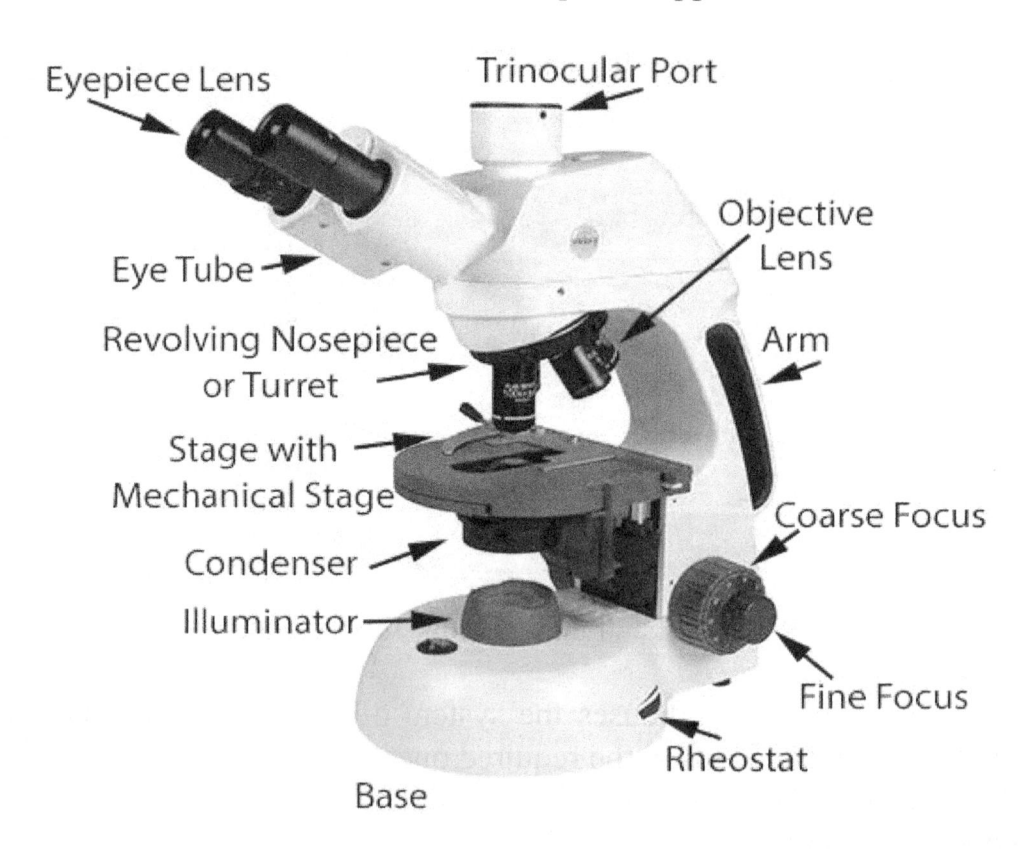

2.3.2 DIFFERENCES BETWEEN LIGHT AND TRANSMISSION ELECTRON MICROSCOPE:

Features	Transmission electron microscope	Light microscope
Radiation source	Electrons	Light
Wavelength	About 0.005nm	400 – 700nm
Maximum resolution	0.5nm	200nm
System of lenses	Electromagnets	Glass
Specimen	Non-living, dehydrated, small and very thin	Living or non-living
Specimen support	A small copper grid	Glass slide
Stains	Contain heavy metals to reflect electrons	Coloured dyes
Image	Black and white	Usually coloured

2.3.3 MICROSCOPIC TECHNIQUES

Preparation of materials for microscopy: investigations into biological specimens can be with a living tissue or dead (preserved) samples. There are two ways specimens can be prepared for light microscopic study.
1. Temporary slide preparation
2. Permanent slide preparation

In both cases, there are series of treatment (processes) the materials need to undergo before the slides can be prepared.

Procedures:

- Collection
- Fixation – FAA (Formalin Acetic Alcohol) mostly recommended for light microscopic study while a combination of primary (formaldehyde /glutaraldehyde) and secondary (Osmium tetroxide) fixatives are used in electron microscopy

- Dehydration in graded series of alcohol and finally in Xylene.
- Critical Point Drying – this is employed during SEM techniques
- Embedding – **wax** is used in light microscopy while **resin** is used for electron microscopy
- Sectioning – metal knives and ordinary microtome are used in light microscope & glass knives for preliminary observation and diamond knives and ultra-microtome are employed for real time electron microscopy
- Staining – series of coloured dyes that could reflect visible light are used for light microscopy while compounds of some heavy metals that are capable of reflecting electrons are employed in the case of electron microscopy
- Mounting – stained specimen are mounted on a glass slide and covered with a cover slip during light
- microscopy, the mounting medium could be glycerin if temporary slide is desire and on Canada balsam used for permanent slide. However, specimen for SEM are gold coated with while those for TEM are placed on copper grids before viewing
- Viewing of specimen

SOME STAINS AND THEIR USES:

STAIN	FINAL COLOUR ON TISSUE	USES
Aniline blue	Blue	Fungal hyphae, spores
Borax	Pink	Nuclei
Eosin	Pink or red	Cytoplasm (pink) cellulose (red)
Light/fast green	Green	Cytoplasm or cellulose
Methylene blue	Blue	Nuclei

Safranin	Red	Lignin, suberin, nuclei, cytoplasm
Iodine	Blue-black	Starch
Phoroglucinol + HCl	Red	Lignin
Aniline HCl or SO_4	Yellow	Lignin
Schultz's solution	Yellow	Lignin, cutin, suberin, protein

CHAPTER 3

BASIC PRINCIPLE OF GENETICS AND HEREDITARY {REPRODUCTION}

3.1 REPRODUCTION

There are different schools of thought or theories on the origin of life on Earth. These theories provide various explanations for how life first appeared and evolved over time. One such theory is the theory of eternity, which suggests that life has always existed in some form and has no specific origin. Another theory is the theory of special creation, which proposes that life was created by a divine being or supernatural force. The theory of catastrophism suggests that major catastrophic events played a significant role in shaping and diversifying life on Earth. Lastly, the theory of spontaneous generation, although dis-proven, once posited that life could spontaneously arise from non-living matter.

Despite the limited lifespan of individual organisms, life continues to persist on Earth through the process of reproduction. Reproduction is a vital mechanism that ensures the continuity of life from one generation to another. While humans may have an average lifespan of 70 to 80 years, life in general thrives and endures through the reproductive abilities of various organisms. For instance, plants that are visible during one season are replaced by younger individuals in the next season, ensuring the survival and propagation of their species.

Reproduction can occur through two main types: sexual and asexual reproduction. Sexual reproduction involves the fusion of genetic material from two parent organisms, typically resulting in offspring that possess a combination of traits from both parents. This method of reproduction promotes genetic diversity and enables the adaptation of species to changing environments.

In contrast, asexual reproduction does not involve the fusion of genetic material from two parents. Instead, it allows organisms to reproduce without the need for a partner. Asexual reproduction methods vary among different organisms and can include processes such as binary fission, budding, and vegetative propagation. Asexual reproduction is advantageous in stable environments as it allows for the rapid production of offspring that are genetically identical to the parent organism.

By employing these diverse reproductive strategies, living organisms have developed mechanisms to maintain the continuity of their existence on Earth. The ability to reproduce ensures the survival and adaptation of species, allowing life to persist and evolve over time. Understanding the intricacies of reproduction is essential for comprehending the vast array of life forms found on our planet.

- Sexual reproduction is a type of reproduction that involves the fusion of male an female gametes. Sexual reproduction allow genetic variability and makes organism adapt better to their environment

- Asexual reproduction is a type of reproduction that does not involve the fusion of gametes but allow an individual organism to reproduce a genetic replica of itself. This type of reproduction does not allow genetic variability

3.2 ASEXUAL REPRODUCTION

Vegetative reproduction is a type of asexual reproduction that involves the generation of a new organism from a vegetative part which could be the stem, leaf, mycelium or specialized propagules like adventitious buds, asexual spores, akinete, homogonia, corms, bulbs and suckers. It may also involve certain human control artificial processes such as layering, grafting, budding and rooting of cuttings of stems and leaves. In any of these methods a portion gets detached from the body of the parent plant which starts a new life in a suitable condition.

3.2.1 BUDDING:- this occurs in yeast, when one or more tiny outgrowths appear on one or more sides of the vegetative cell immersed in a sugar solution, which later get detached from the parent and start to live an independent life. Budding often occur continuously so that finally one more chains, sometimes sub-chains, of cells are formed. The individual cells of the chain separate from one another and form new yeast plants.

Budding, as observed in yeast, is a remarkable process that allows for the rapid expansion of their population. When a vegetative cell is immersed in a sugar solution, one or more tiny outgrowths, known as buds, emerge on one or multiple sides of the cell. Over time, these buds mature and eventually detach from the parent organism, embarking on an independent life of their own. Budding often occurs continuously, resulting in the formation of chains or sub-chains of cells. As the individual cells within the chain separate from one another, they have the capacity to develop into new yeast plants, thus perpetuating the life cycle.

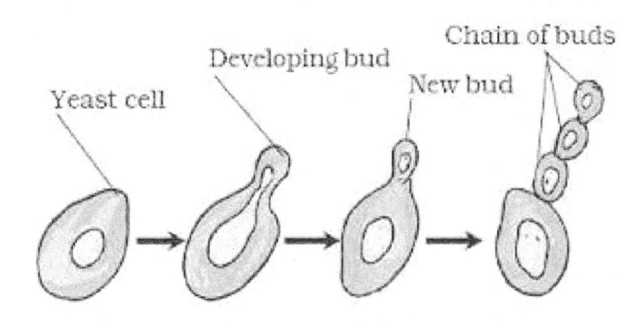

This method of asexual reproduction in yeast holds several advantages. Firstly, budding allows for a rapid increase in population size, as multiple buds can form simultaneously. This efficient mode of reproduction enables yeast to adapt swiftly to changing environments and exploit available resources. Secondly, the ability of yeast cells to detach and exist independently enhances their dispersal potential, allowing them to colonize new habitats. This dispersal mechanism aids in their survival and ensures the continuation of the yeast species.

Budding is not exclusive to yeast but is also observed in various other organisms across different taxonomic groups. In some multicellular organisms, such as hydra budding serves as a means of producing new individuals. During this process, a small outgrowth, or bud, forms on the body of the parent organism. The bud then develops into a miniature replica of the parent, ultimately detaching and establishing itself as an independent organism.

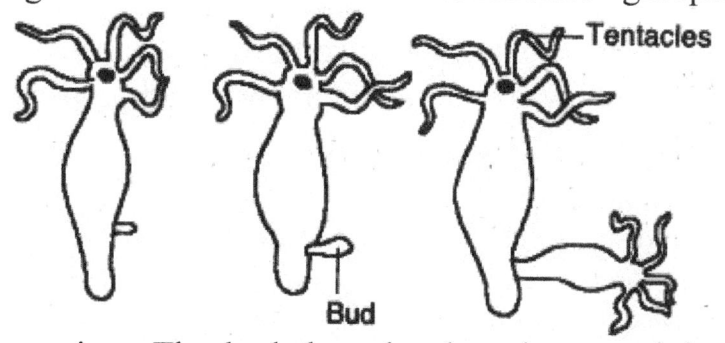

3.2.2 GEMMAE:-

Gemmae, a form of asexual vegetative reproduction, can be observed in certain mosses and liverworts, particularly in species like Marchantia. Gemmae are specialized structures that develop on the leaf branches or thalli of these organisms, serving as a means of propagation. In mosses and liverworts, gemmae are tiny, multicellular structures that contain all the necessary components to develop into a new individual. These structures are often produced in specialized receptacles or gemma cups. The gemmae are typically surrounded by protective structures and are capable of withstanding harsh environmental conditions, ensuring their survival until favorable conditions for growth arise.

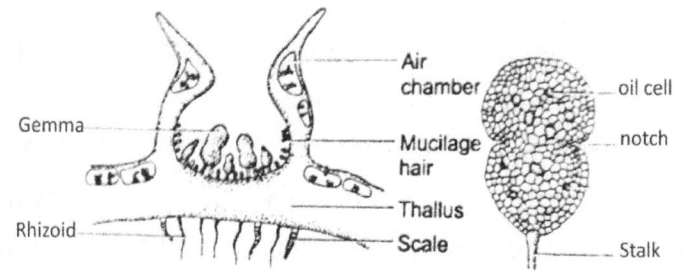

When conditions become favorable, the gemmae detach from the parent plant and are dispersed by various mechanisms such as water, wind, or animal activity. Once they land in a suitable environment, such as moist soil or a suitable substrate, the gemmae germinate and begin to develop into new individuals. The process of gemmae germination involves the growth of a new plant body, including roots, stems, and leaves, from the structures contained within the gemmae.

The ability of mosses and liverworts to reproduce through gemmae offers several advantages. Gemmae allow for the efficient and rapid propagation of these organisms. Since the gemmae are pre-formed structures containing all the necessary components for growth, they can quickly develop into new individuals without the need for extensive developmental processes. This enables mosses and liverworts to colonize new areas and expand their populations effectively. The dispersal mechanisms associated with gemmae contribute to the wide distribution of these organisms. By utilizing external agents such as water or wind, gemmae can be transported over long distances, allowing mosses and liverworts to establish themselves in new habitats and increase their chances of survival.

3.2.3 LEAF:- This occur in certain plant species, including ferns such as Adiantum caudatum, A. Lunulatum, and Polypodium flagelliferum. These ferns possess the ability to propagate using their leaf tips. Also Bryophyllum pinnatum, a succulent plant, which utilizes its leaf margins for vegetative propagation when they come into contact with the ground. As the leaf touches the ground the tip/margin strikes roots and form a bud. The bud grows in to a new plant. Though ferns reproduce vegetatively normally by their rhizomes.

In ferns, the process of leaf propagation involves the formation of new individuals from the tips of their leaves. When the leaf tip of a fern touches the ground, it has the ability to strike roots into the soil and initiate the development of a bud. As the bud grows, it eventually matures into a new plant, establishing an independent existence. Ferns typically reproduce vegetatively through their rhizomes, which are underground horizontal stems. However, the ability of certain fern species to propagate through leaf tips adds another layer of reproductive diversity to their life cycle.

Similarly, Bryophyllum pinnatum, also known as the "Mother of Thousands," employs leaf propagation as a means of asexual reproduction. Once the leaf margins of Bryophyllum pinnatum come into contact with the ground, they have the ability to strike roots and form buds. These buds then develop into new plants, each capable of repeating the process of leaf propagation. This fascinating method allows Bryophyllum pinnatum to spread and colonize new areas efficiently.

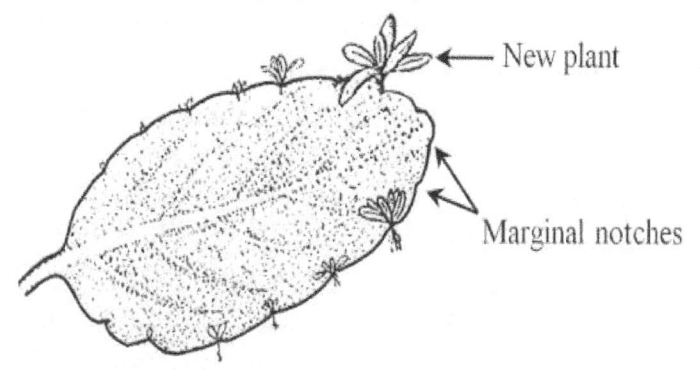

Leaf propagation provides advantages to these plant species. It allows for the rapid multiplication of individuals without the need for specialized reproductive structures or the involvement of sexual reproduction. By utilizing their leaves as propagative structures, ferns and plants like Bryophyllum pinnatum can establish new colonies and expand their populations effectively.

3.2.4 UNDERGROUND STEMS:-
Underground stems play a significant role in the asexual reproduction of numerous flowering plants. These specialized structures, such as Rhizome (ginger), the tuber (potato) bulb (onion),, serve as vital repositories for energy and nutrients, enabling the formation of new buds that can develop into complete plants.

One example of an underground stem used for asexual reproduction is the rhizome, which can be observed in plants like ginger. Rhizomes are horizontal, modified stems that grow beneath the soil surface. Along the length of the rhizome, nodes and internodes are present, and from these nodes, new buds emerge. These buds have the potential to develop into shoots and roots, ultimately giving rise to new plants. The growth of these new shoots and roots from the rhizome ensures the propagation and expansion of the plant population.

Tubers, such as the commonly known potato, is a type of underground stem involved in asexual reproduction. Tubers store nutrients and energy reserves serving as a source of sustenance for the plant. The "eyes" on the surface of a tuber are actually dormant buds. When conditions are favorable, these buds sprout and develop into new shoots, which eventually grow into complete plants. Through the formation of buds on tubers, plants can reproduce asexually, effectively producing clones of the parent plant.

Bulbs, like the onion, are underground stems modified for storage. They consist of layers of specialized leaves known as scales, with each scale containing a bud within it. The buds found in bulbs have the capacity to generate new shoots and roots. As the bulb matures, these buds undergo growth and produce new plants. This process ensures the survival and propagation of the plant species.

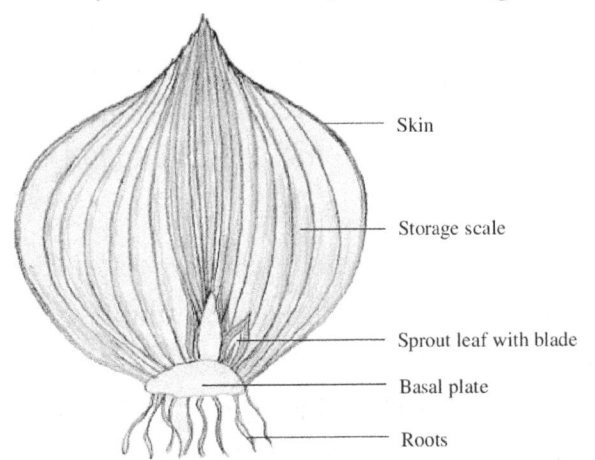

The use of underground stems for asexual reproduction offers numerous advantages to flowering plants. It allows for the efficient production of offspring by utilizing stored nutrients and energy reserves. Additionally, these underground stems aid in vegetative propagation, allowing plants to expand their population without relying on sexual reproduction or the dispersal of seeds.

3.2.5 SUB-AERIAL STEMS- The runner, the stolon, the offset and the suckers are sued by some plant such as *Colocasia* species, water lettuce (*Pistia*) *Chrysanthemum* species, and *Musa* species for vegetative propagation.

Runners are sub-aerial stems that grow horizontally above the ground, typically at or just below the soil surface. Plants like Colocasia species utilize runners for vegetative propagation. Along the length of the runner, nodes and internodes are present, and from these nodes, new roots and shoots emerge. These adventitious roots anchor the runner to the soil, while the shoots develop into new plants, separate from the parent plant. Runners allow for the rapid expansion of the plant population by producing genetically identical clones.

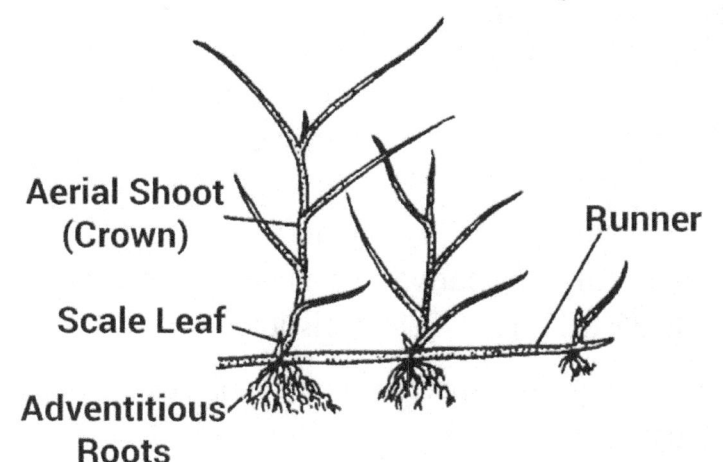

Stolons, similar to runners, are horizontal stems that grow above the ground. Plants like water lettuce (Pistia) employ stolons for vegetative propagation. Stolons extend laterally from the parent plant and produce new adventitious roots and shoots at their nodes. These roots anchor the stolon in the soil, and the shoots develop into independent plants. Stolons contribute to the spread of the plant population by facilitating the establishment of new individuals in suitable habitats.

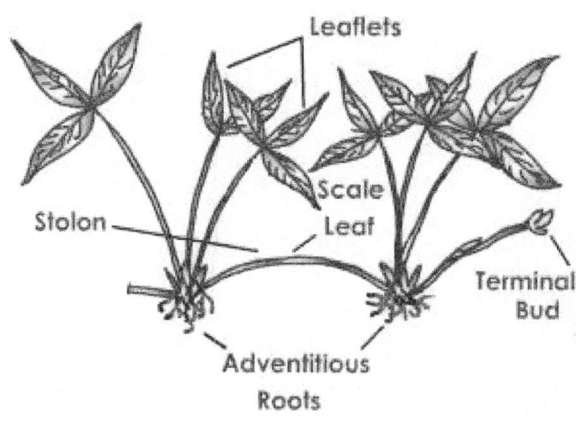

Offsets, also known as offshoots or plantlets, are small, self-contained units that develop along the base of certain plants, such as Chrysanthemum species. These offsets are essentially miniature versions of the parent plant, complete with roots stems, and leaves. Once the offsets have developed sufficiently, they can be detached from the parent plant and grown independently. This method of vegetative propagation ensures the preservation of the parent plant's traits and characteristics.

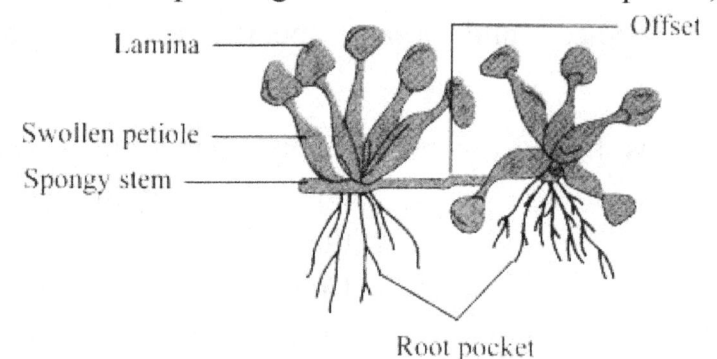

Eichhornia showing offset

Suckers, observed in plants like Musa species (bananas), are shoots that emerge from the base of the main stem or the rhizome. Suckers grow vertically and can develop into fully mature, self-supporting plants. They possess their own root systems, allowing them to draw nutrients and water independently. Suckers contribute to the vegetative propagation of banana plants and play a crucial role in their commercial cultivation.

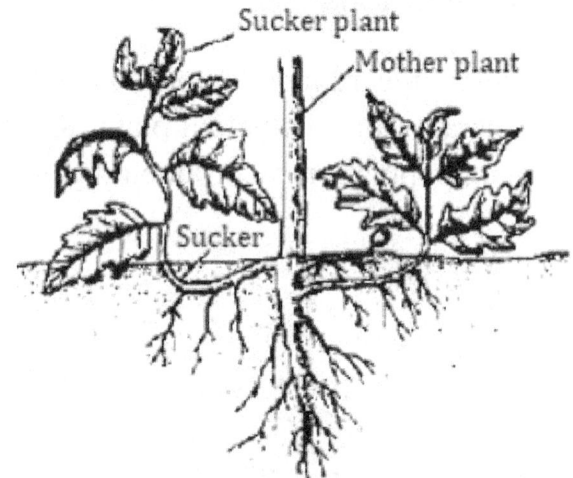

The utilization of sub-aerial stems for vegetative propagation allows for the production of offspring that are genetically identical to the parent plant, preserving desirable traits and characteristics. Sub-aerial stems provide an efficient means of colonization and expansion, enabling plants to occupy new areas and compete for resources effectively.

3.2.6 BULBILS in garlic (*Allium sativum*) some of the lower flowers of the inflorescence become modified into small multicellular bodies known as bulbils, which fall into the ground and grow as a new plant. Sometimes they grow to some extent on the parent plant before falling to the ground. Bulbils are also produce in the leaf axil of wild yam, *Dioscorea bulbifera* and *Lilium bulbiferum*. In pineapple (*Ananas*), the inflorescence generally ends in a reproductive bud, but in some varieties of pineapple the inflorescence becomes surrounded at the base by a whorl of such buds and also crowned by a few of them.

In garlic, some of the lower flowers of the inflorescence undergo modification, transforming into bulbils. These bulbils are small, self-contained structures that resemble miniature bulbs. They possess the necessary components for growth and development, including meristematic cells and stored nutrients. As the bulbils mature, they detach from the parent plant and fall to the ground. There, they have the potential to take root and grow into new garlic plants. In some cases, the bulbils may develop to some extent while still attached to the parent plant before eventually separating and continuing their growth independently.

Wild yam (Dioscorea bulbifera) and Lilium bulbiferum also produce bulbils In these plants, bulbils are formed in the leaf axils, which are the angles between the upper surface of the leaf and the stem. These axillary bulbils possess the capacity to develop into new plants, serving as a means of vegetative propagation. Once detached from the parent plant, the bulbils have the potential to take root in the soil and initiate growth, eventually establishing themselves as independent individuals.

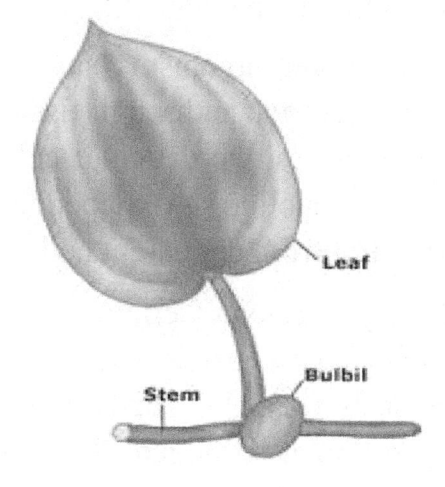

In certain pineapple varieties, the inflorescence terminates with a reproductive bud. However, in some specific pineapple cultivars, the inflorescence undergoes a unique development. It becomes surrounded at the base by a whorl, or cluster, of reproductive buds, and a few buds may also crown the top of the inflorescence. These buds can be considered as bulbils and have the potential to generate new pineapple plants. Through the production of these bulbil-like structures, certain pineapple varieties have an additional means of vegetative propagation, ensuring the continuation of the species.

The presence of bulbils in these plant species offers advantages for vegetative propagation. Bulbils provide a means for the plants to produce genetically identical offspring, allowing for the preservation and perpetuation of favorable traits. Furthermore, bulbils contribute to the colonization and expansion of these plant species, facilitating their survival and establishment in new environments.

3.2.7 FISSION:- Fission is a form of asexual reproduction that involves the parent cell splitting into two new cells. This process results in the formation of two daughter cells, each containing the complete set of genetic material inherited from the parent cell. Over time, these daughter cells grow and develop into independent organisms, resembling the parent cell.

Fission is commonly observed in various unicellular organisms such as algae, fungi, and bacteria. Many unicellular algae reproduce through fission where the parent cell undergoes division, giving rise to two daughter cells. These daughter cells then continue to grow and function as independent organisms, carrying forward the genetic information from the parent cell.

Fungi also exhibit asexual reproduction through fission. Certain fungal species, such as yeast, can reproduce by binary fission. The parent cell undergoes division, leading to the formation of two daughter cells, which can subsequently grow and develop into mature fungal individuals. Fission allows fungi to rapidly multiply their population and colonize new habitats.

Bacteria, being single-celled organisms, commonly reproduce through fission as well. During bacterial fission, the parent cell replicates its genetic material and divides into two daughter cells. These daughter cells inherit the genetic material and cellular components of the parent cell, allowing them to function independently and continue the life cycle of the bacterium. Bacterial fission is a vital mechanism for bacterial proliferation and plays a significant role in their ability to adapt and survive in diverse environments.

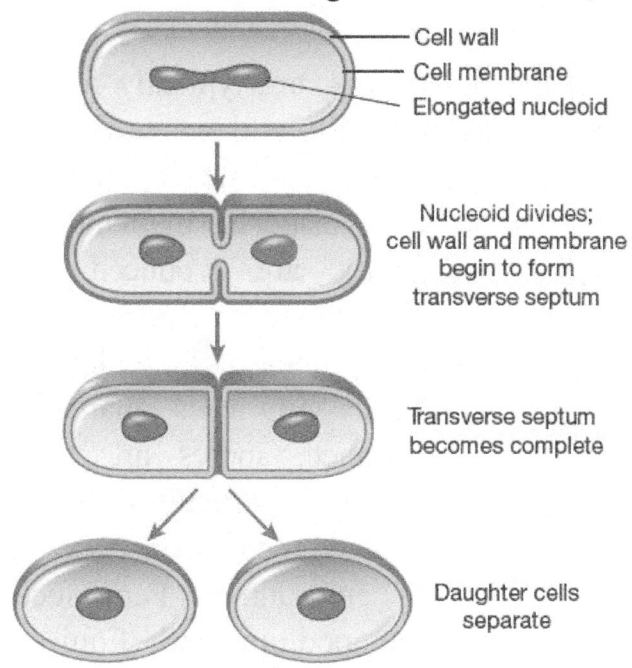

Fission allows for rapid population growth and colonization, as each parent cell can give rise to two or more daughter cells. Additionally, fission enables the inheritance of genetic material without the need for genetic recombination or the involvement of other individuals, ensuring the preservation of favorable traits within the population.

8. SPORE FORMATION:- Spore formation is a common method of asexual reproduction observed in various organisms. Spores are unicellular and microscopic reproductive units that can develop independently without fusing with another unit. They serve as a means of dispersal and can give rise to new individuals. Among spores, there are two main types: motile spores and non-motile spores. Motile spores, also known as zoospores, possess cilia or flagella that allow them to move in a watery environment. Many algae and fungi produce zoospores as part of their asexual reproduction process. These spores are often produced in large numbers and swim in water for a certain period, propelled by their cilia. Eventually, they come to rest and undergo germination, developing into independent individuals. An example of a plant that produces zoospores is Ulothrix, where zoospores are formed abundantly.

Non-ciliate, non-motile spores, on the other hand, are typically found in terrestrial fungi. These spores are lightweight, dry, and possess a tough outer coat, which allows for easy dispersal by wind. True spores are produced by a sporophyte, which is the diploid phase of the plant's life cycle. In moss plants, for instance, spores are formed within structures called sporangia and are released into the environment. Ferns such as Lycopodium and Equisetum also reproduce asexually through spores. These plants are homosporous, meaning they produce only one type of spore.

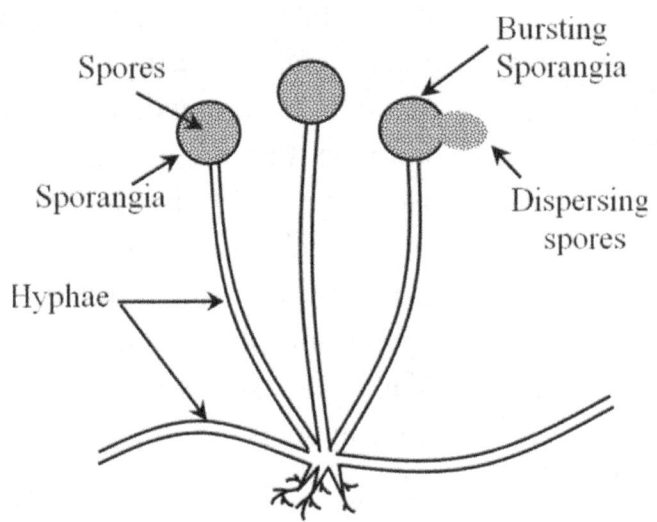

In more advanced plant groups such as Selaginella and flowering plants (gymnosperms and angiosperms), a heterosporous reproductive strategy is observed. These plants produce two types of spores: microspores (male) and megaspores (female). Microspores develop into male gametophytes, while megaspores develop into female gametophytes. This heterospory allows for specialized reproductive structures and enhances the efficiency of sexual reproduction.

The production of spores as a means of asexual reproduction provides several advantages to organisms. Spores are highly resistant structures that can withstand unfavourable conditions, facilitating dispersal to new habitats Spore formation also allows for the preservation and dispersal of genetic material, ensuring the survival and propagation of the species.

3.3 ARTIFICIAL METHODS OF VEGETATIVE PROPAGATION

Artificial methods of vegetative propagation are techniques employed by gardeners and horticulturists to reproduce plants with desired characteristics efficiently. These methods involve the separation or detachment of a portion of the parent plant and promoting its growth into a new, independent individual. These techniques are particularly valuable in flowering plants, where maintaining the traits of the parent plant is crucial for preserving desirable features.

These methods could be through cuttings, layering, grafting, Gootee etc. In most of these methods, a portion of the parent plant is separated by special methods. It is important to note that in flowering plants the methods of vegetative propagation are diverse. The offspring look like the parent plant in all respects so gardeners often use these methods for quick multiplication of flowers in their gardens.

3.3.1 CUTTINGS

Cuttings are a widely used and effective method of vegetative propagation in plants. Stem cuttings involve the removal of a portion of the stem from a parent plant and placing it in a suitable environment to develop roots and grow into a new plant. Many plant species, such as Cassava, Sugar cane, Moringa, Coleus, and many others, can be successfully propagated through stem cuttings.

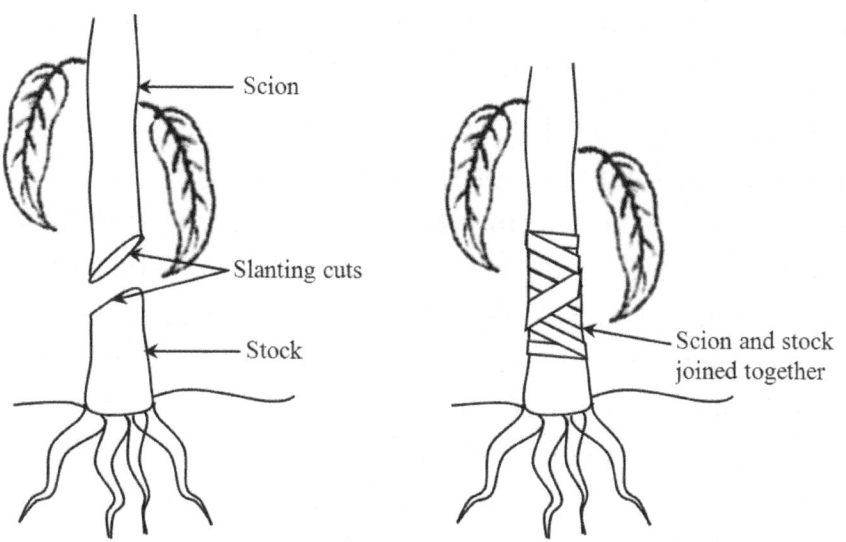

When stem cuttings are placed in moist soil or a rooting medium, they create favourable conditions for root development. Adventitious roots, which are roots that form from non-root tissues, emerge from the base of the stem cutting. These roots enable the cutting to absorb water and nutrients from the soil, supporting its growth. Additionally adventitious buds present on the stem cutting can develop into new shoots and leaves, further contributing to the growth and development of the new plant.

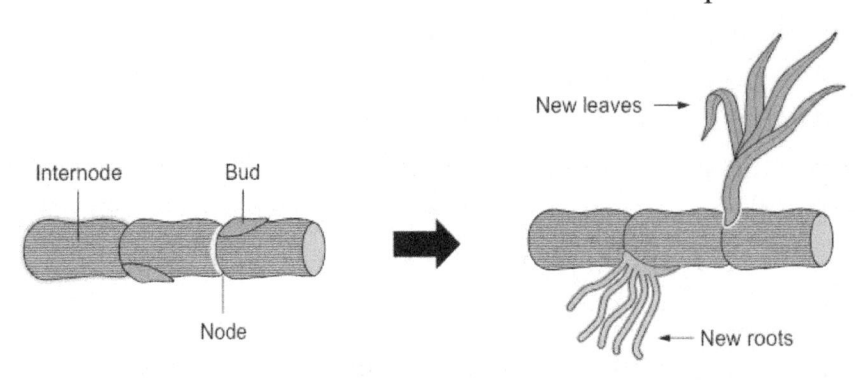

Root cuttings are another method used to propagate certain plants, such as lemon. In this technique, a portion of the root system is cut and planted in moist soil. The root cutting produces both roots and shoots, allowing the development of a new plant. Root cuttings are particularly useful for plants that produce vigorous and extensive root systems, as they can quickly establish themselves and grow into healthy individuals.

The success of stem and root cuttings largely depends on providing the appropriate environmental conditions for rooting and growth. Factors such as moisture, temperature, and hormone treatments can significantly influence the rooting process. Gardeners and horticulturists often employ techniques such as misting, using rooting hormone powders or gels, and maintaining optimal temperature and humidity levels to enhance the chances of successful root formation. Cuttings allow for the preservation of specific traits from the parent plant, ensuring the reproduction of desired characteristics such as flower color, leaf shape, or fruit quality. Cuttings also provide a faster means of propagation compared to growing plants from seeds, as they skip the germination and early seedling stages.

3.3.2 LAYERING

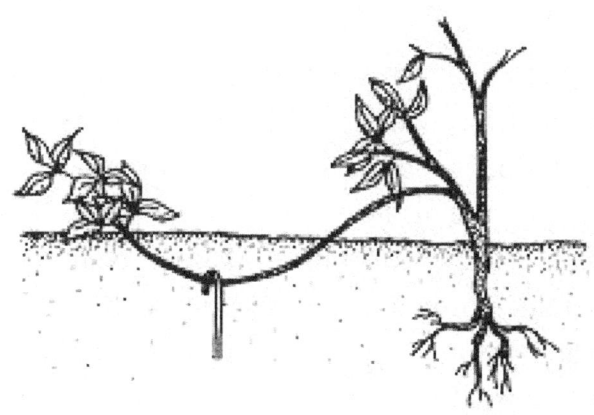

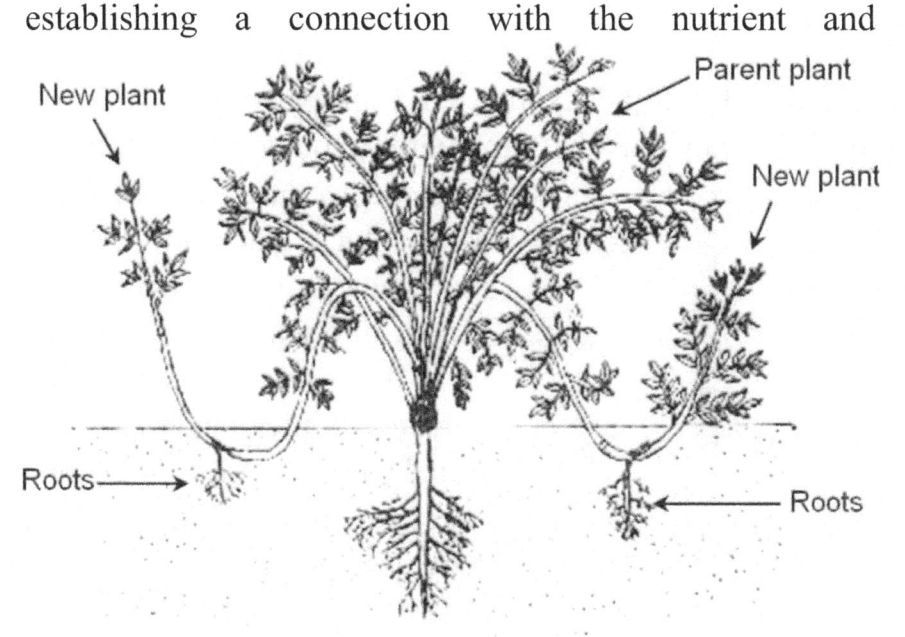

Layering is an effective method of vegetative propagation that involves utilizing the natural ability of plants to develop roots from branches while still attached to the parent plant. In this method, a low-lying branch of the plant is carefully selected, and a section of the bark, approximately 2.5-5cm in length, is removed. This creates a wound that stimulates the formation of roots. The wounded section is then placed in contact with the soil and covered, while the upper part of the branch remains above the ground. As time progresses, the branch gradually develops adventitious roots from the wounded area. These roots penetrate the surrounding soil, establishing a connection with the nutrient and water supply. Simultaneously the upper part of the branch continues to receive nutrients and energy from the parent plant allowing it to grow and develop leaves, shoots, and eventually become an independent plant. After a period of approximately 2-4 months, the rooted portion of the branch, now equipped with its own root system, can be separated from the parent plant and transplanted into a new location. This method of propagation is particularly successful in plants such as Lemon, Grape-vine, Ixora, and Rose.

Layering takes advantage of the plant's own resources, utilizing its natural ability to produce roots from branches, which increases the chances of successful establishment and growth. Second, layering allows for the production of rooted plants while they are still attached to the parent plant, providing them with a continuous supply of nutrients and water. This promotes faster and healthier growth compared to starting with detached cuttings. Layering enables the propagation of specific varieties or cultivars that may be difficult to reproduce accurately through other methods, such as seeds. This ensures the preservation of desirable characteristics, including flower color, fruit quality, or growth habit. Layering also allows for the production of a larger number of plants from a single parent, making it an efficient and cost-effective means of propagation.

3.3.3 GRAFTING

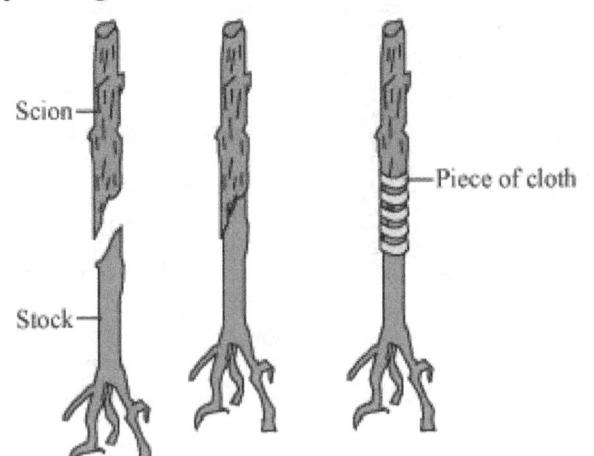

Grafting is a specialized method of vegetative propagation that involves joining a small branch, known as the scion, with a rooted plant of the same or closely related species, known as the stock. The aim is to create an organic union between the two plants, allowing them to grow as one. The scion retains all its unique qualities, such as specific fruit traits or ornamental characteristics, while the stock provides physical strength and a reliable root system. The process of grafting begins by carefully selecting a healthy scion, which is a portion of the plant that possesses the desired qualities. The scion is typically a young shoot or bud from a well-performing plant. The stock, on the other hand, is chosen for its robust root system and ability to support the growth of the scion. It may be a separate plant or a pre-established rootstock.

To achieve successful grafting, the scion and stock are prepared by making precise cuts that match each other, ensuring optimal contact between the two tissues. The scion is then firmly inserted into a corresponding opening or cleft in the stock. As the graft heals, the vascular tissues of the scion and stock begin to connect and fuse together, enabling the flow of water, nutrients, and other essential substances between the two plants.

One of the key advantages of grafting is the ability to propagate plants while maintaining the desired characteristics of the scion. This is especially valuable in fruit and ornamental shrubs and trees, where specific traits like fruit flavor, color, or disease resistance are crucial. Grafting allows growers to reproduce these desirable traits consistently.

Several grafting methods exist, each suited to different plants and purposes. Inarching involves connecting the scion and stock by growing them in close proximity until they naturally fuse together.

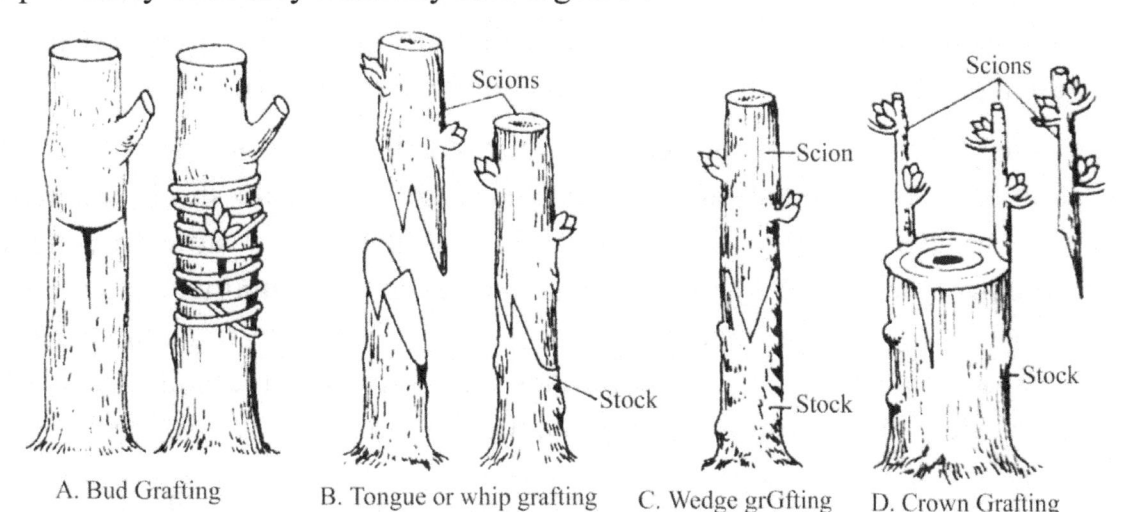

A. Bud Grafting   B. Tongue or whip grafting   C. Wedge grGfting   D. Crown Grafting

Bud grafting involves inserting a single bud from the scion into the stock. Tongue grafting, wedge grafting, and crown grafting are other common techniques that involve more intricate cuts and alignments to promote successful graft union.

Grafting requires precision, patience, and careful consideration of plant compatibility and grafting techniques. It is a skilled technique practiced by horticulturists, gardeners, and farmers to create new plant varieties, rejuvenate older plants, or propagate plants that are challenging to reproduce through other methods.

3.3.4 GOOTEE OR GOOTYING

Gootee is also known as layering and is a useful method for propagating plants that have long and flexible stems. This technique can be done in different ways depending on the type of plant and its growth habit. For instance, in some plants like Wisteria, a single long stem is used for gootee, while in others like Honeysuckle, multiple stems are used to propagate.

Gootee is a simple and effective method that requires little effort. It can be done in the field or in a garden, and it can be used to propagate large numbers of plants. This technique is particularly useful for plants that are difficult to propagate by other methods, such as cuttings or grafting. It is also useful for producing new plants that are genetically identical to the parent plant, which is important for maintaining plant varieties.

Gootee has been used for centuries and is still widely practiced today. In addition to vines and climbers, gootee can also be used for shrubs and trees. It is an ideal method for propagating plants that are slow-growing or difficult to propagate by other means. However, it is important to note that gootee is not always successful, and it may take some time for the new plant to develop. It is also important to ensure that the soil around the buried stem remains moist to promote root growth.

3.4 SEXUAL REPRODUCTION

Sexual reproduction is a fundamental process in the life cycle of angiosperms, or flowering plants. It involves the fusion of male and female gametes, resulting in the formation of a zygote and eventually leading to the development of a new plant. This intricate process takes place within the structures of a flower, specifically in the ovule.

The sexual reproduction in angiosperms is characterized by the alternation of generations, which involves a cycle between the gametophyte phase and the sporophyte phase. The gametophyte phase is haploid, denoted as n, meaning it contains half the number of chromosomes compared to the sporophyte phase, which is diploid and denoted as 2n. The cycle begins with the production of gametes through a process called meiosis, where the number of chromosomes is halved. The male gametophyte, or pollen, contains the male gametes or sperm, while the female gametophyte, located within the ovule, contains the egg.

The sexual reproduction process in angiosperms starts with pollination, where pollen grains are transferred from the anther (the male reproductive organ) to the stigma (the female reproductive organ) of the same or a compatible flower. This can be achieved through various agents such as wind, water, or pollinators like insects, birds, or mammals. Once pollination occurs, the pollen grain germinates and develops a pollen tube that grows down the style, eventually reaching the ovary.

Within the ovary, the pollen tube releases the male gametes, which then travel through the style and reach the embryo sac within the ovule. Fertilization takes place when one of the male gametes fuses with the egg cell, forming a zygote. This process is known as double fertilization and is a unique characteristic of angiosperms. Additionally, the other male gamete fuses with two polar nuclei within the embryo sac, resulting in the formation of endosperm, a nutrient-rich tissue that nourishes the developing embryo.

Following fertilization, the zygote develops into an embryo, and the ovule matures into a seed. The seed contains the embryo, endosperm, and protective seed coat. When the seed germinates under favorable conditions, the embryo resumes growth, and a new plant emerges. This new plant will go through its own life cycle, continuing the process of sexual reproduction and perpetuating the species.

Note that sexual reproduction in angiosperms contributes to genetic diversity. Through the fusion of genetic material from two different parent plants, variations occur in the offspring, allowing for adaptation to changing environmental conditions and evolution over time.

ALTERNATION OF GENERATION

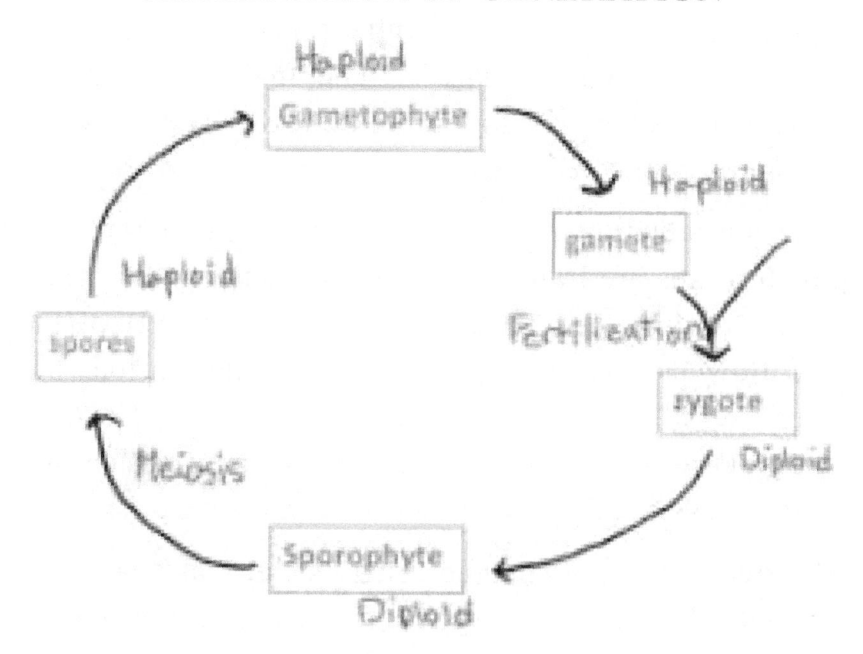

DIAGRAM ON ALTERNATION OF GENERATION

CHAPTER 4

CELL DIVISION

One unique characteristics of all living organisms is that they are all made up of cell. Some are unicellular while others multicellular. In other for living organisms to grow and reproduce their cells need to divide. There are two main types of cell division; mitotic cell division which occurs during growth as well as asexual reproduction and meiotic cell division which occur during gamete formation prior to sexual reproduction.

Characteristics of a cell: a cell consist of two main parts, the cytoplasm and the nucleus. The nucleus contain the genetic material the chromosome which carries the genes which expresses genetic information

4.1 CHROMOSOME DESCRIPTION

The chromosome is a double thread like structure with a constriction called the centromere which divides the chromosome into upper and the lower arms. Each arm is made up of two chromatids. Chromatids of the same chromosome are called sister chromatids while chromatids of different chromosome are called non sister chromatids. The chromosome of all living organisms are made up of the same chemical constituent. There are four main types of chromosomes based on the location of their centromere: metacentric, submetacentric, acrocentric and telocentric chromosome.

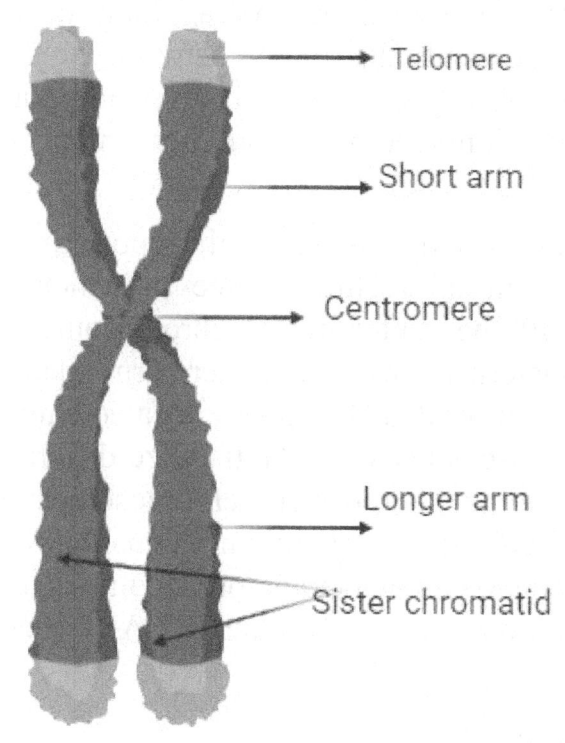

Metacentric chromosome: is a chromosome in which the upper arm is equal to the lower arm in length or chromosome in which the centromere is located at the center of the chromosome. A metacentric chromosome is characterized by having its upper arm equal in length to the lower arm or by having the centromere positioned at the center of the chromosome. This type of chromosome exhibits a balanced distribution of genetic material between its arms. The centromere, which plays a crucial role in chromosome segregation during cell division, is located at a central position enabling the arms to maintain relative symmetry. Metacentric chromosomes are commonly found in many species, including humans, and contribute to the overall stability and functionality of the genome

Submetacentric chromosome: is a chromosome in which the centromere is slightly off the center or chromosome in which the upper arm is a little shorter than the lower arm. A submetacentric chromosome is a type of chromosome that possesses a centromere slightly off-center or exhibits an upper arm that is slightly shorter than the lower arm. This slight asymmetry in arm length or centromere position distinguishes submetacentric chromosomes from their metacentric counterparts. As a result of this asymmetry, the genetic material is not evenly distributed between the arms of the chromosome. Submetacentric chromosomes are observed in various organisms and can impact gene expression, recombination rates, and overall chromosomal structure

Acrocentric chromosome: is a chromosome in which the upper arm is extremely shorter than the lower arm or a chromosome in which the centromere is located completely off the center of the chromosome. An acrocentric chromosome is characterized by an upper arm that is significantly shorter than the lower arm or by a centromere located entirely off-center. This structural arrangement creates a substantial size difference between the arms of the chromosome. Acrocentric chromosomes are typically found in the genomes of certain species, including humans, where they play a crucial role in the formation of nucleolar organizing regions (NORs). These NORs are regions responsible for ribosomal RNA synthesis, making acrocentric chromosomes vital for protein synthesis and cellular function

Telocentric chromosome: is a chromosome in which the centromere is located at one end of the chromosome i.e. the chromosome has only one arm. A telocentric chromosome is a type of chromosome that possesses a centromere located at one end, resulting in the chromosome having only one arm. The centromere's positioning at the end of the chromosome creates a distinct morphology, where one arm is absent. Telocentric chromosomes are relatively rare in most organisms, but they can be found in certain species. Although they lack the symmetrical structure seen in metacentric and submetacentric chromosomes, telocentric chromosomes still carry genetic information and contribute to the overall genomic makeup of an organism

CLASSIFICATION OF CHROMOSOMES
BASED ON THE POSITION OF CENTROMERE

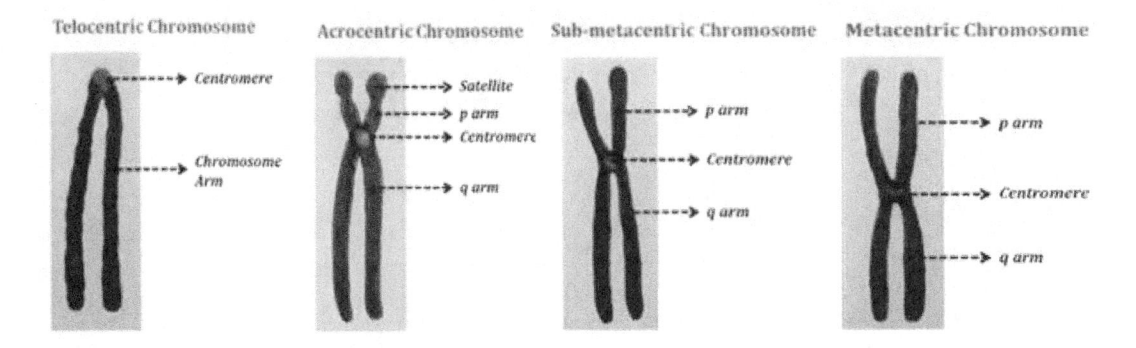

4.2 MITOTIC CELL DIVISION (MITOSIS)

This is a type of cell division that occurs during growth in higher organism immediately after fertilization (mitosis start). A single zygote divides in to 2, then 4, then 8, then 16 and so on, the cell in the organism start to multiply and the organism start to grow.

4.2.1 CHARACTERISTICS OF MITOSIS

Two daughter cells are formed at the end of the division.
Each of the daughter cell is genetically similar to the parent cell.
Each daughter cell is a diploid.
It is an equational division.
It involves one karyokinesis and one cytokinesis.
It occurs during growth and asexual reproduction.

The process of mitosis include the following stages: Prophase, metaphase, anaphase and telophase. Before a cell start to divide it goes through an interphase stage during which the cell get ready for the division.

INTERPHASE: Before a cell initiates the process of division, it undergoes a crucial stage called interphase, during which it prepares itself for division and carries out various essential activities. Interphase is the longest phase of the cell cycle and consists of three distinct stages: G1 (Gap 1), S (Synthesis) and G2 (Gap 2).

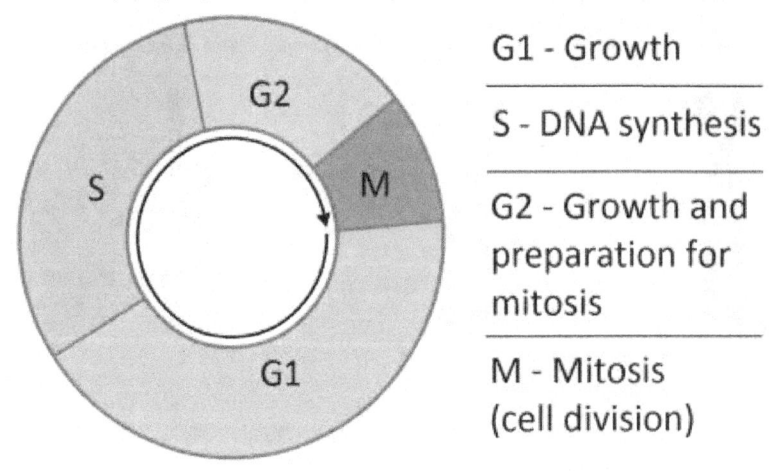

G1 - Growth

S - DNA synthesis

G2 - Growth and preparation for mitosis

M - Mitosis (cell division)

During G1, the cell undergoes a period of growth and performs necessary metabolic activities. Enzymes and proteins are synthesized to support the cell's functions and energy activation processes take place to provide the necessary energy for cellular activities. The cell prepares itself for DNA synthesis in the subsequent S phase.

The S phase of interphase is characterized by DNA synthesis. During this phase, the cell's chromosomes are replicated, resulting in the formation of identical copies of the DNA molecules. Each chromosome becomes double-stranded as the genetic material is duplicated, ensuring that each daughter cell receives an accurate and complete set of genetic information during division.

Following the S phase, the cell enters the G2 phase. In G2, the cell undergoes further growth and prepares for the actual process of cell division It checks and verifies the accuracy of DNA replication, repairs any DNA damage, and ensures that all necessary cellular components are present for division to occur successfully.

Note that during interphase, the cell is not actively dividing but rather undergoing essential preparations for division. Interphase is a critical period for the cell's growth, metabolism, and DNA replication, and it plays a vital role in maintaining the integrity and functionality of the cell. Cells spend a significant amount of time in interphase compared to the actual process of division, as this preparation phase is crucial for the successful progression of cell division.

PROPHASE

At prophase which is the first stage of mitosis. Chromosomes appear as a thin thread like structure and each of the chromosomes cannot be easily distinguished by observation. The nuclear membrane is intact but later get disorganized. The chromosome has two chromatids at this stage and DNA content remain at 4C.

During the prophase stage of mitosis, the cell undergoes significant changes in preparation for chromosome segregation and nuclear division. Key features of prophase:

Chromosome Condensation: The chromatin, which normally appears as a diffuse and thread-like structure in the nucleus during interphase, undergoes extensive condensation. As a result, the chromosomes become more compact and visible under a microscope. At this stage, each chromosome consists of two identical sister chromatids held together by a specialized region called the centromere.

Chromosome Distinguishability: In early prophase, the individual chromosomes may not be easily distinguished as discrete structures due to their condensed and tangled nature. However, as prophase progresses, the distinct chromosomes become more apparent and can be identified based on their size, shape, and position within the nucleus.

Nuclear Membrane Disorganization: Initially, the nuclear envelope or membrane remains intact during early prophase. However, as the prophase continues, the nuclear membrane starts to disintegrate and break down into small vesicles. This breakdown allows the spindle apparatus to access the chromosomes during subsequent stages of mitosis.

Formation of Spindle Apparatus: As the nuclear membrane disorganizes, the spindle apparatus begins to assemble. The spindle fibers, composed of microtubules, extend from structures called centrosomes located at opposite poles of the cell. The spindle fibers play a critical role in capturing and aligning the chromosomes during metaphase.

DNA Content: The DNA content in each chromosome remains the same as in the previous interphase, known as 4C DNA content. This indicates that DNA replication has already occurred during the preceding S phase, and each chromosome now consists of two identical sister chromatids.

Prophase marks the initiation of the mitotic process and prepares the cell for the subsequent stages of mitosis. The condensation of chromosomes, disorganization of the nuclear membrane, and assembly of the spindle apparatus are vital steps that ensure proper chromosome alignment and distribution during the later stages of cell division.

METAPHASE

After prophase the cell move to metaphase. At this stage the chromosome are maximally contracted, nuclear membrane completely disappears, spindle fibers appear and chromosomes line up at the equator. The DNA content remain at 4C.

During the metaphase stage of mitosis, the cell continues to undergo changes in preparation for the separation of sister chromatids. key features of metaphase:

Chromosome Alignment: The chromosomes, which have undergone maximum condensation during prophase, now align themselves along the equatorial plane of the cell, also known as the metaphase plate. The alignment occurs due to the interaction between the chromosomes' centromeres and the spindle fibers.

Disappearance of the Nuclear Membrane: By metaphase, the nuclear membrane completely disappears. This allows the spindle fibers to interact directly with the chromosomes, facilitating their proper alignment and subsequent separation.

Formation of Spindle Fibers: The spindle fibers, composed of microtubules, continue to grow and extend from the centrosomes located at opposite poles of the cell. These fibers attach to the chromosomes at their kinetochores, which are specialized protein structures located at the centromeres.

Chromosome Maximal Contraction: During metaphase, the chromosomes reach their maximum level of condensation and become highly visible under a microscope. This maximal contraction ensures that the sister chromatids are closely associated and can be properly separated during the subsequent stage of mitosis.

DNA Content: The DNA content of the chromosomes remains unchanged from the previous stages of mitosis, maintaining a 4C DNA content. This indicates that the DNA has already been replicated during the preceding S phase, and each chromosome now consists of two identical sister chromatids.

Metaphase ensures that the chromosomes are properly aligned at the metaphase plate before their subsequent separation during anaphase. The disappearance of the nuclear membrane and the alignment of the chromosomes along the equator are essential for the accurate distribution of genetic material to the daughter cells.

ANAPHASE

At this stage each individual chromosome breaks open at the centromere and move to opposite poles through the pulling by the spindle fibers. Each chromosome become single stranded and the DNA content become 2C. The breakage of the chromosome at the centromere is called karyokinesis i.e. the division of the nucleus. Key features of anaphase:

Separation of Sister Chromatids: In anaphase, the cohesion proteins that hold the sister chromatids together at the centromere are cleaved. This results in the separation of each chromosome into two individual chromatids. Once separated, the chromatids are now considered independent daughter chromosomes.

Movement of Chromosomes: The separated daughter chromosomes, now single-stranded, are pulled towards opposite poles of the cell. This movement is facilitated by the spindle fibers attached to the kinetochores of the chromosomes. The shortening of the spindle fibers exerts a force that pulls the chromosomes towards their respective poles.

DNA Content: As the sister chromatids separate and become independent chromosomes, the DNA content of each chromosome is halved. This reduction in DNA content occurs because each chromosome now consists of only one copy of the DNA molecule, resulting in a 2C DNA content.

Karyokinesis: The breakage of the centromere and the subsequent separation of sister chromatids is referred to as karyokinesis. It is the division of the nucleus, ensuring that each daughter cell receives a complete set of chromosomes during cell division.

Anaphase ensures the proper distribution of genetic material to each daughter cell. The movement of chromosomes to opposite poles of the cell sets the stage for the subsequent stage of mitosis, telophase, where the cell prepares for the physical separation of the cytoplasm and the formation of two distinct daughter cells.

TELOPHASE:

After anaphase the cell move into telophase. At this stage movement of chromosomes stops. Chromosomes resolve into thin threadlike structure again and the nuclear membrane reappears. The division of the cytoplasm (cytokinesis) occur and two daughter cells are produced each with the same chromosome number as in the parent cell. Key features of telophase:

Cessation of Chromosome Movement: In telophase, the movement of chromosomes towards the opposite poles of the cell comes to a halt. The chromosomes reach their respective poles and begin to decondense.

Chromosome Decondensation: As telophase progresses, the chromosomes gradually resolve from their condensed state into thin, threadlike structures. They revert to their less visible and diffuse form, resembling the chromatin seen during interphase.

Reformation of Nuclear Membrane: The nuclear membrane, which had disintegrated during earlier stages of mitosis, starts to reappear during telophase. It reforms around the decondensing chromosomes, separating them from the cytoplasm and restoring the distinct nuclear compartments in each daughter cell.

Cytokinesis: Concurrently with telophase, cytokinesis occurs, which is the division of the cytoplasm. Cytokinesis involves the physical separation of the cell into two distinct daughter cells. In animal cells, a cleavage furrow forms, pinching the cell membrane inward until it eventually divides the cytoplasm. In plant cells, a cell plate forms at the center, which gradually develops into a cell wall to separate the daughter cells.

Chromosome Number in Daughter Cells: At the end of telophase and cytokinesis, two daughter cells are formed, each containing the same chromosome number as the parent cell. The genetic material is evenly distributed to ensure that each daughter cell receives a complete set of chromosomes.

Telophase marks the final stage of mitosis, where the cell completes the process of division and transitions into the interphase of the cell cycle. The reformation of the nuclear membrane and the decondensation of chromosomes allow the daughter cells to resume their normal cellular functions and prepare for subsequent rounds of division or carry out specialized functions in multicellular organisms.

The diagram below shows the stages of mitotic cell division in an organism with chromosome number of 2n=2

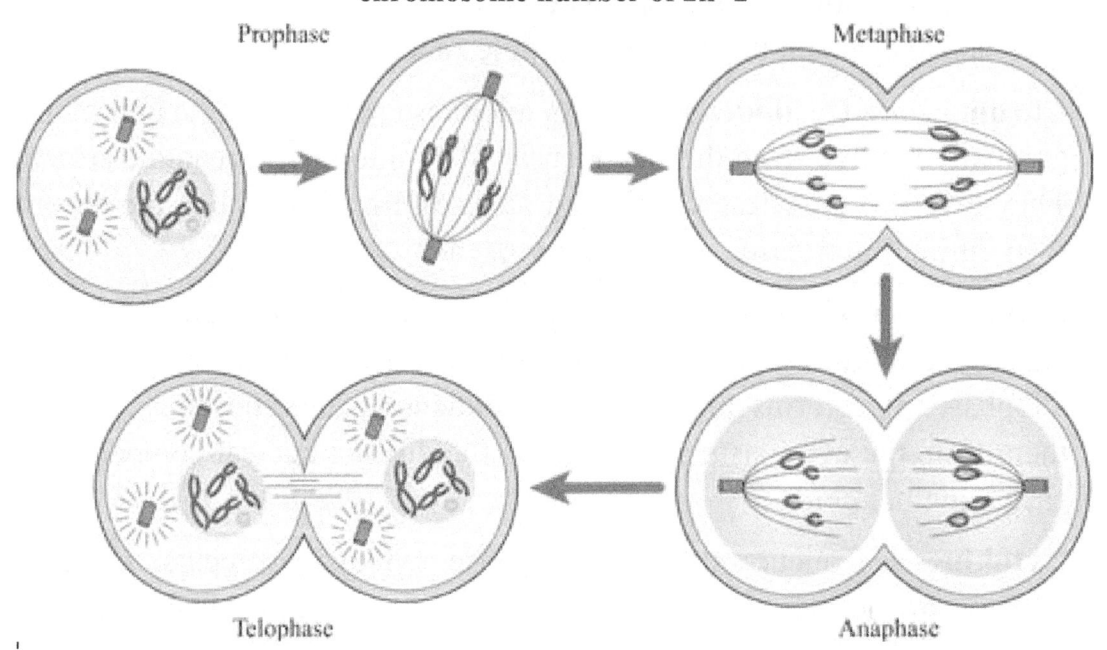

4.3 MEIOTIC CELL DIVISION (MEIOSIS)

Meiosis unlike mitosis occur during gamete formation prior to sexual reproduction. It is a reductional and equational division.it involves two divisions, the first meiotic division which is a reductional division and the second meiotic division which is an equational division and similar to mitosis. The importance of meiosis is to ensure constant and stable chromosome number within a breeding population from one generation to another and allow room for variability which enable organisms to adapt better to their changing environment.

4.3.1 CHARACTERISTICS OF MEIOSIS

It is an equational and reductional division
Four daughter cells are produced at the end of the divisions
Each daughter cell contain haploid number of chromosomes and different from the parent cell
It involves two divisions
It involves one karyokinesis and two cytokinesis
It occur during gamete formation

The **first meiotic division** consist of **prophase I, Metaphase I, Anaphase I and Telophase I**. Prophase I is subdivided into five sub stages: Leptonema, Zygonema, Pachynema, Diplonema and Diakinesis. Just as in mitosis a diploid cell go into interphase. When all processes of G1, S and G2 are completed the cell move directly into prophase I.

PROPHASE I

LEPTONEMA: During this stage, the chromosomes are still in their diffuse and thread-like form. They are thin and randomly distributed within the nucleus.

ZYGONEMA: In zygotene, the pairing of homologous chromosomes, which are identical pairs of chromosomes, begins. This pairing process is called synapsis and involves the formation of protein structures called the synaptonemal complex. Synapsis allows for the exchange of genetic material between non-sister chromatids of homologous chromosomes through a process called crossing over. This crossing over contributes to genetic variability in sexually reproducing organisms. The paired homologous chromosomes are referred to as a bivalent or a tetrad.

PACHYNEMA:

Pachytene is characterized by the further thickening of the chromosomes. They appear shorter and thicker than in previous stages. Pairing of homologous chromosomes is completed during zygotene, and at pachytene, this pairing stops. Chromosomes that are not paired at this stage remain unpaired throughout the rest of the division process. The physical evidence of crossing over, known as chiasmata, can be observed, indicating the exchange of genetic material between homologous chromosomes.

DIPLONEMA:

In diplotene, the chromosomes continue to thicken and become even more visible. The paired homologous chromosomes start to separate from each other, but they remain connected at specific points called chiasmata. The chiasmata can still be visibly seen during diplotene, although they gradually start to disappear.

DIAKINESIS:

During diakinesis, the crossing over becomes much less visible, and the process of terminalization occurs. Terminalization refers to the movement of the chiasmata towards the terminal ends of the chromosomes. The homologous chromosomes can be observed joined at the tips during diakinesis.

Prophase I allows for the pairing of homologous chromosomes and the exchange of genetic material through crossing over. This process contributes to genetic diversity and plays a significant role in the formation of genetically distinct gametes during sexual reproduction.

Note: all the stages of prophase I can be called the names in the diagram below.

Diagram showing different stages of prophase I

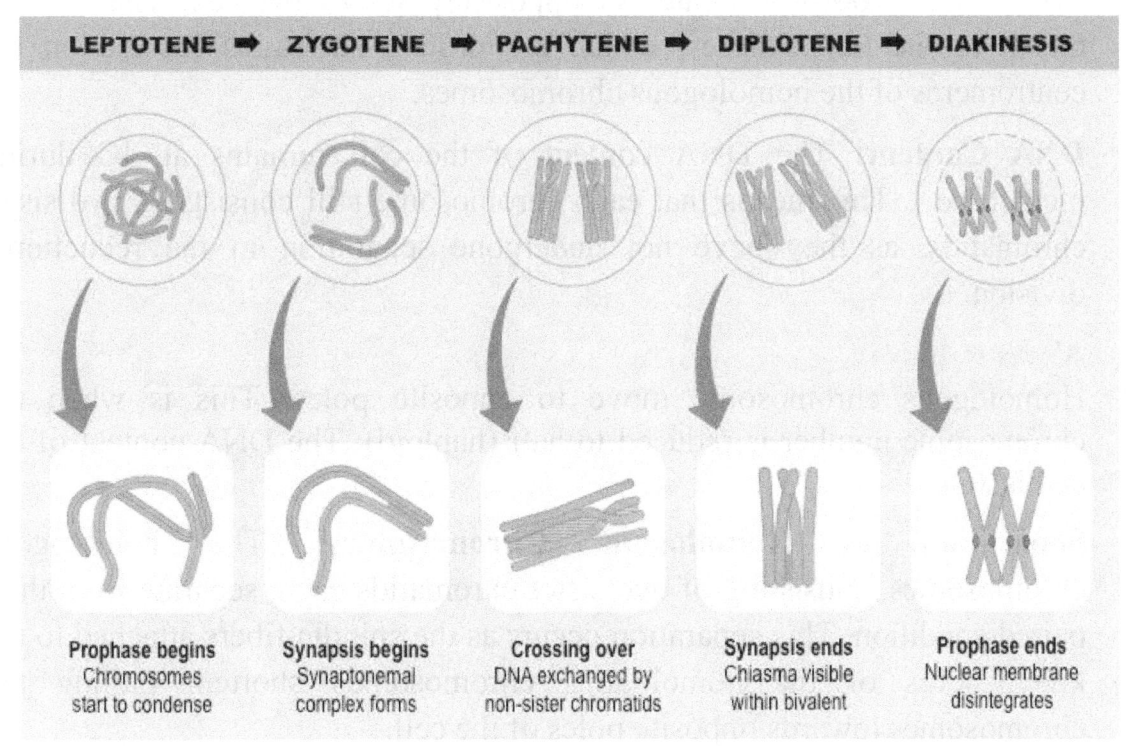

LEPTOTENE ➡ ZYGOTENE ➡ PACHYTENE ➡ DIPLOTENE ➡ DIAKINESIS

Prophase begins	Synapsis begins	Crossing over	Synapsis ends	Prophase ends
Chromosomes start to condense	Synaptonemal complex forms	DNA exchanged by non-sister chromatids	Chiasma visible within bivalent	Nuclear membrane disintegrates

METAPASE I

The chromosomes line up at the equator. Nuclear membrane disappear and spindle fibers appear. .The DNA content of the cell is 4C at this stage.

Chromosome Alignment: The homologous chromosomes, which were paired during prophase I and have undergone crossing over, now line up side by side at the metaphase plate. The alignment is not in a single file, as in metaphase of mitosis, but rather as homologous pairs.

Disappearance of the Nuclear Membrane: As the cell progresses into metaphase I, the nuclear membrane completely disappears. This allows the spindle fibers, which are microtubules emanating from the centrosomes, to interact directly with the chromosomes.

Spindle Fiber Formation: The spindle fibers begin to form and extend from the centrosomes located at opposite poles of the cell. These fibers attach to the kinetochores, which are protein structures located on the centromeres of the homologous chromosomes.

DNA Content: The DNA content of the cell remains at 4C during metaphase I. This means that each chromosome still consists of two sister chromatids, as they have not undergone separation in the reductional division.

ANAPHASE I

Homologous chromosome move to opposite poles. This is when the chromosome number is reduced to half (haploid). The DNA content of the cell is 2C.

Separation of Homologous Chromosomes: The homologous chromosomes, consisting of two sister chromatids each, separate from their paired condition. This separation occurs as the spindle fibers attached to the kinetochores of the homologous chromosomes shorten, pulling the chromosomes towards opposite poles of the cell.

Movement to Opposite Poles: The separated homologous chromosomes migrate towards the poles of the cell. The movement is facilitated by the contraction of the spindle fibers, which exert a force on the chromosomes to guide their migration. This movement ensures that each resulting daughter cell receives one member of each homologous pair.

Reduction of Chromosome Number: Anaphase I is a reductional division because it reduces the chromosome number by half. The separation of homologous chromosomes ensures that each daughter cell will contain one set of chromosomes, rather than the two sets present in the parent cell. As a result, the DNA content of the cell is reduced to 2C, as each chromosome consists of a single chromatid.

TELOPHASE I

The movement of chromosomes stops. Nuclear membrane reappear, chromosome become thin threadlike again. The division of the cytoplasm (cytokinesis) occur and two daughter cells are produced each containing haploid number of chromosomes. The DNA content of the cell is still 2C. Each of the cell produced go into the second meiotic division. In some organisms the cell move directly from anaphase I to prophase II The second meiotic division is like a mitotic division. It consist of prophase II, Metaphase II, Anaphase II and telophase II.

Chromosome Movement Ceases: The homologous chromosomes reach their respective poles and come to a complete stop. This marks the end of chromosome movement during telophase I.

Reappearance of Nuclear Membrane: The nuclear membrane, which had disappeared during earlier stages, reassembles around the separated homologous chromosomes. This results in the formation of two distinct nuclei within the cell.

Chromosome Decondensation: The chromosomes, which were condensed and visible during earlier stages, start to decondense. They return to their threadlike, diffuse form, resembling the chromatin observed during interphase.

Cytokinesis: The division of the cytoplasm, known as cytokinesis, occurs during telophase I. This physical separation of the cell results in the formation of two daughter cells, each containing one set of chromosomes, which is the haploid number for that particular organism. These daughter cells are now ready to enter the second meiotic division, also known as meiosis II.

It is important to note that in some organisms, such as certain fungi and algae, the cell may bypass a separate telophase I and enter directly into prophase II. In these cases, the processes of nuclear membrane reformation and chromosome decondensation occur in a condensed form.

The second meiotic division, meiosis II, is similar to a mitotic division and consists of prophase II, metaphase II, anaphase II, and telophase II. These stages mirror the corresponding stages of mitosis, with the main difference being that the starting cells in meiosis II already contain a haploid number of chromosomes. Meiosis II results in the separation of sister chromatids, leading to the formation of four haploid daughter cells, each containing one set of chromosomes.

PROPHASE II

The chromosome appear as thin thread like structure and the nuclear membrane is still intact.

Chromosome Condensation: The chromosomes, which decondensed during telophase I, start to condense again. They become visible as distinct thread-like structures within the nucleus.

Nuclear Membrane: The nuclear membrane, which reformed during telophase I, remains intact during prophase II. It continues to enclose the chromosomes within the nucleus.

Spindle Fiber Formation: Similar to prophase in mitosis, the spindle fibers begin to assemble and extend from the centrosomes located at opposite poles of the cell. These fibers will facilitate the movement and separation of the sister chromatids during subsequent stages.

Prophase II sets the stage for the further separation of sister chromatids in the second meiotic division. The condensation of chromosomes prepares them for precise alignment and separation in the subsequent stages of meiosis II. The intact nuclear membrane helps maintain the organization of the genetic material and provides a boundary for the cellular processes occurring within the nucleus.

METAPHASE II

The nuclear membrane disappear, while the spindle fibers appear. The chromosome line up at the equator and are maximally contracted.

Nuclear Membrane Dissolution: The nuclear membrane, which remained intact during prophase II, dissolves, allowing the spindle fibers to interact directly with the chromosomes.

Spindle Fiber Formation: As the nuclear membrane disappears, the spindle fibers, which have been organizing and extending from the centrosomes, become more prominent. These fibers attach to the kinetochores on the sister chromatids.

Chromosome Alignment: The sister chromatids, each containing one copy of the replicated chromosome, align themselves along the equatorial plane of the cell. This alignment occurs with the help of the spindle fibers, which exert tension on the kinetochores.

Maximum Chromosome Contraction: The chromosomes reach their maximum level of condensation during metaphase II. They become highly visible and appear as maximally contracted structures, facilitating their precise alignment and separation in subsequent stages.

The proper alignment of sister chromatids in metaphase II is essential for ensuring accurate separation during anaphase II. The disappearance of the nuclear membrane allows for direct interaction between the spindle fibers and the chromosomes, ensuring proper attachment and alignment of the chromatids along the equatorial plane.

ANAPHASE II

Each individual chromosome breaks open at the centromere. This is when karyokinesis occur in meiosis. At this stage the DNA content of the cell become 1C and the haploid number of chromosome maintained.

Separation of Sister Chromatids: Each individual chromosome, consisting of two sister chromatids, breaks open at the centromere. The spindle fibers attached to the kinetochores of the sister chromatids shorten, pulling the chromatids apart.

Movement to Opposite Poles: The separated sister chromatids migrate towards the opposite poles of the cell. This movement is driven by the contraction of the spindle fibers, which exert force on the chromatids, guiding their migration.

Division of Genetic Material: Anaphase II is the stage of meiosis where the division of genetic material occurs. The separation of sister chromatids ensures that each resulting daughter cell receives one chromatid from each chromosome.

DNA Content: As the sister chromatids separate, the DNA content of the cell becomes 1C. This means that each chromosome is represented by a single chromatid, and the cell maintains the haploid number of chromosomes.

TELOPHASE II

Movement of chromosome stops, the chromosomes resolve into thin threadlike structure, nuclear membrane reappear. Cytokinesis occur in each of the cell; each producing two daughter cells. Four haploid daughter cell are eventually produced.

Chromosome Movement Ceases: The movement of chromosomes comes to a halt during telophase II. The separated sister chromatids reach their respective poles and remain in position.

Chromosome Decondensation: The chromosomes, which were condensed and visible during earlier stages, start to decondense. They unwind and resolve into thin threadlike structures, resembling chromatin.

Reappearance of Nuclear Membrane: The nuclear membrane, which had dissolved during earlier stages, reassembles around the separated sister chromatids. This leads to the formation of distinct nuclei within each of the daughter cells.

Cytokinesis: Cytokinesis, the division of the cytoplasm, occurs in each of the daughter cells. This results in the physical separation of the cytoplasmic contents and the formation of two daughter cells from each parent cell. Ultimately, four haploid daughter cells are produced.

Telophase II marks the conclusion of meiosis, resulting in the formation of four haploid daughter cells. These daughter cells are genetically distinct and carry a unique combination of alleles, contributing to genetic diversity in sexually reproducing organisms. The reformation of the nuclear membrane and the decondensation of chromosomes allow the daughter cells to resume their normal cellular functions.

Diagram showing different stages of meiosis in organism with chromosome number of 2n=4

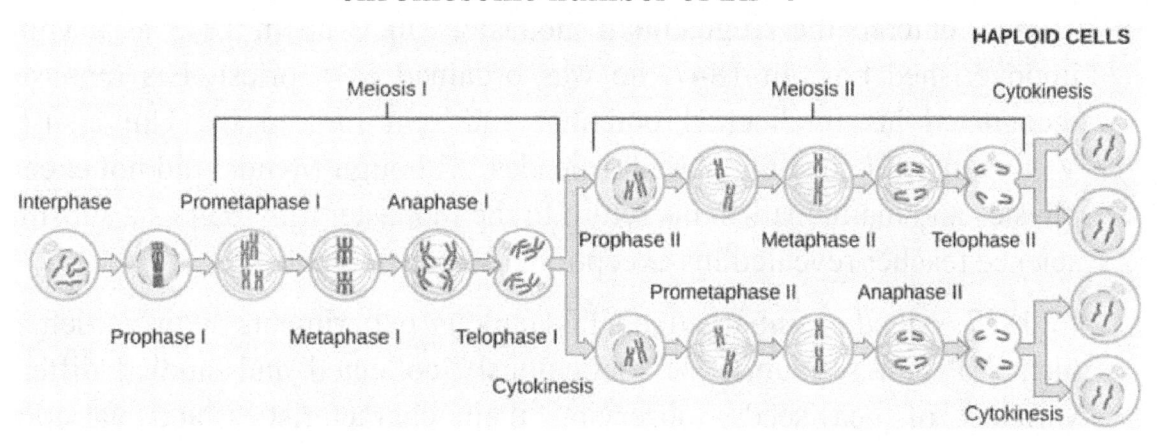

CHAPTER 5

MENDELIAN GENETICS

5.1 SHORT HISTORY ON GREGORY MENDEL

Gregor Mendel, often referred to as the "father of modern genetics," was born on July 20, 1822, in what is now the Czech Republic. He is renowned for his pioneering work on inheritance and the principles of heredity. Despite not being a biologist by profession, Mendel's contributions to the field revolutionized our understanding of genetics.

Mendel entered the Augustinian monastery in Brno in 1843 as a young, impoverished boy. In 1847, he was ordained as a priest. His monastery recognized his intellectual potential and sent him to the University of Vienna in 1851 to study natural sciences. Although Mendel did not excel in physics and mathematics, his return to the monastery in 1854 as a substitute science teacher revealed his exceptional scientific mind.

In 1857, Mendel embarked on his famous experiments with garden pea plants (Pisum sativum). He meticulously collected and studied different varieties of pea seeds, observing their characteristics and variations. Through rigorous experimentation in the monastery garden, Mendel investigated the patterns of inheritance and developed his groundbreaking laws of heredity.

After seven years of dedicated work, Mendel presented his findings in 1866 publishing his paper "Experiments on Plant Hybridization." Unfortunately, his work received little attention at the time, and Mendel's revolutionary ideas went largely unrecognized.

It was not until 1900, long after Mendel's death in 1884, that his work gained widespread recognition. Three different scientists independently rediscovered Mendel's laws of inheritance, including the law of segregation, which explained the passing on of traits from parent to offspring. This rediscovery brought Mendel's work into the limelight, cementing his place as a pivotal figure in the field of genetics.

Mendel's experiments and principles formed the foundation of modern genetics. His concepts of dominant and recessive traits, as well as his laws of segregation and independent assortment, provided the framework for understanding how traits are inherited and passed down through generations. Mendel's work laid the groundwork for future scientists to unravel the complexities of genetic inheritance and paved the way for significant advancements in the field of genetics and the understanding of heredity.

5.2 MENDEL'S WORK ON GARDEN PEA (PISUM SATIVUM)

Gregor Mendel's choice to work with garden pea plants (Pisum sativum) was a key factor in the success of his experiments on inheritance. Unlike his predecessors, who often focused on complex traits influenced by multiple genes, Mendel deliberately selected simple characters for his hybridization experiments.

Mendel recognized that studying complex traits would make it difficult to identify clear patterns of inheritance. Complex traits are often influenced by multiple genes and can be affected by environmental factors, making it challenging to discern the underlying principles of inheritance.

Instead, Mendel chose to work with garden pea plants that exhibited distinct and easily observable traits, such as flower color (purple or white), seed shape (round or wrinkled), and seed color (yellow or green). These traits are controlled by a single gene with clear dominant and recessive alleles.

By focusing on these simple characters, Mendel was able to carefully track the inheritance patterns and ratios across generations. He performed controlled crosses between plants with contrasting traits, ensuring that only one characteristic was studied at a time. This allowed him to establish clear patterns of inheritance and make precise observations.

Mendel's choice of garden pea plants as his experimental organism, with their well-defined and easily distinguishable traits, enabled him to formulate his laws of inheritance with clarity and precision. It provided him with a solid foundation to establish the principles of dominance, segregation, and independent assortment.

Mendel's decision to study simple characters in his garden pea experiments was a deliberate and strategic choice that laid the groundwork for his ground breaking discoveries in genetics. By focusing on the inheritance of straightforward traits, he was able to unravel the fundamental laws that govern genetic inheritance, setting the stage for the modern understanding of genetics and heredity.

The characters Mendel investigated are listed below.

S/N	Characters	Character states
1.	Plant height	Tall and dwarf
2.	Flower position	Axillary and terminal
3.	Flower colour	Red and white Red and white
4.	Seed coat colour	Grey and white
5.	Seed coat texture	Smooth and wrinkled
6.	Endosperm colour	Green and yellow
7.	Colour of pod	Green and yellow
8.	Texture of pod	Full and wrinkled

5.3 SOME IMPORTANT GENETIC TERMS

- **Allele:** An allele refers to the alternative forms or states of a gene. For example, in a gene with two possible states, T and t, T and t are considered alleles of that gene.
- **Dominant Allele:** A dominant allele is an allele that expresses its phenotypic trait in both the homozygous (having two copies of the same allele) and heterozygous (having two different alleles) states. The presence of a dominant allele masks the expression of a recessive allele.

- **Recessive Allele:** A recessive allele is an allele that expresses its phenotypic trait only in the homozygous state. In the presence of a dominant allele, the recessive allele's trait is not manifested.
- **Locus:** The locus refers to the specific location or position of a gene on a chromosome. Each gene occupies a specific locus on a chromosome.
- **Homozygous:** Homozygous refers to the condition in which an individual possesses two identical alleles for a particular gene at a given locus. For example, if an individual has two copies of the allele T at a specific gene locus, they are said to be homozygous for that gene.
- **Heterozygous:** Heterozygous refers to the condition in which an individual possesses two different alleles for a particular gene at a given locus. For example, if an individual has one copy of the allele T and one copy of the allele t at a specific gene locus, they are said to be heterozygous for that gene.
- **Homologous Chromosomes:** Homologous chromosomes are a pair of chromosomes that are similar in size, shape, and carry genes for the same traits. One chromosome in the pair is inherited from the mother, and the other is inherited from the father.
- **Genotype:** The genotype refers to the genetic constitution of an individual, specifically the combination of alleles present at a given locus or across multiple loci. It represents the genetic information carried by an organism.
- **Phenotype:** The phenotype refers to the observable physical or biochemical characteristics of an organism, which are the result of the interaction between the genotype and the environment. It represents the expression of the genetic information carried by an organism.
- **Backcross:** A backcross involves crossing an F1 hybrid (offspring of a cross between two different parental strains) with one of its parental strains. This is done to reintroduce the genetic traits of one parent back into the offspring population.

- **Test Cross:** A test cross is performed by crossing an individual showing a dominant phenotype (but with an unknown genotype) with a homozygous recessive individual. This cross helps determine the genotype of the individual showing the dominant phenotype.

5.4 SIMPLE MONOHYBRID EXPERIMENT

In Mendel's simple monohybrid experiment, he focused on the inheritance of a single trait, specifically plant height. He conducted a cross between a homozygous tall plant (genotype TT) and a homozygous dwarf plant (genotype tt).

In the first filial generation (F1), Mendel observed that all the offspring plants were tall. This led him to conclude that the tall character state is dominant over the dwarf character state. The genotype of the F1 plants was heterozygous (Tt) since they inherited one allele for tallness from the tall parent (T) and one allele for dwarfness from the dwarf parent (t).

To study the inheritance pattern further, Mendel allowed the F1 plants to self-pollinate and produce the second filial generation (F2). In the F2 generation, Mendel observed a phenotypic ratio of 3 tall plants to 1 dwarf plant (3T-:1tt). This ratio indicated that the dwarf trait reappeared in the offspring, even though it was absent in the F1 generation.

Upon examining the genotypes of the F2 plants, Mendel discovered a genotypic ratio of 1 TT:2 Tt:1 tt. This ratio revealed that among the tall plants, there were individuals with two dominant alleles (TT), individuals with one dominant and one recessive allele (Tt), and individuals with two recessive alleles (tt).

Mendel's experiment demonstrated the principles of dominance and segregation in inheritance. The dominance of the tall trait over the dwarf trait and the segregation of alleles during gamete formation were key findings that laid the foundation for Mendel's laws of inheritance and the field of modern genetics.

cross   TT x tt
↓

Tt (F1 tall plant)
↓

Tt x Tt (selfing of F1 plant)

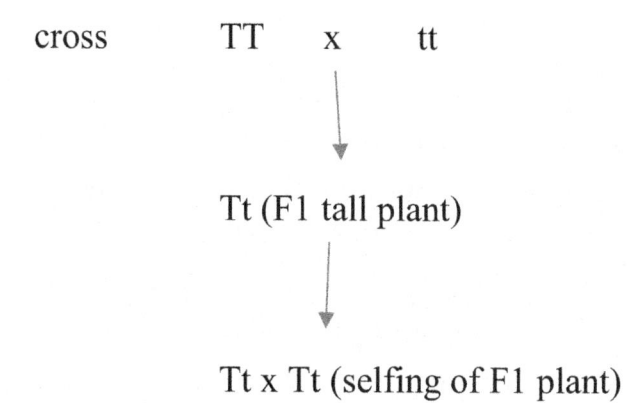

Punnett's table showing selfing of F1 plant

	T	T
T	TT (homozygous tall)	Tt (heterozygous tall)
T	Tt (heterozygous tall)	tt (homozygous dwarf)

5.5 DIHYBRID INHERITANCE

In his dihybrid inheritance experiment, Mendel crossed a true-breeding tall plant (with both alleles for tallness, TT) and a true-breeding dwarf plant (with both alleles for dwarfness, tt). Additionally, he considered another pair of traits: seed texture, specifically smooth (SS) and wrinkled (ss). By studying the inheritance of both height and seed texture simultaneously, Mendel aimed to understand how these traits are inherited together.

By crossing the tall, smooth-seeded plant with the dwarf, wrinkled-seeded plant, Mendel created a dihybrid cross. This means that he was simultaneously examining the inheritance of two different traits, height and seed texture.

Through his experiments, Mendel observed the phenotypes of the offspring in the F1 generation, which were all tall and smooth-seeded. This suggested that both the tall and smooth-seeded traits were dominant over their respective alternatives (dwarf and wrinkled).

To further explore the inheritance patterns, Mendel allowed the F1 plants to self-fertilize, giving rise to the F2 generation. In the F2 generation, Mendel observed a phenotypic ratio of 9 tall, smooth-seeded plants: 3 tall, wrinkled-seeded plants: 3 dwarf, smooth-seeded plants: 1 dwarf, wrinkled-seeded plant (9TSS:3TWS:3DSS:1DWS).

Mendel's observations indicated that the traits for height and seed texture were segregating independently, meaning that the inheritance of one trait did not influence the inheritance of the other. This led to the formulation of Mendel's Law of Independent Assortment, which states that different pairs of alleles segregate independently during gamete formation.

Mendel's dihybrid inheritance experiment provided crucial insights into the principles of inheritance and the concept of independent assortment, furthering our understanding of how traits are passed from one generation to the next.

Tall, smooth seeded (TTSS) x dwarf, wrinkled seeded (ttss)

TTSS X ttss

↓

TtSs F1 (tall smooth seeded)

All the F1 were tall and smooth seeded
The F1 plants were advanced to F2 generation by selfing the FI plant (TtSs X TtSs)

Gametes formed by each sex. Male (TS, Ts, tS, ts). Female (TS, Ts, tS, ts) Four different types of gametes will be formed due to independent assortment of genes on chromosomes.

Punnett's table showing the gametes and genotypes formed in dihybrid inheritance by selfing an F1 individual

Male gametes	Female gametes	TS	Ts	tS	ts
TS		TTSS	TTSs	TtSS	TtSs
Ts		TTSs	TTss	TtSs	Ttss
tS		TtSS	TtSs	ttSS	ttSs
Ts		TtSs	Ttss	ttSs	ttss

SUMMARY OF PHENOTYPES AND GENOTYPES PRODUCED BY SELFING AN F1 IN THE DIHYBRID CROSS ABOVE.

PHENOTYPIC RATIO	GENOTYPIC RATIO	PHENOTYPES
9T-s- plant	(1TTSS, 2TtSS, 2TTSs, 4TsSs)	Tall smooth seeded
3T-ss plant	(2Ttss, 1TTss)	Tall wrinkled seeded
3ttS- plant	(2ttSs 1ttSS)	Dwarf smooth seeded
1ttss	(ttss)	Dwarf wrinkled seeded

This indicates that if 1000 plants were produced in the F2

9/16=563 will be tall smooth seed
3/16 =188 will be tall wrinkled seeded
3/16= 188 will be dwarf smooth seeded
1/16= 63 will be dwarf wrinkled seeded

The segregation pattern observed in F2 generation of dihybrid inheritance was due to the independent assortment of genes/alleles on chromosomes which means each type of gamete have equal chance of being formed Based on the monohybrid inheritance and dihybrid inheritance two laws were produced called Mendel's law

The First Law (Law Of Segregation): States that genes are responsible for the development of an individual and are independently transmitted from one generation to the other without any alteration

The Second Law (Law Of Independent Assortment): States that the segregation of one pair of allele is independent of the segregation of the other pair of allele

CHAPTER 6
MOLECULAR GENETICS

The molecule called deoxyribonucleic acid (DNA) was discovered by Friedrich miescher in 1869. In 1868 Miescher demonstrated that the nuclear material could be isolated from the cytoplasm. This nuclear material was later shown to be composed of protein and nucleic acids. Because of the repetitive nature of nucleic acids and the apparent similarity in chemical structure irrespective of source, it was thought not to play an important role in heredity. Amino acids were thought to specify genetic information.

The significance of DNA in heredity was not fully understood until the mid-20th century. In the early 1950s, the groundbreaking work of James Watson and Francis Crick, along with contributions from Rosalind Franklin and Maurice Wilkins, elucidated the structure of DNA and proposed its role in carrying genetic information.

Watson and Crick's model of DNA, known as the double helix structure, revealed that DNA is composed of two complementary strands that are intertwined. The strands are held together by hydrogen bonds between specific nucleotide base pairs: adenine (A) with thymine (T), and guanine (G) with cytosine (C). This complementary base pairing provides the basis for DNA replication and the transmission of genetic information.

The discovery of the structure of DNA revolutionized the field of molecular genetics. It established DNA as the molecule responsible for storing and transmitting genetic information from one generation to the next. The central dogma of molecular biology, proposed by Francis Crick in 1958, states that DNA is transcribed into RNA, and RNA is translated into proteins, thus forming the basis for the flow of genetic information.

Further advancements in molecular genetics, such as the development of techniques like polymerase chain reaction (PCR) and DNA sequencing, have allowed scientists to study and manipulate DNA in greater detail. These tools have opened up new avenues of research, leading to breakthroughs in areas such as gene expression, genetic engineering, and the understanding of genetic diseases.

6.1 TRANSFORMATION

Transformation is the evidence to identify the DNA as the genetic material or uptake of foreign piece of DNA by an organism occasioning the alteration of genotype. In *Diplococcus pneumoniae* bacteria the thin capsule enclosing the cell wall consists of substances called polysaccharides, which are of specific types. Type II capsules for example elicit the formation of antibodies in the bloodstream of rabbits that are different from antibodies formed with type III capsules. These capsules types are distinct properties of bacterial strains and bacteria of one kind do not appear in pure cultures of another kind.

On the other hand the capsules themselves are subject to variability in respect to their presence or absence. That is, any particular type of *D. pneumoniae* may occasionally give rise to bacteria that do not have capsule at all (mutation) such non-capsulated bacteria have rough (R) appearance when grown on culture plates and relatively harmless in contrast to the smooth (S) appearance and virulence of encapsulated bacteria. When cultured separately, R and S colonies will transmit their respective characters to future generations except for rare mutations. These bacteria are sensitive to heat and if temperature is raised sufficiently high the bacteria are heat-killed and can no longer divide.

Griffith, in 1928, showed that heat-killed bacteria of one type could have a hereditary influence on bacteria of another type. In one experiment he injected heat killed bacteria of type IIIS into a mouse that carry type IIR non heat killed and obtained virulent live cultures which were of the type IIIS variety. Because the injection of heat killed type IIIS bacteria by itself does not result in any live bacteria culture, nor does type II mutate in to type III. A change or transformation of type IIR into type IIIS must have occurred through the transfer of some active substance.

Grifffith's results could be duplicated by mixing different types of heat killed and live-bacterial strain in mouse (*in vivo*) as well by mixing item in test tubes (*in vitro*). The search for the specific agent responsible for transformation, the transforming principle continued until 1944. In this year Avery, Macleod and Mccarty showed that the transforming principles consist entirely of DNA.

In their experiment they extracted the DNA from the heat-killed cells of type IIIS, and mix this DNA extract directly with *in vitro* cultures of type IIR. A serum was added whose antibodies react with the R cell and cause them to precipitate to the bottom. When transformation occurs, type IIIS cells, not being precipitated, now grow diffusely throughout the medium.

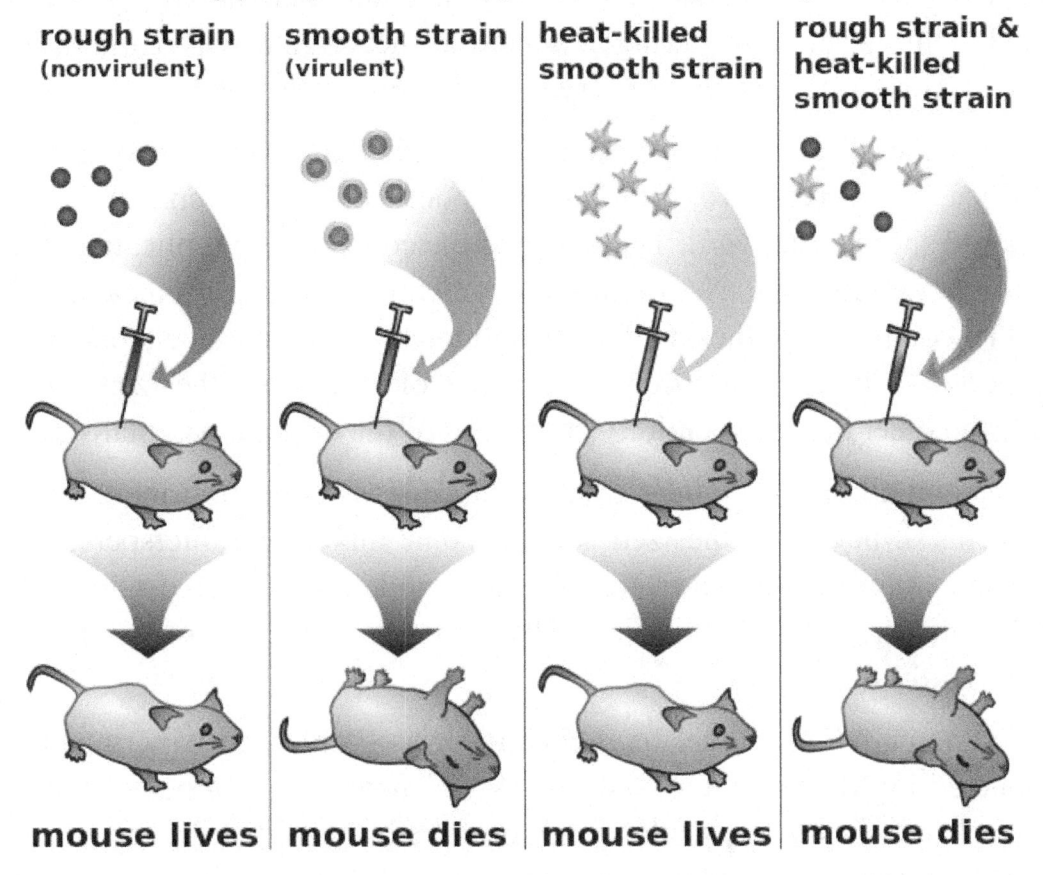

rough strain (nonvirulent)	smooth strain (virulent)	heat-killed smooth strain	rough strain & heat-killed smooth strain
mouse lives	mouse dies	mouse lives	mouse dies

6.2 SUMMARY OF GRIFFITH'S EXPERIMENT

Rough Bacteria in Mouse Mouse does not die

Smooth Bacteria injected in mouse Mouse die

Heat killed smooth bacteria injected in mouse Mouse does not die

Heat killed smooth bacteria +non heat killed Rough bacteria injected in mouse Mouse die.

6.3 SEMI-CONSERVATIVE DNA REPLICATION

Semi-conservative DNA replication is the specific mode of DNA replication proposed by Watson and Crick. According to this model, each strand of the double-stranded DNA molecule acts as a template for the synthesis of a new complementary strand. As a result, the newly formed DNA molecules are composed of one original (parental) strand and one newly synthesized strand. This means that during replication, the original DNA molecule is partially conserved or semi-conserved in the new molecules.

Initially, the idea of semi-conservative replication lacked experimental evidence. However, two alternative models of DNA replication were proposed shortly afterward. The first was the conservative replication model in which the original DNA duplex remains intact and serves as a template for the complete synthesis of a new duplex. In this model, one newly synthesized duplex is entirely composed of newly synthesized strands, while the other duplex is entirely composed of the original parental strands.

The second alternative model was the dispersive replication model, where fragments of newly synthesized DNA become mixed and interspersed with fragments of the original DNA. In this model, both the original DNA duplex and the newly synthesized DNA contain a mixture of parental and newly synthesized segments.

To determine which model of replication was correct, an experiment was conducted by Meselson and Stahl in 1958. They used isotopes of nitrogen to label the DNA molecules and observed their distribution through subsequent rounds of replication. The results of their experiment supported the semi-conservative replication model, as they observed DNA molecules with hybrid densities (a mix of light and heavy isotopes), which indicated the presence of both parental and newly synthesized DNA strands.

The discovery of semi-conservative DNA replication provided a crucial understanding of how genetic information is faithfully transmitted from one generation to the next. It has since become a foundational principle in molecular biology.

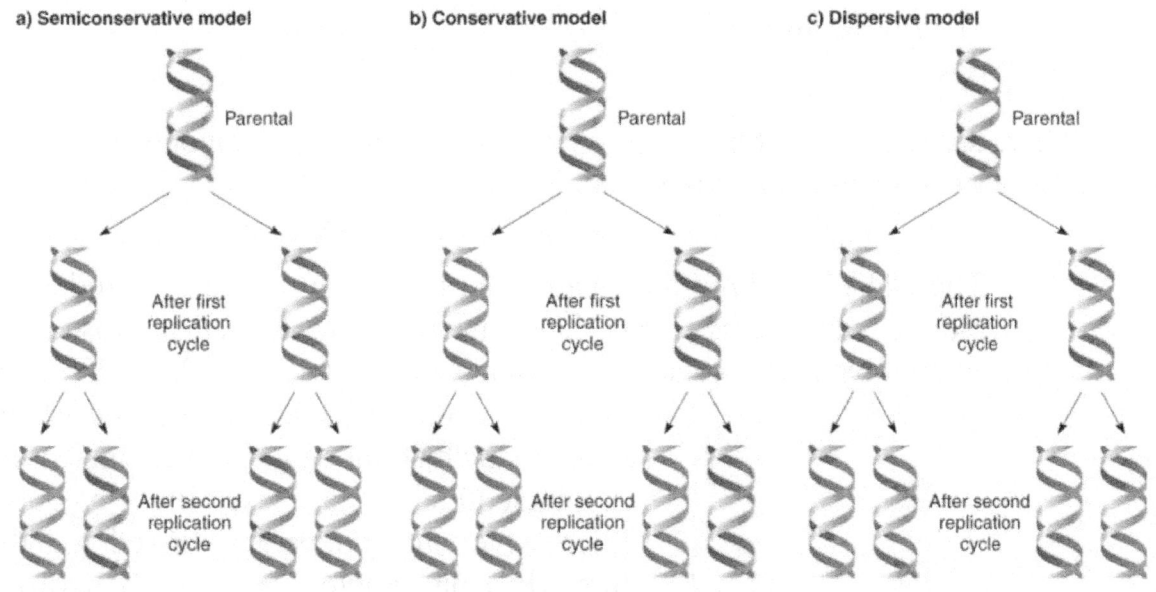

Diagram showing the three possible modes of duplex DNA replication. The predicted distribution of original parental in purple color and newly synthesized strand in blue color for two rounds of replication.

6.4 THE NATURE OF DNA

After the isolation and purification of the DNA molecule it was soon found that it molecular weight varied widely. However it was established that DNA from all organisms are chemically similar.

The DNA from any species consists of three chemical groups.
1. Phosphate
2. Deoxyribose sugar
3. Bases

There are two types of bases the pyrimidines (one ring compounds) and the purines (two ring compounds). The pyrimidines are thymine (T) and Cytosine (C) while the purines are Adenine (A) are guanine (G)

The DNA is a three-dimensional structure that is accepted today and for which a noble prize was later awarded, was first proposed and demonstrated by J. D. Watson and Francis crick in 1953 (1953-1955). The DNA is a long molecule consisting of two strands. Each strand is a chain of nucleotides A nucleotide is a base plus a sugar. Successive nucleotides are linked together through a phosphate group and a hydroxyl group on the sugar component. The two strands are weekly associated by hydrogen bonds. The width of the two strands is always the same. This is so because a purine (two ting structure) always pairs with a pyrimidine (one ring structure). Specifically, adenine always pair with thymine and Cytosine always pair with guanine.

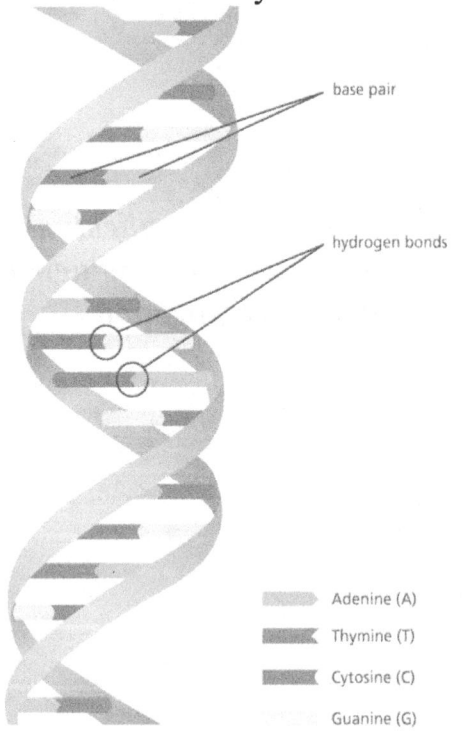

In any organism DNA molecule A=T and C=G. However the ratio varies widely for different species of DNA molecules

Since the above pairing is specific, if you

know the nucleotide sequence of one strand you can tell the sequence of the other strand. This phenomenon is referred to as complementarity.

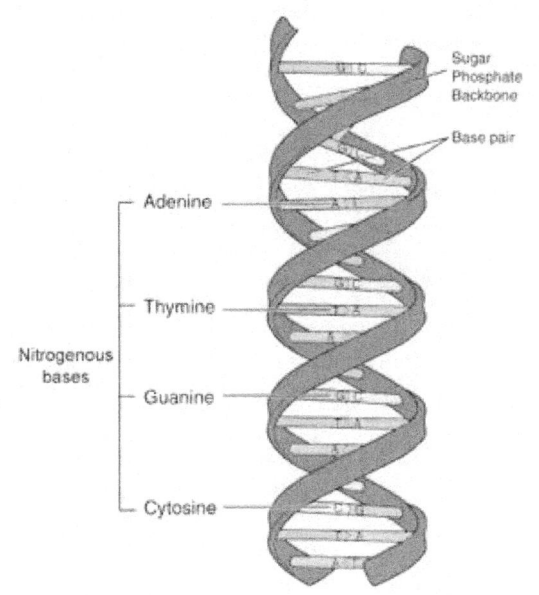

Diagram showing Structure of the DNA

6.5   RIBONUCLEIC ACID (RNA)

RNA is also found as a component of the chromatin fibre (Chromosome). Although both the DNA and RNA are nucleic acids. RNA differs from DNA in a number of ways. The DNA is double stranded and has deoxyribose sugar. The RNA is always single stranded with a ribose sugar. Both DNA and RNA contain the same bases except that uracil is present in RNA in place of thymine which is found only in DNA. RNA is synthesized in the nucleus on the DNA template transported into the cytoplasm where it is more prevalent and associated with ribosomes. DNA on the other hand is self-duplicating and exclusively located in the nucleus. Several classes of RNA exist based on size and function. Heterogeneous nuclear RNA are very large RNA molecules found in the nucleus. They undergo further processing before being transported through the nuclear membrane into the cytoplasm as messenger RNA (mRNA) The mRNA molecules carry the information from the gene in the nucleus to the ribosomes where such information is translated into protein synthesis. Other classes of RNA include the smaller transfer RNA (tRNA) and the ribosomal RNA (rRNA), which is found as integral parts of the ribosomes. Both tRNA and rRNA play important roles in protein synthesis.

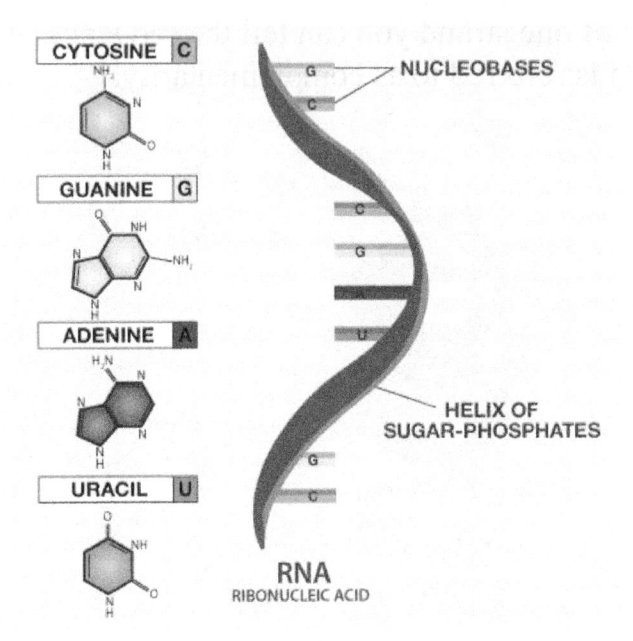

6.6 THE GENETIC CODE

One definition of the gene is that it specifies a peptide chain, through a complex procedure which involves the transfer of information in a gene from the nucleus through a messenger (mRNA) to the cytoplasm where the message is read during protein synthesis to make peptide chain.

The genetic code refers to the set of rules that determines how the sequence of nucleotides in DNA or mRNA is translated into the sequence of amino acids in a protein during protein synthesis. The genetic code is essentially a correspondence between specific sequences of three nucleotides, called codons, and specific amino acids.

There are four different nucleotides in RNA: adenine (A), cytosine (C), guanine (G), and uracil (U). With three nucleotides per codon, there are 64 possible codons (4^3) combinations. However, there are only 20 different amino acids used to build proteins. This means that multiple codons can code for the same amino acid, creating redundancy in the genetic code. For example, the amino acid alanine is specified by the codons GCU, GCC, GCA, and GCG.

The genetic code was deciphered through a combination of experimental and theoretical approaches. Scientists conducted experiments to determine which codons corresponded to which amino acids by studying the effects of mutations and genetic alterations. In the early 1960s, Marshall Nirenberg and his colleagues used synthetic RNA sequences to decipher specific codon assignments, which earned Nirenberg a Nobel Prize.

Diagram for the base combinations (triplet codes) that specify different amino acids.

The assignment of codons to amino acids follows a universal code, meaning that it is largely the same across different organisms. This universality allows genetic information to be accurately translated into proteins regardless of the species. However, some codons can have additional meanings beyond specifying amino acids. For instance, certain codons can serve as start or stop signals for protein synthesis.

6.7   CHARACTERISTICS OF THE GENETIC CODES

1. The genetic code is Universal (the same in all living organisms).
2. The code is degenerate since we have sixty four codes for only twenty amino-acids. This means that one amino acids is coded for by more than one codon. The significance of this is thought be a kind of stabilizer of species so that degenerate code reduces the effect of base change mutations as the third of the codons are less important than the first two.
3. Codons are linear with no overlaps and only one base is part of only one codon.
4. The codon is directional and its is always read from a fixed point
5. In *E. coli* there is an initiation codon usually AUG at the beginning of all genes. There is some evidence in support of AUG as initiation codon in higher organisms.
6. Termination codons in all systems are UAA, UAG and UGA.

CHAPTER 7

MOVEMENT OF MATERIALS IN AND OUT OF CELLS

7.1 CELLULAR TRANSPORT

In order to ensure the survival and proper functioning of a cell, it is crucial to have a well-regulated system for the movement of nutrients, products, and wastes in and out of the cell. This cellular transport plays a vital role in maintaining the cell's internal environment and facilitating the exchange of substances with the external environment.

Cells are surrounded by a watery fluid known as the extracellular fluid or interstitial fluid. This fluid serves as a medium through which various substances can move around inside and outside the cell. It also enables the transportation of molecules across the cell membrane, which acts as a selective barrier regulating the passage of substances.

The ability of a cell to gather and conserve resources from its environment is essential for the success of plants. Unlike animals, plants are immobile and rely on their surroundings for essential nutrients, water, and other substances. Efficient transport mechanisms allow plants to absorb water and minerals from the soil, transport sugars produced during photosynthesis to different parts of the plant, and eliminate waste products.

The transport of materials is not limited to individual cells but is also crucial for the integrated functioning of the whole plant. Plants have specialized tissues and structures dedicated to transporting substances throughout their body. Xylem vessels, for example, are responsible for transporting water and minerals from the roots to the leaves, while phloem tissues transport sugars and other organic compounds to various parts of the plant.

Several mechanisms contribute to cellular transport. Diffusion is a passive process where substances move from an area of higher concentration to an area of lower concentration. Active transport, on the other hand, requires energy and enables cells to move substances against their concentration gradient, allowing them to accumulate specific molecules or ions as needed. Bulk flow involves the mass movement of fluids, such as water, in response to pressure differences, facilitating the long-distance transport of substances.

The coordination and regulation of these transport processes are crucial for the survival and proper functioning of plants. They ensure that nutrients reach the cells that need them, waste products are efficiently eliminated, and the plant maintains its overall physiological balance.

7.2 CELL MEMBRANE/PLASMA MEMBRANE

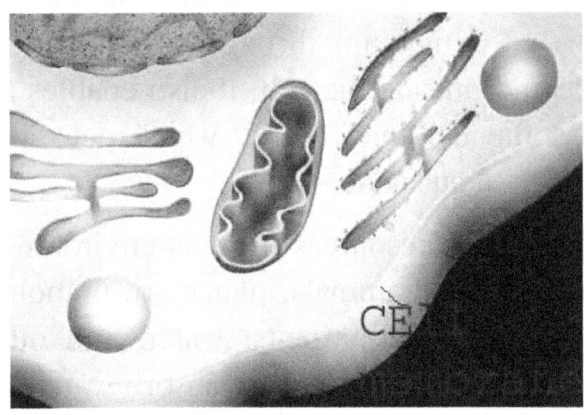

The cell membrane controls what moves in and out of the cell. The plasma membrane of a cell is a network of lipids and proteins that forms the boundary between a cell's contents and the outside of the cell. It is also simply called the cell membrane. The cells of all living things have plasma membranes. The cell membrane, also known as the plasma membrane, is a vital structure that encloses the cell and acts as a selective barrier. It plays a crucial role in controlling the movement of substances in and out of the cell, thereby maintaining the internal environment necessary for the cell's proper functioning.

CELL MEMBRANE

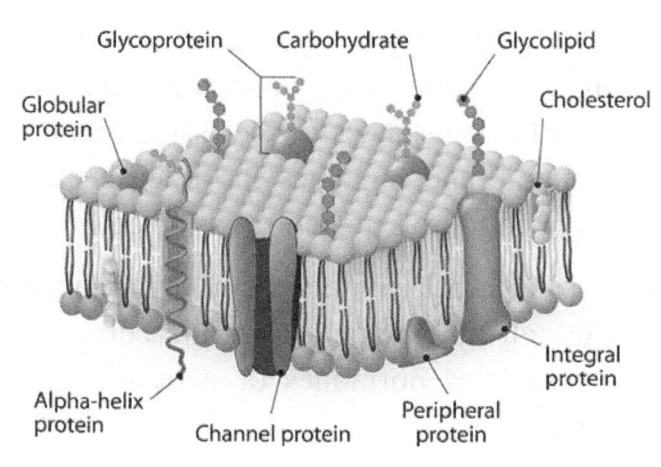

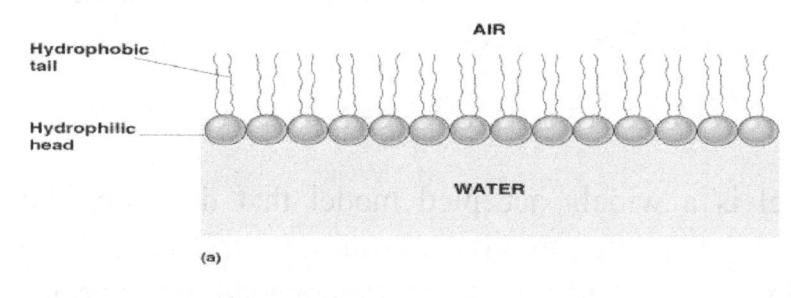

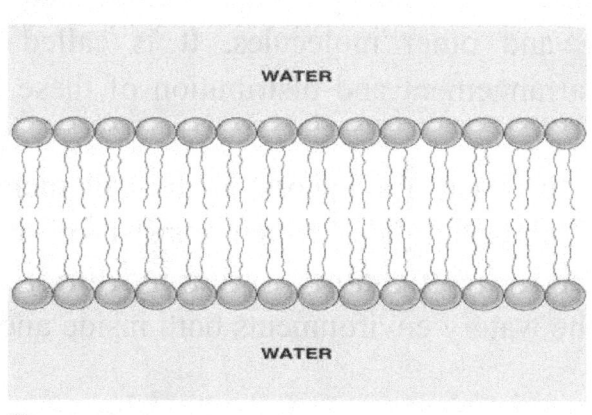

Composed of a phospholipid bilayer embedded with proteins and other molecules, the cell membrane exhibits a semi-permeable nature. This means that while some substances can freely pass through the membrane, others require specific mechanisms or assistance to cross it. The phospholipid bilayer consists of two layers of phospholipids, with their hydrophilic (water-loving) heads facing outward and their hydrophobic (water-repelling) tails facing inward. This arrangement creates a barrier that prevents the free diffusion of hydrophilic and large molecules across the membrane. Embedded within the phospholipid bilayer are various proteins that serve as transporters, channels, and receptors. These proteins play a key role in facilitating the movement of specific substances across the membrane. Transport proteins, such as carrier proteins and ion channels, provide pathways for ions and molecules to traverse the membrane, either through passive processes like diffusion or through active processes that require energy.

The cell membrane's ability to control what moves in and out of the cell is crucial for maintaining the cell's internal environment. It allows the cell to selectively take in essential nutrients, such as glucose and amino acids, while preventing the entry of harmful substances or excessive amounts of certain ions. Likewise, the cell membrane regulates the release of waste products and other molecules that need to be expelled from the cell.

The cell membrane also plays a role in cell signaling and communication. The presence of receptors on the cell membrane allows the cell to detect and respond to external signals, such as hormones or neurotransmitters, triggering specific cellular responses.

The main function of the plasma membrane is to
- protect the cell from its surrounding environment.
- It is semi-permeable and regulates the materials that enter and exit the cell.
- Another important function of the membrane is to facilitate communication and signaling between cells.

7.3 FLUID MOSAIC MODEL

The fluid mosaic model is a widely accepted model that describes the structure and behavior of the cell membrane. According to this model, the cell membrane is composed of a fluid lipid bilayer embedded with various proteins, carbohydrates, and other molecules. It is called a "mosaic" because of the diverse arrangement and distribution of these components within the membrane.

The lipid bilayer forms the basic framework of the cell membrane. It is primarily composed of phospholipids, which consist of a hydrophilic (water-loving) head and hydrophobic (water-repelling) tails. The hydrophilic heads face the watery environments both inside and outside the cell, while the hydrophobic tails align with each other in the interior of the bilayer, shielding themselves from the surrounding water.

This lipid bilayer provides a fluid matrix in which various proteins, carbohydrates, and other molecules are embedded. The proteins within the membrane can have different functions, such as transporters, receptors, enzymes, and structural components. Carbohydrates are often found attached to lipids or proteins on the outer surface of the membrane, forming glycolipids or glycoproteins.

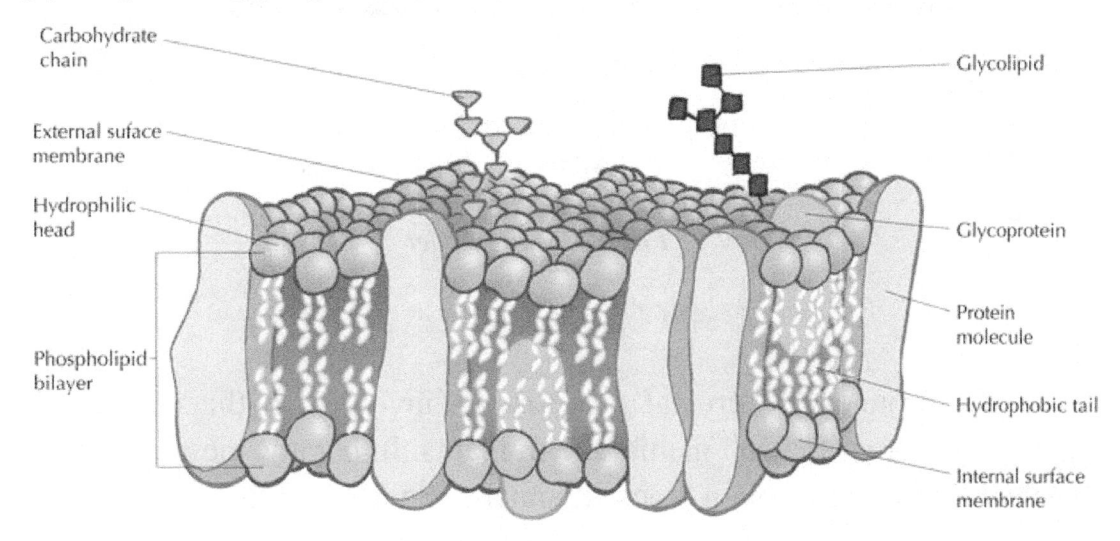

One of the key features of the fluid mosaic model is the concept of fluidity. The cell membrane is not a rigid structure but rather behaves like a fluid. The lipid bilayer allows for lateral movement of individual lipid molecules, resulting in the dynamic nature of the membrane. The proteins and other molecules within the membrane can also move and diffuse within the lipid bilayer.

This fluidity and the ability of lipids, proteins, and carbohydrates to freely diffuse within the membrane contribute to the diverse functions of the cell membrane. It allows for the flexibility of the membrane, facilitating processes such as cell growth, division, and the movement of molecules in and out of the cell. The fluid nature of the cell membrane also enables the formation of specialized structures, such as lipid rafts, which play a role in organizing specific proteins and signaling processes.

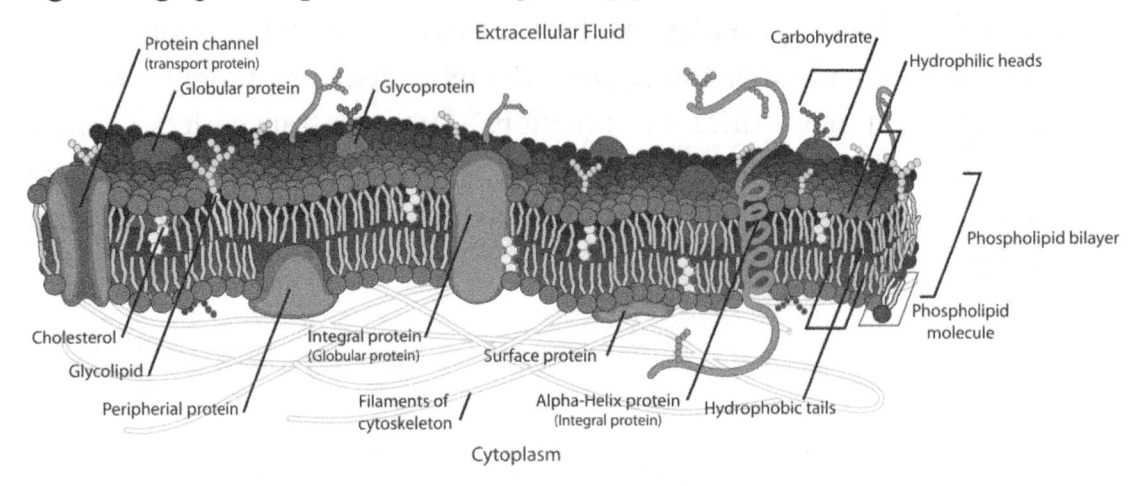

Fluid Mosaic Model

7.4 TYPES OF TRANSPORT PROTEIN

Transport proteins are crucial components involved in the movement of substances across the cell membrane in plants. In botany, these proteins can be classified into two main types: channel proteins and carrier proteins.

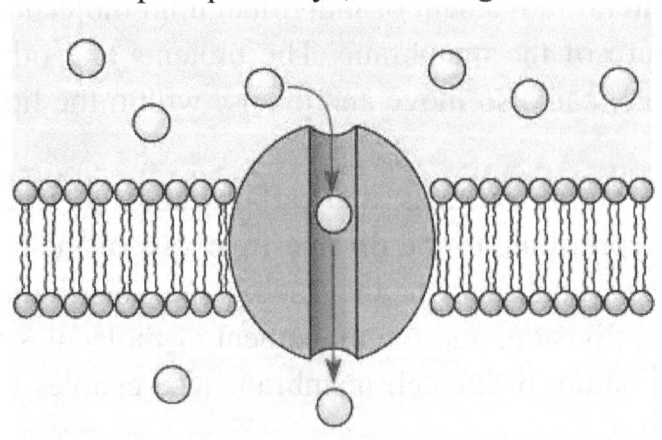

Channel proteins are embedded within the cell membrane and form pore-like structures that allow specific ions or molecules to pass through. They create open pathways, enabling the facilitated movement of substances across the membrane. Channel proteins possess selectivity, meaning they only allow certain substances to pass based on factors such as size, charge, or chemical properties. Through these channels, plants are able to regulate the transport of ions and small molecules, maintaining ion concentrations and facilitating the exchange of nutrients and signaling molecules.

Carrier proteins, also known as transporters, helps in transporting specific molecules or ions across the cell membrane. These proteins undergo conformational changes when they bind to a specific substance, allowing it to be transported across the membrane. Carrier proteins exhibit specificity in their binding sites, ensuring the selective transport of particular molecules. They can transport a wide range of substances, including sugars, amino acids, and ions. Carrier proteins are essential for the uptake of nutrients from the environment, the distribution of metabolites within the plant, and the removal of waste products.

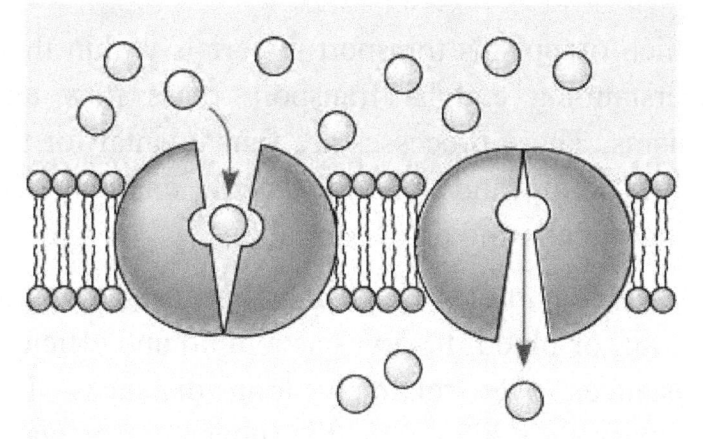

The presence of these transport proteins in plant cells is vital for the successful functioning and survival of plants. They enable plants to acquire essential nutrients from the soil, transport water and minerals through roots and stems, and distribute sugars produced during photosynthesis to various parts of the plant. The regulated movement of these substances is critical for plant growth, development, and overall metabolism.

CHAPTER 8

PHYSIOLOGY

Two primary methods by which organisms transport materials within their bodies are crucial for understanding cellular transport: mass flow and membrane transport mechanisms. These processes are fundamental for the movement of substances over short and long distances, enabling the survival and proper functioning of cells and organisms.

Mass flow is a mechanism in which particles are physically carried along in a fluid stream, such as water, air, or blood. It serves as a rapid and efficient means of transporting substances over relatively long distances. For example, in vascular plants, mass flow allows for the movement of water and nutrients from the roots to the leaves through the xylem vessels. In animals, mass flow is observed in the circulatory system, where blood carries oxygen, nutrients, and waste products to and from various tissues and organs.

On the other hand, within cells and across cellular membranes, diffusion, osmosis, and active transport are crucial mechanisms for moving single molecules or small structures. Diffusion is the passive movement of molecules from an area of higher concentration to an area of lower concentration, driven by the natural tendency of particles to spread out and achieve equilibrium. Osmosis is a specific type of diffusion that refers to the movement of water across a selectively permeable membrane to equalize the concentration of solutes on both sides. Active transport, in contrast to passive processes, requires energy expenditure to move substances against their concentration gradient, often using specialized transport proteins.

The transport of substances in and out of cells is of utmost importance in biology, as it is essential for cellular function and the survival of organisms. Nutrients must enter cells, while waste products and toxins must be eliminated. The cell membrane acts as a selective barrier, allowing only certain substances to pass through. Diffusion, osmosis, and active transport are the primary means by which substances cross the cell membrane. Mass flow, although crucial at higher organizational levels, such as organs, tissues, and whole organisms, cannot directly transport substances across cellular membranes.

8.1 SIMPLE DIFFUSION

Simple diffusion is a passive process that does not require the input of energy. It allows molecules to move from an area of high concentration to an area of low concentration. In this process, substances naturally tend to move down their concentration gradient, which is the difference in concentration between two regions.

When there is a higher concentration of a particular molecule in one area compared to another, the molecules will diffuse or spread out to achieve an equilibrium state. The driving force behind this movement is the inherent tendency of molecules to distribute evenly. As a result, molecules move freely through the spaces between other molecules until their concentration is equalized throughout the system.

Simple diffusion occurs across cell membranes when molecules are small and non-polar or when there is a concentration gradient present. The cell membrane acts as a selectively permeable barrier, allowing certain molecules to pass through freely. For example, small lipophilic molecules, such as oxygen and carbon dioxide, can readily diffuse through the lipid bilayer of the cell membrane.

Simple Diffusion

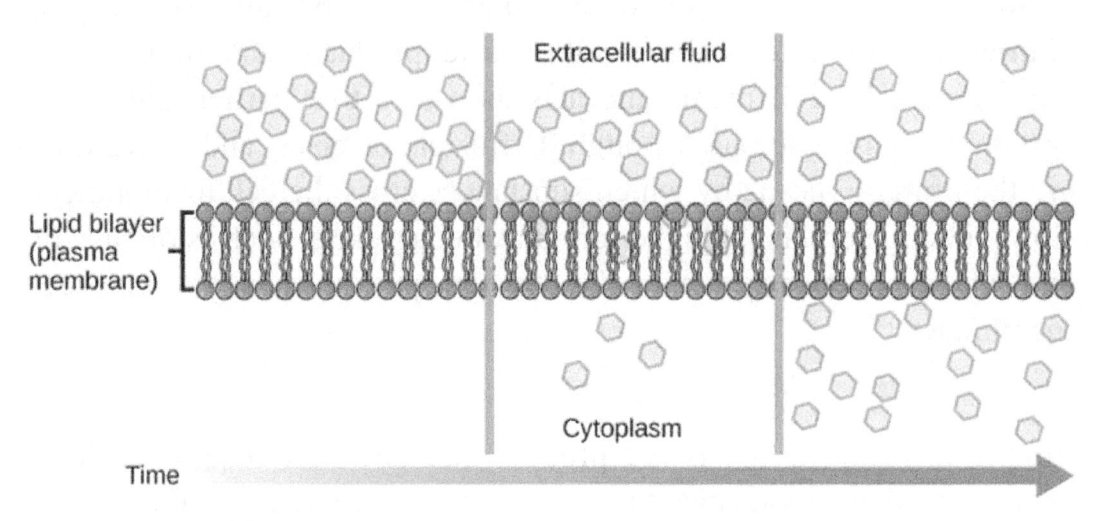

Lipid bilayer (plasma membrane)

Extracellular fluid

Cytoplasm

Time

The rate of diffusion depends on factors such as the concentration gradient, the size of the molecules, and the permeability of the membrane. The larger the concentration gradient, the faster the rate of diffusion. Additionally, smaller molecules diffuse more rapidly than larger ones. The permeability of the membrane to a specific molecule also influences the rate of diffusion, as some molecules may require the assistance of specialized transport proteins.

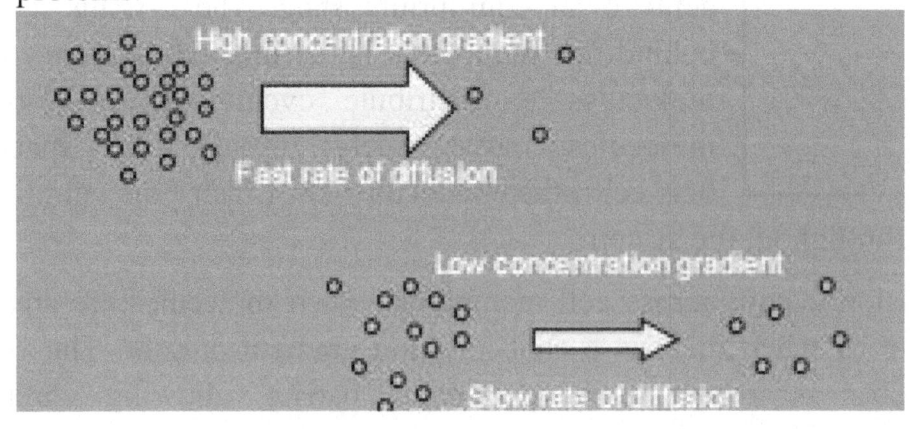

The *difference* between the regions of high concentration and low concentration is called the *concentration gradient*. The *steeper* the concentration gradient, the *faster* diffusion takes place

What do you see here?

What is diffusing into what?

8.2 DIFFUSION

Diffusion and simple diffusion are often used interchangeably, but there is a subtle distinction between the two terms.

Diffusion refers to the process by which molecules or particles move from an area of higher concentration to an area of lower concentration. It is a fundamental concept in physics and biology that describes the spontaneous movement of substances to achieve equilibrium.

Simple diffusion, on the other hand, specifically refers to the passive movement of molecules or particles directly through a membrane or across a concentration gradient without the involvement of any transport proteins or energy expenditure. In simple diffusion, molecules move freely through the spaces within the lipid bilayer of the cell membrane or across a permeable barrier. This process occurs when the molecules are small, non-polar, and able to dissolve in the lipid portion of the membrane.

Diffusion is a general term describing the movement of substances from high to low concentration, while simple diffusion is a specific type of diffusion that occurs directly through a membrane without the involvement of transport proteins or energy. Simple diffusion is a subset of diffusion that is applicable to certain molecules and membranes that meet the criteria for passive movement.

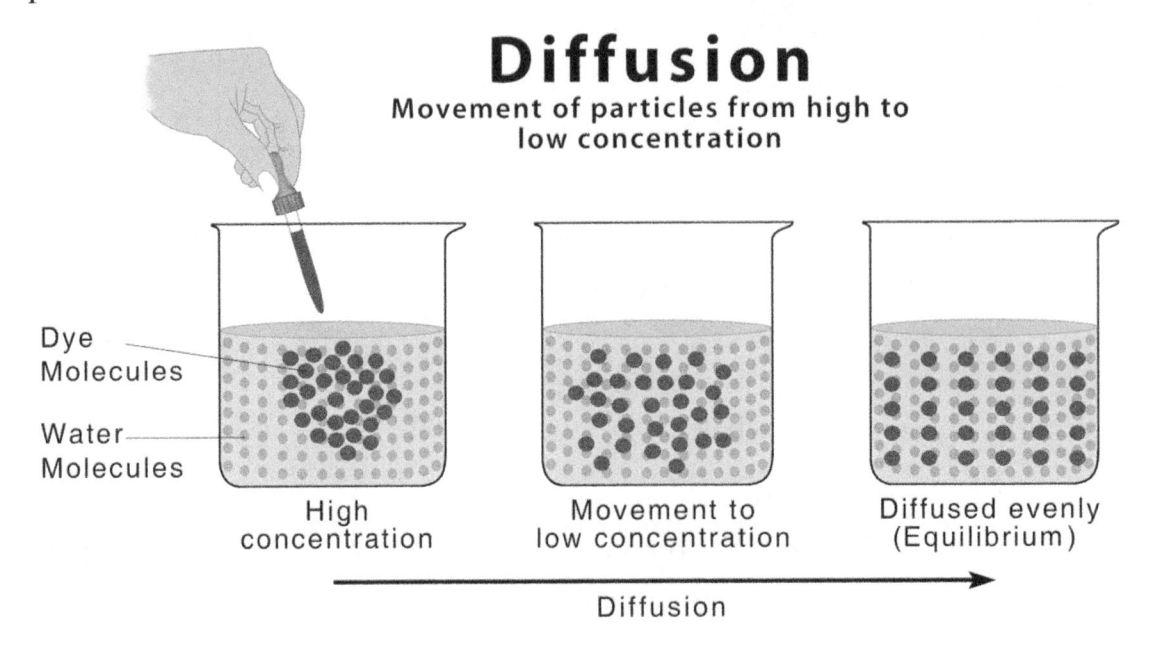

Diffusion
Movement of particles from high to low concentration

Dye Molecules
Water Molecules

High concentration — Movement to low concentration — Diffused evenly (Equilibrium)

Diffusion →

8.3 PASSIVE TRANSPORT: SIMPLE DIFFUSION AND FACILITATED DIFFUSION.

8.3.1 FACILITATED DIFFUSION

Facilitated diffusion is a significant process involved in the transport of specific molecules across the cell membrane in plants. It is a form of passive transport that does not require the input of energy. Facilitated diffusion relies on specialized transport proteins to facilitate the movement of molecules from an area of higher concentration to an area of lower concentration.

Certain molecules, such as glucose or amino acids, are unable to freely diffuse through the lipid bilayer of the cell membrane due to their size or charge. Instead, they rely on specific transport proteins called carriers or channels to aid their movement. These transport proteins act as selective pathways, allowing the molecules to cross the membrane. Channel proteins are integral membrane proteins that form pores or channels in the cell membrane. These channels are selective and permit the passage of specific molecules based on their size or charge. They enable the movement of ions and small molecules across the membrane.

Carrier proteins, on the other hand, bind to the molecules they transport and undergo conformational changes to facilitate their passage across the membrane. When a molecule binds to the carrier protein, it triggers a change in the protein's shape, allowing the molecule to be transported to the other side of the membrane. Carrier proteins exhibit high selectivity and transport specific molecules or closely related groups of molecules.

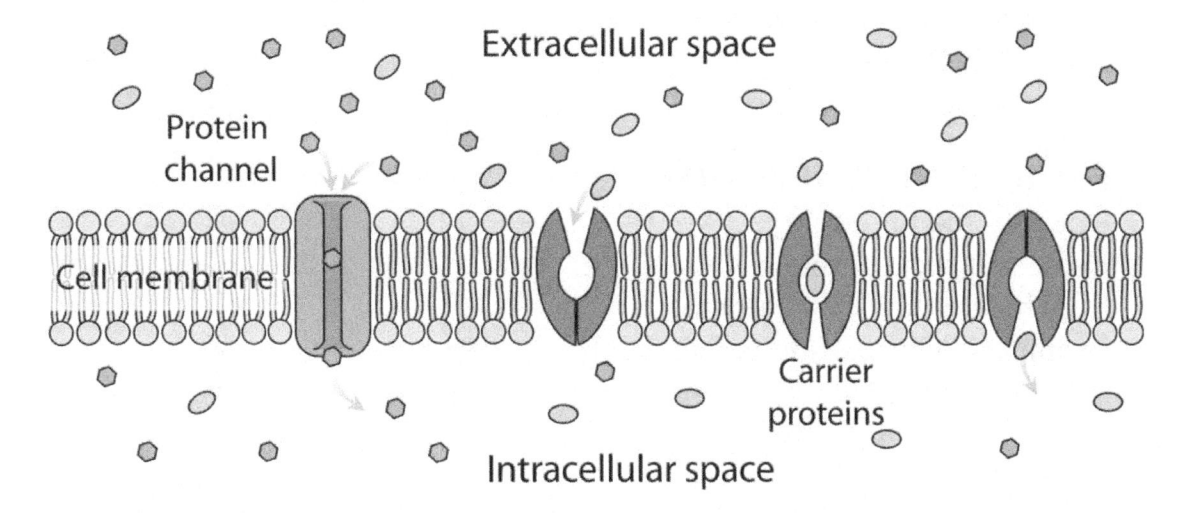

Facilitated diffusion is particularly important for the uptake of essential molecules by plant cells. For instance, glucose, a vital energy source for cells, is transported into the cells through facilitated diffusion. The presence of transport proteins on the cell membrane enables the movement of glucose from areas of higher concentration, such as the surrounding environment or the vascular system, into the cells. Its provides insights into how plants acquire and distribute essential molecules. It contributes to the efficient uptake of nutrients from the environment, ensuring the proper functioning and growth of plants. The coordinated action of channel proteins and carrier proteins in facilitated diffusion is crucial for maintaining the balance of molecules within plant cells.

8.3.2 SUMMARY: PASSIVE TRANSPORT

DIFFUSION

0Diffusion is a PASSIVE process which means no energy is used to make the molecules move, they have a natural KINETIC ENERGY.

SIMPLE DIFFUSION
Doesn't require energy
Moves from high to low concentration
Example: Oxygen or water diffusing into a cell and carbon dioxide diffusing out.

FACILITATED DIFFUSION
Does not require energy.
Uses transport proteins to move from high to low concentration
Examples: Glucose or amino acids moving from blood into a cell.

Passive transport

simple diffusion

facilitated diffusion

Materials move down their concentration gradient through the phospholipid bilayer.

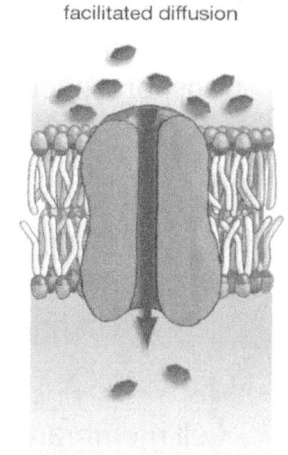

The passage of materials is aided both by a concentration gradient and by a transport protein.

8.4　ACTIVE TRANSPORT

Active transport is an essential cellular process that enables the movement of molecules or ions against their concentration gradient, from an area of lower concentration to an area of higher concentration. Unlike passive transport mechanisms such as diffusion and facilitated diffusion, active transport requires the expenditure of energy in the form of adenosine triphosphate (ATP).

One prominent example of active transport is the Sodium-Potassium Pump. This specialized transport protein, found in the cell membrane of many cells, actively pumps sodium ions (Na+) out of the cell and potassium ions (K+) into the cell. The pump uses the energy derived from ATP hydrolysis to carry out this uphill transport against their respective concentration gradients.

The Sodium-Potassium Pump maintains a higher concentration of potassium ions inside the cell and a higher concentration of sodium ions outside the cell. This concentration gradient is vital for various cellular processes such as nerve impulse transmission, muscle contraction, and nutrient uptake.

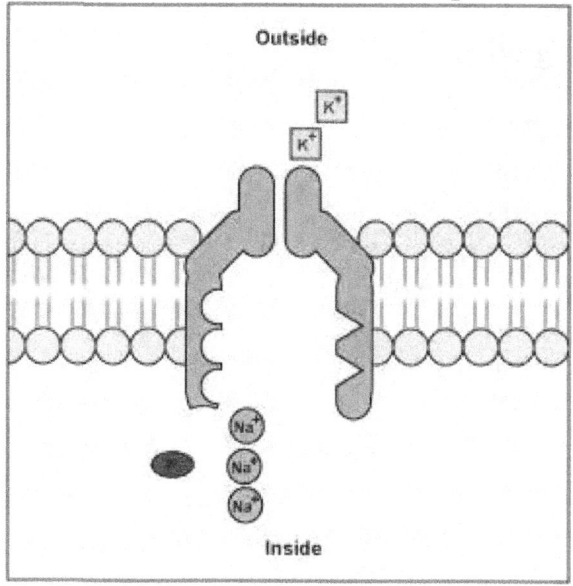

The pump operates by binding three sodium ions from the intracellular environment. This binding stimulates the phosphorylation of the pump by transferring a phosphate group from ATP to the pump. The phosphorylation induces a conformational change in the pump, leading to the release of the sodium ions outside the cell. Subsequently, two potassium ions from the extracellular space bind to the pump, triggering the dephosphorylation of the pump and returning it to its original conformation. This releases the potassium ions into the intracellular space.

The Sodium-Potassium Pump is an example of primary active transport, where the energy for transport directly comes from ATP hydrolysis. This process is crucial for maintaining the electrochemical balance across the cell membrane and ensuring proper cell function.

8.4.1 SUMMARY: TYPES OF TRANSPORT

Passive Transport
- Molecules move from area of high concentration to area of low concentration.
- Movement is down the concentration gradient.
- No energy needed.
- Small molecules such as H_2O, O_2 and CO_2.

Active Transport
- Molecules move from area of low concentration to area of high concentration.
- Movement is up the concentration gradient.
- Energy is required (ATP).
- Large molecules, ions.

8.5 OSMOSIS

Osmosis is a process that describes the diffusion of water molecules across a selectively permeable membrane. It occurs from an area of higher water concentration to an area of lower water concentration. The direction of water movement is influenced by the concentration of solutes, such as salts or sugars, on either side of the membrane.

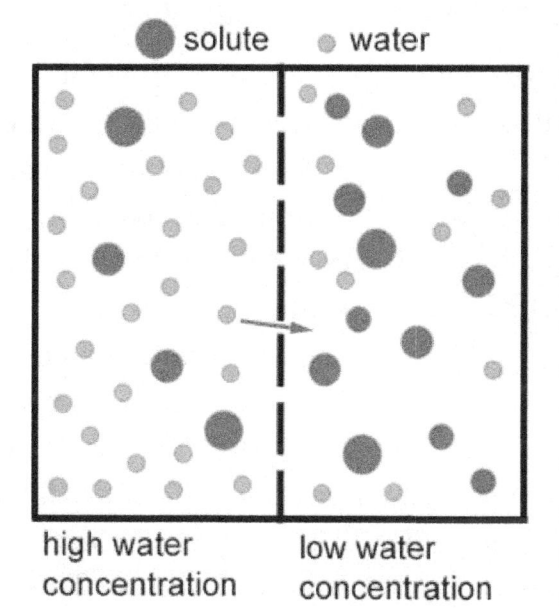

Water has a natural tendency to move from an area of higher water concentration (lower solute concentration) to an area of lower water concentration (higher solute concentration) in order to equalize the solute concentration on both sides of the membrane. This movement of water helps maintain osmotic balance and enables the transport of essential substances within cells and across different compartments in plants.

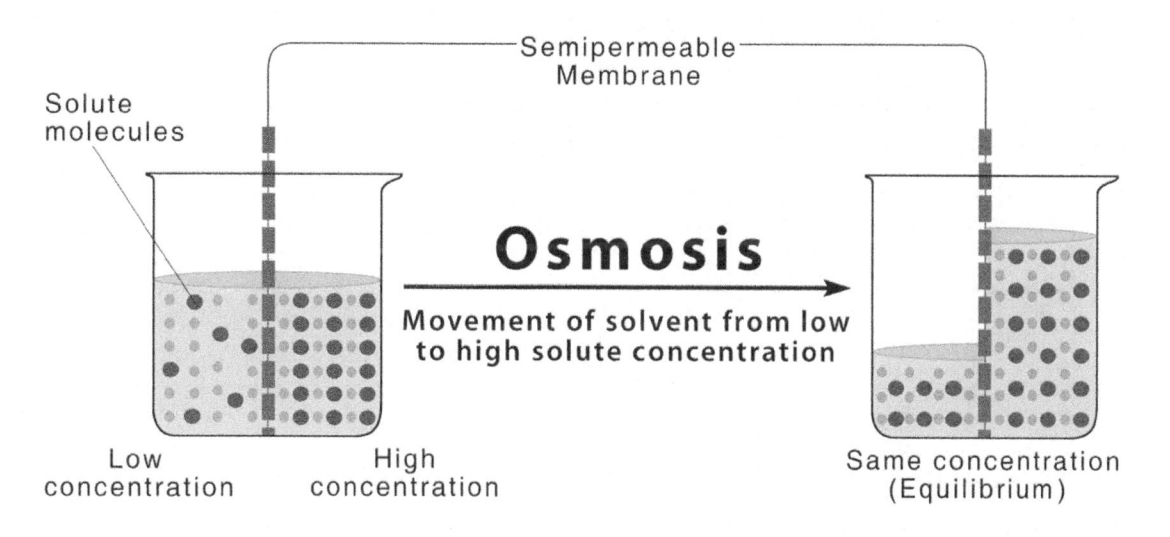

The driving force behind osmosis is the attraction between water molecules and solute particles. Water molecules are attracted to solutes and tend to move towards areas of higher solute concentration to dilute the solute concentration. Consequently, water molecules will also move to areas of lower solute concentration to balance the concentration gradient.

Partially-permeable membranes, also known as semi-permeable or selectively permeable membranes, allow certain molecules, such as water, to pass through while restricting the passage of larger solute molecules. The permeability of the membrane is determined by its structure and the specific characteristics of the molecules involved. In plant cells, the cell membrane acts as a partially-permeable membrane. It allows water molecules to freely pass through via osmosis while selectively controlling the movement of solutes. This is important for regulating the water balance within cells and maintaining cell turgor pressure, which is essential for plant structure and function.

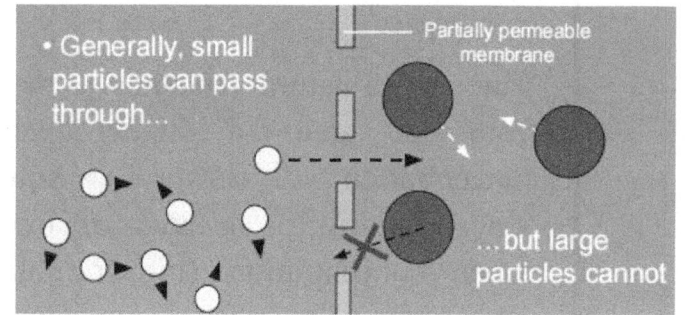

DIFFUSION OF WATER ACROSS A MEMBRANE

High water concentration ⟶ Low water concentration
Low solute concentration ⟶ High solute concentration

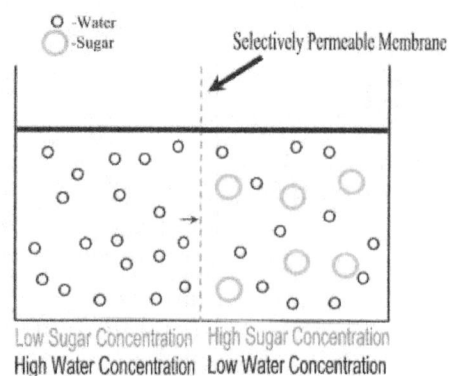

8.5.1 WHAT CONTROLS OSMOSIS?

One important factor that regulates osmosis is the presence of a concentration gradient, which refers to the unequal distribution of particles. In osmosis, water molecules move across a partially-permeable membrane from an area of higher water concentration to an area of lower water concentration. This movement is driven by the concentration gradient established by the difference in solute concentrations on either side of the membrane.

The concentration gradient plays a significant role in determining the direction and intensity of osmotic flow. When one side of the membrane has a higher concentration of solute particles, water molecules are drawn towards that region, resulting in the movement of water to equalize the concentration on both sides of the membrane.

Controlled manipulation of the concentration gradient is essential for regulating osmosis in plant cells. Cells and organisms possess various mechanisms to maintain or alter solute concentrations, thereby influencing the movement of water. For example, active transport processes can actively move solute particles across the cell membrane, creating concentration gradients that drive osmotic water movement.

Environmental factors, such as temperature and pressure, can also impact osmosis. Changes in temperature affect the kinetic energy of water molecules and can influence the rate of osmotic movement. Pressure, including hydrostatic and osmotic pressure, can counteract or enhance water movement, thereby affecting the overall osmotic process.

In summary, osmosis is controlled by factors such as the concentration gradient, solute characteristics, and environmental conditions. Managing the concentration gradients are important in regulating osmosis, which is vital for water regulation, nutrient uptake, and maintaining cellular homeostasis in plants.

CONCENTRATIONS AFFECT OSMOSIS

Water moves across membrane (in/out) at equal rates.

Water moves out of the cell

Water moves into the cell

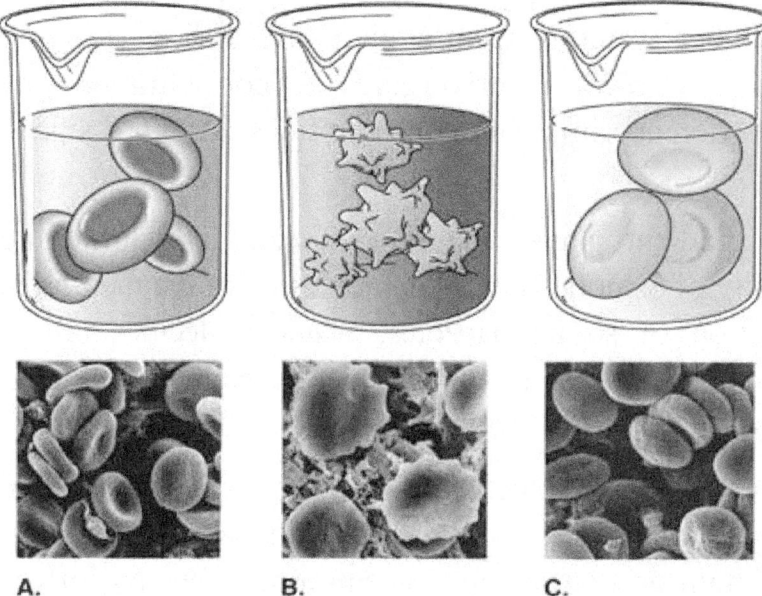

A.
Isotonic solution (equal concentration of ions in solution and cell)

B.
Hypertonic solution (higher concentration of ions in solution than in cell)

C.
Hypotonic solution (lower concentration of ions in solution than in cell)

8.6   CELLS IN SOLUTIONS

ISOTONIC

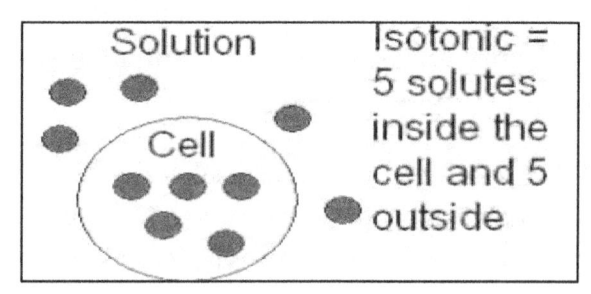

Isotonic = 5 solutes inside the cell and 5 outside

A solution whose solute concentration is the same as the solute concentration inside the cell.

HYPOTONIC

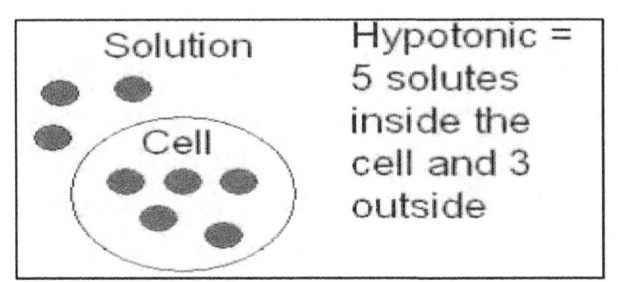

Hypotonic = 5 solutes inside the cell and 3 outside

A solution whose solute concentration is lower than the solute concentration inside a cell.

HYPERTONIC

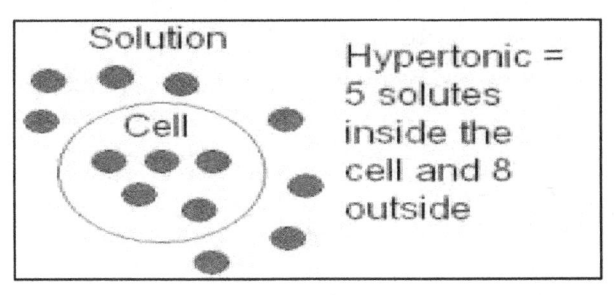

Hypertonic = 5 solutes inside the cell and 8 outside

A solution whose solute concentration is higher than the solute concentration inside a cell.

8.6.1   CELL IN ISOTONIC SOLUTION

When a cell is in an isotonic solution, the direction of water movement is balanced. The solute and water concentrations are the same inside and outside the cell, resulting in no net movement of water. The cell is in a state of equilibrium, and water will flow in both directions, both outside and inside the cell.

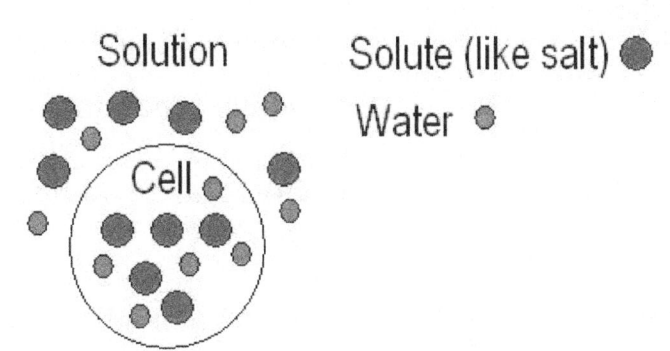

Solute (like salt) ●
Water ○

The solute and water concentrations are the same inside and outside the cell.

8.6.2 CELL IN HYPOTONIC SOLUTION

In a hypotonic solution, the direction of water movement is towards the inside of the cell. The external solution has a lower solute concentration compared to the cell's interior. As a result, water is attracted to the higher solute concentration inside the cell, causing it to flow into the cell This influx of water can lead to cell swelling or even bursting in extreme cases.

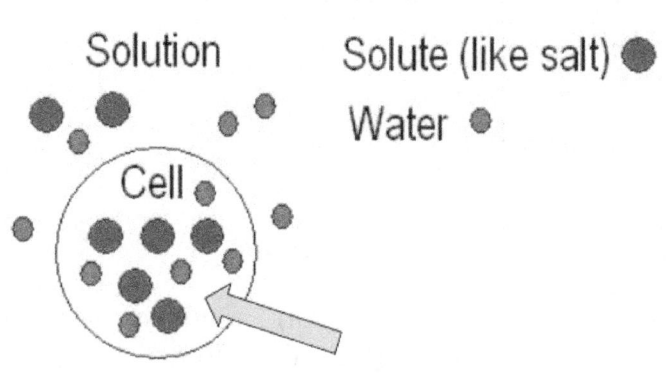

8.6.3 CELL IN HYPERTONIC SOLUTION

Conversely, in a hypertonic solution, the water movement is directed out of the cell. The external solution has a higher solute concentration than the cell's interior. Water molecules are attracted to the higher solute concentration outside the cell, resulting in a net flow of water from inside the cell to the external solution. This loss of water can lead to cell shrinkage and potential cell damage.

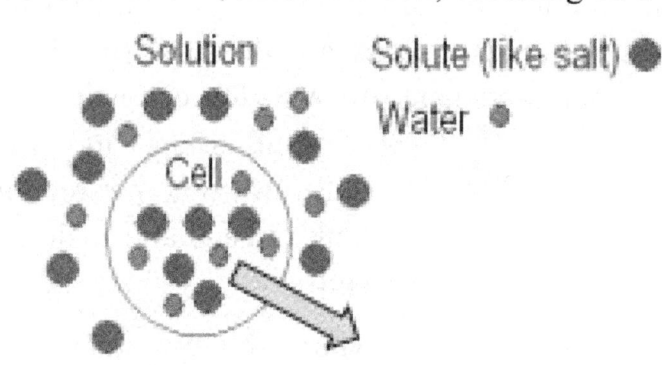

The solute concentration is greater outside the cell, therefore water will flow outside the cell.

The direction of water movement in cells is dependent on the relative concentrations of solutes inside and outside the cell.

8.7 ENDOCYTOSIS AND EXOCYTOSIS

8.7.1 ENDOCYTOSIS

Endocytosis is a cellular process used by plant cells to internalize materials from their external environment. It involves the formation of specialized vesicles, which are small portions of the cell membrane that surround and engulf the external substances. This allows the cell to bring these substances into its interior for various purposes, such as nutrient uptake, receptor-mediated signaling, and the removal of waste materials.

There are three primary types of endocytosis that occur in plant cells:

1. **Phagocytosis:** This type of endocytosis involves the engulfment of large solid particles, such as food particles or foreign pathogens. Specialized cells, including certain immune cells, are particularly involved in phagocytosis as part of the plant's defense mechanisms.

2. **Pinocytosis:** Known as "cell drinking," pinocytosis is the process by which cells internalize fluid droplets or dissolved substances from their surroundings. It allows the cell to take in extracellular fluid and dissolved molecules through the formation of small vesicles.

3. **Receptor-mediated endocytosis:** This type of endocytosis relies on specific interactions between molecules, called ligands, and their corresponding receptors on the cell surface. It enables the selective internalization of specific substances, such as hormones or growth factors, through the formation of vesicles.

8.7.2 EXOCYTOSIS

Exocytosis is a reverse process to endocytosis, where plant cells release materials from their interior into the extracellular space. It is a crucial mechanism for cellular secretion, intercellular communication, and the removal of cellular waste.

During exocytosis, secretory vesicles or granules, which are membrane-bound structures within the cell, fuse with the cell membrane. This fusion enables the contents of these vesicles to be released outside the cell. The fusion process is mediated by specific proteins and is triggered by various cellular signals, such as changes in calcium concentration or specific signaling pathways.

Exocytosis function in the release of neurotransmitters in nerve cells, the secretion of hormones from endocrine cells, and the export of enzymes and other proteins from cells. It facilitates inter-cellular communication and helps maintain the proper functioning of plant cells.

CELLULAR TRANSPORT MECHANISMS

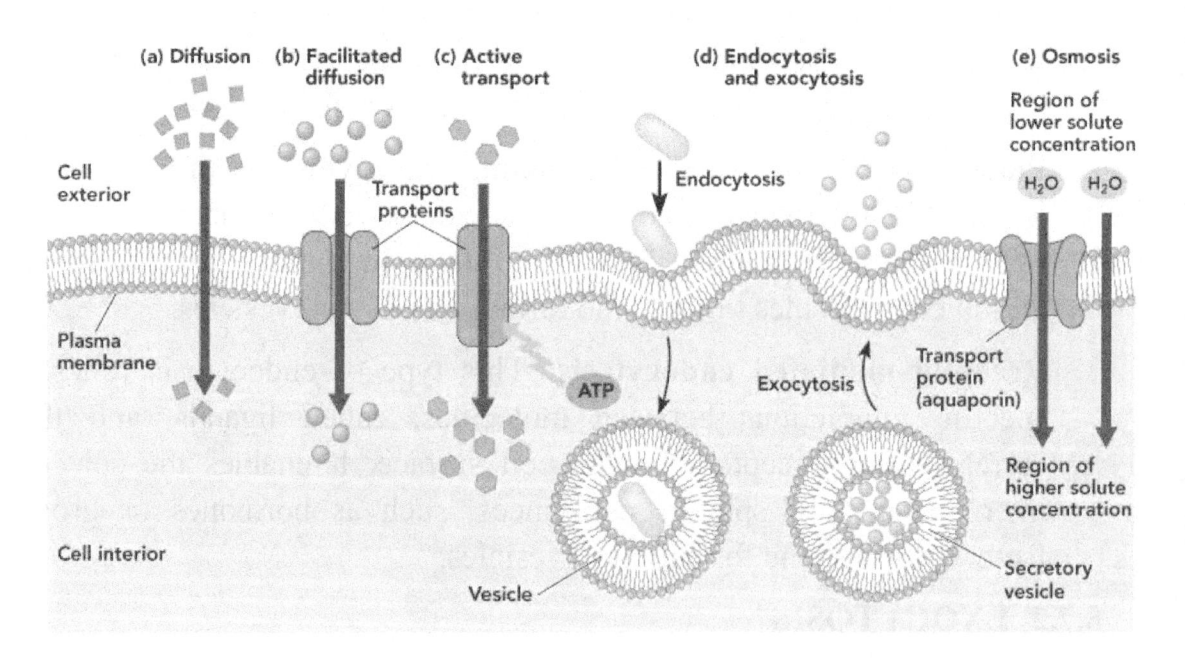

OSMOSIS IN PLANT AND ANIMAL CELLS

The drawings below show the appearance of a red blood cell and a plant cell in isotonic, hypotonic, and hypertonic environments. Label each environment in the spaces provided.

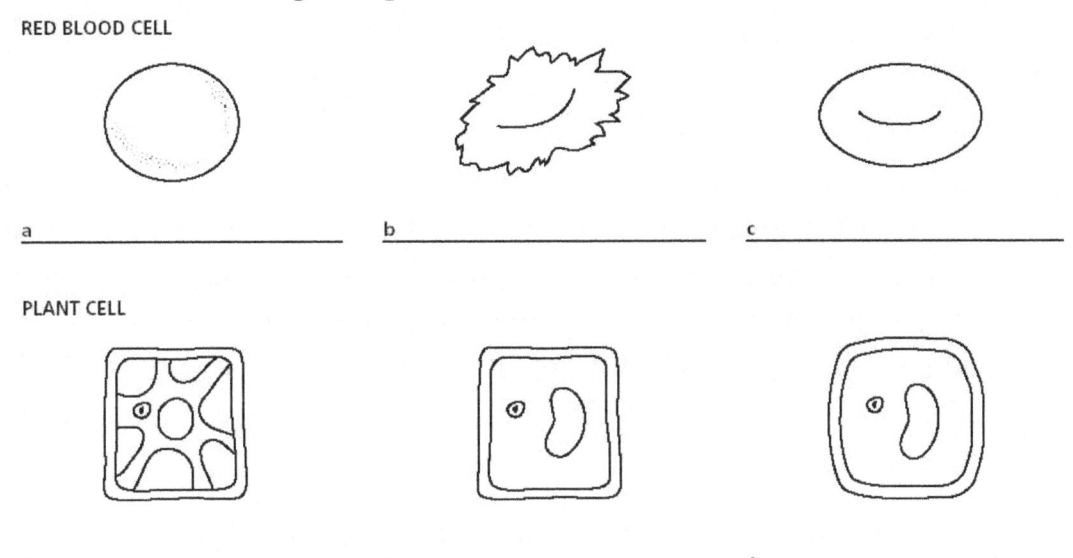

RED BLOOD CELL

a _____ b _____ c _____

PLANT CELL

d _____ e _____ f _____

CHAPTER 9

PLANT HORMONES AND GROWTH REGULATION

9.1 INTRODUCTION TO PLANT HORMONES

Plant hormones, also known as phytohormones, are essential chemical messengers that regulate and coordinate various aspects of plant growth, development, and responses to environmental stimuli. They play a pivotal role in ensuring the proper functioning and adaptation of plants in their dynamic surroundings. The study of plant hormones has greatly advanced our understanding of plant biology and has significant implications for agriculture, horticulture, and biotechnology.

HISTORICAL DEVELOPMENT AND KEY SCIENTISTS IN PLANT HORMONE RESEARCH

The exploration of plant hormones has a rich historical background, characterized by landmark discoveries made by eminent scientists in the field. Charles Darwin and his son Francis Darwin made pioneering contributions in the late 19th century by observing the growth patterns of coleoptiles in canary grass. Their investigations led to the recognition of an unknown substance that influenced plant growth and initiated the path to the discovery of plant hormones.

During the early 20th century, Fritz Went demonstrated the existence of a mobile chemical substance, which he named "auxin," originating from the tip of oat seedlings' coleoptiles. This pivotal finding laid the foundation for understanding the role of auxins in regulating diverse processes associated with plant growth and development.

Further advancements in plant hormone research came through the efforts of exceptional scientists. Kenneth V. Thimann made significant contributions to elucidate the properties and functions of plant hormones, while Johannes van Overbeek contributed to the discovery of cytokinins and their involvement in cell division and differentiation. Carl Djerassi, renowned for synthesizing plant hormones, made notable strides in understanding their chemical structures and functions. The identification and characterization of additional plant hormones, such as gibberellins, abscisic acid (ABA), and ethylene, expanded our understanding of the complexity and versatility of plant hormonal regulation. These discoveries, combined with modern techniques in molecular biology and genetic engineering, have propelled plant hormone research to new frontiers, revealing intricate signaling networks and interactions among different hormones.

The study of plant hormones provides a framework for comprehending the mechanisms through which plants perceive and respond to internal and external cues. It equips students with a profound understanding of the intricate processes governing plant growth, development, and adaptation. It fosters the application of this knowledge to address practical challenges in agriculture, horticulture, and the development of sustainable plant-based industries.

9.2 AUXINS

Auxins are a class of plant hormones that play a fundamental role in the regulation of growth and development in higher plants. Discovered and extensively studied by Charles Darwin and his son Francis, auxins have since garnered significant attention in the field of plant physiology. These hormones exert their effects on various physiological processes, including cell elongation, tissue differentiation, apical dominance, phototropism, gravitropism, and root development.

CHEMICAL STRUCTURE OF AUXINS: Auxins, including the primary hormone IAA, are organic compounds derived from the amino acid tryptophan. The core structure of auxins consists of an indole ring fused with a carboxylic acid group. The carboxylic acid group can undergo various modifications, such as esterification or amidation, which affect auxin activity and stability. Additionally, auxins can exist in different forms, including free acids, conjugated forms with other molecules, or as inactive precursors. The structural diversity of auxins allows for distinct physiological effects and regulatory mechanisms.

BIOSYNTHESIS OF AUXINS: The biosynthesis of auxins occurs primarily in young developing tissues, such as shoot apices, root tips, and developing fruits. It involves a multistep process that begins with the conversion of tryptophan into indole-3-pyruvic acid (IPA). Subsequent enzymatic reactions catalyze the conversion of IPA to IAA, the primary active form of auxins. Further modifications, such as conjugation with sugars or amino acids, can occur to regulate auxin transport and distribution within the plant. The biosynthetic pathway of auxins is tightly regulated, allowing for precise control of hormone levels and responses.

TRANSPORT OF AUXINS: Auxins exhibit polar transport within plants, enabling them to move directionally from their site of synthesis to target tissues. This polar transport is facilitated by specialized transport proteins, including PIN-FORMED (PIN) proteins, AUXIN RESISTANT1/LIKE AUX1 (AUX/LAX) influx carriers, and ATP-binding cassette (ABC) transporters. The movement of auxins can occur through both cell-to-cell transport and long-distance transport within the plant's vascular system. Polar auxin transport ensures the spatial distribution of auxins, enabling them to exert their effects on various plant tissues and organs.

RECEPTORS AND SIGNALING PATHWAYS: Auxins exert their physiological effects by binding to specific receptors and initiating signaling cascades within plant cells. The perception of auxins occurs through a family of auxin receptors known as the TIR1/AFB (TRANSPORT INHIBITOR RESPONSE1/AUXIN SIGNALING F-BOX) proteins. Upon auxin binding, the TIR1/AFB receptors form a complex with AUXIN/INDOLE-3-ACETIC ACID (Aux/IAA) repressor proteins, leading to their degradation. This degradation releases the transcriptional repressor function of Aux/IAA proteins, allowing for the activation of auxin-responsive genes, including those involved in growth and developmental processes.

ROLES AND FUNCTIONS OF AUXINS

Cell Elongation: Auxins act as potent stimulators of cell elongation, promoting growth in the shoot and root systems of plants. This process is predominantly mediated by the activation of proton pumps located at the plasma membrane of target cells. Through the establishment of a proton gradient, auxins lower the pH of the cell wall, triggering the enzymatic activity of expansins. Expansins loosen the cell wall structure, enabling the uptake of water and facilitating the elongation of cells. This mechanism allows auxins to contribute to the growth of plant organs and the overall architecture of the plant.

Tissue Differentiation: Auxins play a crucial role in the process of tissue differentiation, enabling the formation of specialized cell types. By influencing the expression of specific genes, auxins regulate the development of vascular tissues, such as xylem and phloem. Additionally, auxins contribute to the formation of meristematic tissues, which are responsible for continuous cell division and the generation of new organs. The precise spatial and temporal distribution of auxins is crucial for proper tissue differentiation and the establishment of plant architecture.

Apical Dominance: Auxins exert control over the phenomenon of apical dominance, wherein the growth of the main shoot is favored over the lateral branches. This regulatory mechanism ensures that the plant allocates its resources efficiently and maintains an optimal balance between upward growth and lateral branching. Auxins are synthesized in the apical meristem and transported downwards, inhibiting the growth of lateral buds. This inhibition occurs through the suppression of cytokinin synthesis in the lateral buds, limiting their growth potential. By regulating apical dominance auxins contribute to the overall shape and form of the plant.

Phototropism and Gravitropism: Auxins are critical mediators of plant responses to light (phototropism) and gravity (gravitropism). In phototropism, auxins accumulate on the shaded side of a plant, promoting cell elongation and bending towards the light source. This directional growth is achieved through the redistribution of auxins via polar auxin transport. Gravitropism, on the other hand, involves the redistribution of auxins in response to gravity. In shoots, auxins move towards the lower side inhibiting elongation, while in roots, auxins accumulate on the lower side, promoting growth. These tropic responses allow plants to optimize their growth and orientation in relation to environmental stimuli.

Root Development: Auxins are vital for root development, influencing both primary and lateral root growth. In the primary root, auxins stimulate cell elongation and promote root hair formation, enabling nutrient uptake and anchorage in the soil. Additionally, auxins play a role in the initiation and elongation of lateral roots, contributing to the overall root system architecture. The distribution and concentration gradients of auxins within the root system are carefully regulated, ensuring proper root growth and establishment.

9.3 GIBBERELLINS

Gibberellins are a class of plant hormones that work in regulating various aspects of plant growth and development. Discovered in the 20th century when studying the pathogen-induced elongation of rice plants, gibberellins have since been extensively studied and found to have diverse physiological functions. This section will explore the biosynthesis, mode of action, and effects of gibberellins on plant growth and development.

BIOSYNTHESIS OF GIBBERELLINS: Gibberellins are synthesized via the terpenoid pathway, starting from the precursor molecule geranylgeranyl diphosphate (GGDP). A series of enzymatic reactions catalyzed by specific enzymes, including copalyl diphosphate synthase (CPS) and ent-kaurene synthase (KS), lead to the formation of an intermediate called ent-kaurene. This intermediate is subsequently modified through oxidation and cyclization steps to produce various gibberellin forms, including the active gibberellic acid (GA3). The biosynthesis of gibberellins is tightly regulated and can be influenced by environmental and internal cues.

MODE OF ACTION: The mode of action of gibberellins involves their interaction with specific receptors located in the plant cells. Gibberellin receptors belong to the GIBBERELLIN INSENSITIVE DWARF1 (GID1) family, which consists of soluble proteins capable of binding to gibberellins. Upon gibberellin binding, the GID1 receptor undergoes a conformational change, allowing it to interact with and target DELLA proteins for degradation. DELLA proteins are negative regulators of gibberellin responses. Their degradation releases the repression on downstream gibberellin-responsive genes, allowing for the activation of growth-promoting processes.

ROLES AND FUNCTIONS OF GIBBERELLINS

Seed Germination: Gibberellins are involved in breaking seed dormancy and promoting seed germination. They stimulate the synthesis of enzymes, such as α-amylase, which degrade stored starch into sugars that fuel seedling growth. Gibberellins also promote the synthesis of hydrolytic enzymes that break down stored proteins, lipids, and nucleic acids, providing essential nutrients for seedling establishment.

Stem Elongation: One of the most well-known functions of gibberellins is their role in stem elongation. They promote cell elongation by stimulating cell division, enlargement, and cell wall loosening. Gibberellins induce the synthesis of enzymes involved in cell expansion, such as expansins, which facilitate cell wall loosening, allowing cells to elongate. This growth-promoting effect of gibberellins is particularly evident in internode elongation, leading to increased plant height.

Flowering: Gibberellins are involved in regulating flowering time and flower development. They interact with other flowering regulatory pathways, such as photoperiod, vernalization, and autonomous pathways, to promote the transition from vegetative growth to reproductive development. Gibberellins also play a role in flower induction, ensuring the timely initiation of floral organs.

Fruit Development: Gibberellins influence fruit development by promoting cell division, cell enlargement, and fruit growth. They stimulate the growth of fruit tissues, such as the pericarp, contributing to fruit size and overall fruit quality. Gibberellins also affect fruit ripening processes, including the production of enzymes that break down cell walls and promote fruit softening.

Delay of Senescence: Gibberellins can delay senescence, the process of aging and deterioration in plants. They inhibit the breakdown of chlorophyll and other photosynthetic pigments, thus maintaining photosynthetic activity and delaying leaf yellowing. Gibberellins also influence the synthesis and activity of enzymes involved in nutrient mobilization during senescence.

Induction of Parthenocarpy: Parthenocarpy refers to the development of fruit without fertilization. Gibberellins can induce parthenocarpy in certain plant species, promoting the formation and growth of seedless fruits. This has economic significance in horticulture, as it allows for the production of seedless fruits, such as seedless grapes or seedless watermelons.

9.4 CYTOKININS

Cytokinins are a class of plant hormones that play a fundamental role in regulating various physiological processes, including cell division, shoot formation, and plant growth and development. Their involvement in cell division and shoot formation makes cytokinins crucial for the proper growth and maintenance of plant tissues.

ROLES AND FUNCTIONS OF CYTOKININS

Role of Cytokinins in Cell Division: Cytokinins are known for their ability to promote cell division in plants. They stimulate the cell cycle progression from the G1 phase to the S phase, triggering DNA replication and subsequent cell division. Cytokinins achieve this by interacting with specific receptors, which are predominantly histidine kinases, located on the plant cell surface. The binding of cytokinins to their receptors initiates a signaling cascade that leads to the activation of cyclin-dependent kinases (CDKs) and other cell cycle regulatory proteins. This activation results in the progression of the cell cycle and the promotion of cell division.

Interaction with Auxins in Shoot Formation: Cytokinins also play a crucial role in shoot formation, including the initiation and outgrowth of buds and the formation of lateral shoots. The interaction between cytokinins and another class of plant hormones, auxins, is particularly important in this process. Auxins, which are primarily produced in the apical meristem and transported towards the basal parts of the plant, promote cell elongation and differentiation. Cytokinins, on the other hand, are involved in stimulating cell division and promoting the formation of new shoot meristems. The balance between cytokinins and auxins determines the formation and growth of shoots, with higher cytokinin-to-auxin ratios promoting bud and shoot initiation.

Regulation of Apical Dominance: Cytokinins also regulate apical dominance, which refers to the inhibition of lateral bud outgrowth by the apical bud. The apical bud produces higher levels of cytokinins compared to lateral buds, creating a cytokinin gradient in the plant. This gradient suppresses the growth and development of lateral buds by inhibiting their cell division and shoot formation. When the apical bud is removed or its cytokinin production is reduced, the lateral buds are released from apical dominance and can grow and develop into branches. This process is crucial for shaping the plant's architecture and determining the distribution of growth throughout the plant.

Synergistic Interactions with Other Growth Regulators: Cytokinins interact synergistically with other growth regulators to modulate cell division and shoot formation. For example, cytokinins work in conjunction with gibberellins, another class of plant hormones, to promote stem elongation and internode development. Cytokinins also interact with other growth regulators, such as ethylene and abscisic acid, to coordinate various aspects of plant growth and development. These interactions highlight the complexity of hormonal regulation in plants and the integrated nature of multiple signaling pathways.

9.5 ABSCISIC ACID (ABA)

Abscisic acid (ABA) is a plant hormone that plays a pivotal role in regulating various physiological processes, particularly in response to environmental stresses and the regulation of dormancy. ABA functions as a signaling molecule that helps plants adapt and survive unfavorable conditions.

ROLES AND FUNCTIONS OF ABSCISIC ACID (ABA)

Stress Responses: ABA is known for its involvement in the regulation of plant responses to various environmental stresses, including drought, salinity, cold, and heat. When plants experience stress, ABA levels increase, triggering a series of responses aimed at minimizing water loss and enhancing stress tolerance. ABA acts as a stress signal by modulating stomatal closure, reducing water loss through transpiration. It promotes the synthesis of proteins and other molecules that protect cells from damage caused by stress, such as antioxidants and osmoprotectants. ABA also regulates gene expression, activating stress-responsive genes that help plants adapt to adverse conditions.

Regulation of Seed Dormancy: ABA plays a critical role in the regulation of seed dormancy, a state of suspended growth and development that enables seeds to remain viable under unfavorable conditions until suitable germination conditions are encountered. During seed development, ABA accumulates, promoting seed maturation and inducing dormancy. ABA inhibits germination by preventing the activation of cell cycle processes and growth-related genes in the embryo. It also interacts with other hormones, such as gibberellins, to maintain seed dormancy. When conditions become favorable, a decrease in ABA levels, often in response to environmental cues like light and temperature changes, relieves dormancy and allows seed germination to occur.

Stomatal Regulation: ABA is a key regulator of stomatal behavior, which influences gas exchange and water loss in plants. In response to water deficit or drought conditions, ABA is synthesized and accumulates in leaves. It triggers the closure of stomata, the small pores on leaf surfaces, to reduce transpirational water loss. ABA induces the efflux of potassium ions from guard cells surrounding the stomatal pore, leading to osmotic water loss and subsequent stomatal closure. By regulating stomatal aperture, ABA helps plants conserve water and maintain cellular hydration under water-limiting conditions.

Developmental Processes: Apart from stress responses and dormancy regulation, ABA also participates in various developmental processes in plants. It influences seed development, controlling aspects such as embryo growth and maturation. ABA is involved in the regulation of leaf and bud development, promoting bud dormancy during winter months. ABA also influences the growth and differentiation of root systems, particularly under drought conditions, by regulating root architecture and elongation.

9.6 ETHYLENE

Ethylene is a plant hormone that function in the regulation of fruit ripening and senescence. It is a gaseous hormone that is involved in various physiological processes throughout a plant's life cycle.

ROLES AND FUNCTIONS OF ETHYLENE

Initiation of Fruit Ripening: Ethylene is a key regulator of fruit ripening, initiating the process that transforms mature, green fruits into fully ripe and flavorful ones. As fruits reach their physiological maturity, they start producing ethylene. The production of ethylene can be induced by various factors such as hormonal signals, changes in internal metabolic processes, and external stimuli like wounding or exposure to other ripe fruits. Once ethylene is produced, it acts as a signal to trigger a cascade of biochemical and physiological changes associated with fruit ripening.

Ethylene Signaling Pathway: Ethylene exerts its effects on fruit ripening through a complex signaling pathway. The perception of ethylene occurs through ethylene receptors, which are located on the cell surface. Upon ethylene binding, the receptors activate a signaling cascade that leads to changes in gene expression and cellular responses. The key component in this pathway is the transcription factor EIN3 (Ethylene Insensitive 3), which regulates the expression of genes involved in fruit ripening. These genes are responsible for the production of enzymes that degrade cell wall components, the accumulation of sugars, color development, flavor synthesis, and aroma production.

Softening of Fruit Tissues: One of the prominent changes that occur during fruit ripening, regulated by ethylene, is the softening of fruit tissues. Ethylene promotes the breakdown of cell wall components, such as pectin, cellulose, and hemicellulose, leading to the loss of cell wall integrity. This enzymatic degradation is facilitated by the upregulation of genes encoding cell wall-modifying enzymes, including polygalacturonases and cellulases. As a result, the fruit becomes softer and more palatable.

Color Changes: Ethylene also influences the color changes that occur during fruit ripening. It stimulates the production of pigments, such as anthocyanins, carotenoids, and chlorophyll degradation products. These pigments contribute to the characteristic coloration of ripe fruits, such as the red color of tomatoes or the orange hue of citrus fruits. Ethylene regulates the expression of genes involved in pigment biosynthesis and degradation pathways, thus influencing the development of vibrant fruit colors.

Flavor and Aroma Development: The production of volatile compounds responsible for fruit flavor and aroma is regulated by ethylene. Ethylene promotes the synthesis of enzymes involved in the production of esters, alcohols, and other volatile compounds that contribute to the characteristic flavors and aromas of ripe fruits. The expression of genes encoding these enzymes is upregulated during fruit ripening in response to ethylene signaling.

Senescence and Shelf Life: In addition to fruit ripening, ethylene also plays a role in senescence, which is the natural aging and deterioration of plant tissues. Ethylene accelerates the senescence process by promoting the degradation of chlorophyll, proteins, and nucleic acids. This leads to changes in tissue texture, color fading, and a decline in overall fruit quality. The control of ethylene levels is crucial for extending the shelf life of harvested fruits, as inhibiting ethylene production or its perception can delay senescence and maintain fruit quality during storage and transportation.

9.7 PLANT HORMONE INTERACTIONS AND SIGNAL TRANSDUCTION

Plant growth and development are intricately regulated by the interplay of various plant hormones. These hormones act as chemical messengers, coordinating physiological processes and responses to internal and external signals. The interactions between different plant hormones and their signal transduction pathways play a crucial role in shaping plant growth, development, and adaptive responses.

Crosstalk between Plant Hormones: Plant hormones do not act in isolation; rather, they interact and influence each other's activities through a process known as crosstalk. Crosstalk occurs at multiple levels, including hormone synthesis, signaling, and response pathways. The interactions between hormones can be synergistic, where they work together to amplify a specific response, or antagonistic, where one hormone counteracts the effect of another. For example, auxins and cytokinins often exhibit synergistic interactions in promoting cell division and shoot formation, while auxins and abscisic acid (ABA) may display antagonistic effects in regulating stomatal closure.

Hormone Signal Perception and Receptors: Plant hormones exert their effects by binding to specific receptors, which are typically located on the cell surface or inside the cell. These receptors can be receptor-like kinases, G-protein-coupled receptors, or other types of protein receptors. Upon hormone binding, the receptors undergo conformational changes, leading to the activation of downstream signaling pathways. The specificity of hormone signaling is determined by the presence of specific receptors that recognize and bind to each hormone.

Intracellular Signal Transduction: Once hormone receptors are activated, they initiate intracellular signal transduction pathways that transmit the hormonal signal to various parts of the cell. These pathways often involve second messengers, such as calcium ions ($Ca2+$), cyclic adenosine monophosphate (cAMP), or reactive oxygen species (ROS), which amplify the signal and trigger a cascade of cellular responses. The activated receptors may also activate protein kinases or transcription factors, leading to changes in gene expression and protein synthesis.

Hormone-Responsive Genes and Transcription Factors: Hormones regulate gene expression by modulating the activity of transcription factors, which are proteins that bind to specific DNA sequences and control the transcription of target genes. Different hormones can activate or suppress the activity of specific transcription factors, thereby regulating the expression of hormone-responsive genes. These genes encode proteins involved in various processes, such as cell division, differentiation, elongation, senescence, and stress responses.

Hormone Signaling Integration: Plant hormone signaling pathways are highly interconnected and integrated to ensure coordinated responses to different stimuli. Integration occurs at multiple levels, including hormone biosynthesis, receptor activation, and downstream signal transduction. The integration of hormone signals allows plants to fine-tune their responses to changing environmental conditions and developmental cues. For example, the interaction between gibberellins and DELLA proteins regulates plant growth in response to varying levels of gibberellins and other hormones.

Environmental and Developmental Regulation: Plant hormone interactions and signal transduction are influenced by both environmental factors and developmental cues. Environmental signals, such as light, temperature, humidity, and nutrient availability, can modulate hormone synthesis, receptor activity, and signal transduction pathways. Similarly, developmental processes, such as seed germination, flowering, fruit ripening, and senescence, involve intricate hormonal regulation and cross-talk. These dynamic interactions allow plants to adapt and respond to changing conditions throughout their life cycle.

9.8 APPLICATIONS OF PLANT HORMONES IN AGRICULTURE AND HORTICULTURE

Plant Growth Promotion: Plant hormones such as auxins, gibberellins, and cytokinins are widely used to promote plant growth and development. They can stimulate cell division, elongation, and differentiation, leading to increased biomass, larger leaves, longer stems, and overall improved plant architecture. Plant growth regulators derived from these hormones, such as indole-3-acetic acid (IAA) and gibberellic acid (GA), are commonly applied as foliar sprays or root drenches to enhance crop yields and promote vigorous plant growth.

Seed Germination and Root Development: Plant hormones play a vital role in seed germination and early seedling establishment. Gibberellins are often used to break seed dormancy and promote uniform germination in certain crops. Additionally, auxins, such as indole-3-butyric acid (IBA), can be applied as root-inducing agents to stimulate root development in cuttings and promote the establishment of new plants.

Fruit Ripening and Quality: Ethylene, the hormone responsible for fruit ripening, is widely utilized in horticulture to control and manage fruit quality. Ethylene gas is commonly applied to promote fruit ripening in climacteric fruits, such as bananas and tomatoes, by inducing ethylene synthesis and accelerating the natural ripening process. Conversely, inhibitors of ethylene synthesis or perception, such as 1-methylcyclopropene (1-MCP), can be used to delay fruit ripening and extend post-harvest shelf life.

Flowering Control: Plant hormones play a crucial role in regulating flowering processes. Gibberellins and cytokinins can be applied to induce flowering in certain plant species and promote flower bud initiation and development. Conversely, the use of plant growth regulators such as growth retardants can delay flowering, allowing for better synchronization of flowering time and improved management of flowering plants in commercial settings.

Plant Stress Management: Plant hormones are valuable tools for managing plant responses to various environmental stresses. Abscisic acid (ABA), for example, is involved in regulating plant responses to drought, salinity, and other abiotic stresses. Exogenous application of ABA can enhance plant stress tolerance and improve survival rates under adverse conditions. Jasmonic acid (JA) and salicylic acid (SA) play roles in plant defense against pathogens and pests, and their application can help activate plant defense mechanisms and enhance resistance to diseases and insect infestations.

Fruit Setting and Fruit Thinning: Plant hormones can be used to promote fruit setting in crops where fruit development is challenging. Synthetic auxins, such as naphthaleneacetic acid (NAA) or 2,4-dichlorophenoxyacetic acid (2,4-D), can be applied to enhance fruit set by stimulating the growth of fruiting structures and preventing premature fruit drop. Conversely, plant growth regulators like gibberellins can be used for fruit thinning by promoting fruit abscission, thereby achieving a balanced fruit load and improving fruit size and quality.

9.9 FUTURE PERSPECTIVES AND RESEARCH FRONTIERS IN PLANT HORMONES

Plant hormones play critical roles in regulating various aspects of plant growth, development, and responses to environmental cues. As our understanding of plant hormones deepens, new research frontiers and technological advancements are emerging, offering exciting possibilities for sustainable agriculture and crop improvement.

Unraveling Hormonal Crosstalk and Integration: One area of future research in plant hormones involves unraveling the complex crosstalk and integration among different hormone signaling pathways. Scientists are investigating how plant hormones interact and modulate each other's activities to orchestrate precise responses and developmental programs. Understanding these intricate interactions will provide insights into the coordination of growth and development in plants and open avenues for targeted manipulation of hormone signaling for desired agricultural traits.

Advancements in Hormone Sensing and Imaging: Technological advancements have facilitated the development of novel tools for hormone sensing and imaging. Researchers are utilizing genetically encoded sensors and advanced imaging techniques to monitor hormone dynamics with high spatial and temporal resolution. These tools enable the visualization of hormone distribution and signaling in real-time, providing valuable information about hormone transport, localization, and responses in different plant tissues and organs.

Hormone-Responsive Gene Expression Networks: Plant hormones regulate gene expression networks to orchestrate specific physiological responses. Future research aims to decipher the intricate regulatory networks underlying hormone-responsive gene expression. This includes identifying key transcription factors, cis-regulatory elements, and epigenetic modifications involved in hormone-mediated gene regulation. Unraveling these networks will enhance our understanding of hormone-mediated developmental processes and pave the way for targeted manipulation of gene expression for crop improvement.

Harnessing Hormonal Signaling for Stress Tolerance: Plant hormones play a crucial role in plant responses to various environmental stresses. Future research is focused on harnessing hormonal signaling to enhance stress tolerance in crops. Scientists are exploring how hormone signaling pathways can be modulated to improve plant resilience to abiotic stresses such as drought, heat, and salinity, as well as biotic stresses including pathogen and pest attacks. Understanding the intricate interplay between hormones and stress responses will aid in the development of stress-tolerant crop varieties.

Application of Hormones in Precision Agriculture: Advancements in technology and data analytics have paved the way for precision agriculture, which involves the targeted application of agricultural inputs based on precise plant needs. Plant hormones can be utilized as tools in precision agriculture to optimize plant growth, nutrient uptake, and resource utilization. By integrating real-time hormone sensing, data analysis, and automated systems, it becomes possible to deliver precise hormone-based treatments, leading to enhanced crop productivity, reduced resource wastage, and improved environmental sustainability.

CHAPTER 10

ELEMENTS OF ECOLOGY, TYPES OF HABITAT

10.1 ECOLOGY

Ecology is a scientific discipline that focuses on the study of organisms and their interactions with the physical and chemical environment. It encompasses the examination of spatial distribution patterns of organisms and the relationships between the abundance and diversity of plants and animals with environmental factors. These relationships are explored not only in terms of space but also over time.

In simpler terms, ecology can be understood as the investigation of the interconnections between organisms and their surroundings, including soil, weather, climate, and the dynamics of the entire ecosystem. It involves understanding how organisms interact with each other and with their abiotic environment.

The environment can be divided into two components: abiotic and biotic. The abiotic environment consists of non-living factors such as soil, water, light, and temperature. These factors play crucial roles in shaping the structure and functioning of ecosystems. On the other hand, the biotic environment includes the living organisms themselves, including plants, animals, and microorganisms.

Ecology encompasses a wide range of research areas, from studying the relationships between individuals within a population, the interactions between different species in a community, to the examination of entire ecosystems and their response to environmental changes. By investigating ecological principles, scientists gain valuable insights into the functioning of ecosystems, conservation of biodiversity, and the sustainable management of natural resources.

10.2 UNITS FOR THE STUDY OF ECOLOGY.

Ecology, the scientific study of organisms and their interactions with the environment, employs various units of analysis to examine ecological phenomena. These units provide a hierarchical framework for understanding the complexity of ecological systems.

Organism: The individual organism serves as the fundamental unit of ecological observation. Ecologists study the characteristics, behavior, and adaptations of organisms in relation to their environment. This branch of ecology is known as autecology, which focuses on the ecology of a single organism.

Population: When it is impractical to study individuals individually, ecologists often study groups of related organisms called populations. A population comprises individuals of the same species living in a specific area at a given time. It is a dynamic unit that is capable of interbreeding and producing fertile offspring. Populations can be analyzed to understand their size, density, distribution, and interactions with the environment.

Community: A community encompasses multiple populations of different species coexisting and interacting within a particular area. It represents the assemblage of plants, animals, and microorganisms and their intricate ecological relationships. For example, a forest community consists of all the trees, shrubs, animals, and microorganisms that interact and depend on one another within the forest ecosystem.

Ecosystem: An ecosystem is a broader unit that includes both the biotic (living) and abiotic (non-living) components of a defined area. It comprises the community of organisms interacting with their physical environment, such as soil, water, and climate. Ecosystems exhibit the flow of energy and cycling of nutrients, enabling the interdependence and functioning of the biotic and abiotic components.

Within ecosystems, organisms can be classified into two main groups based on their carbon source:

Producers (Autotrophs): These are typically green plants or other photosynthetic organisms capable of converting light energy into organic compounds. They are self-sustaining and form the foundation of the food web by producing their own food through photosynthesis.

Consumers and Decomposers (Heterotrophs): Consumers are organisms that obtain energy by consuming other organisms or organic matter. They include herbivores, carnivores, and omnivores. Decomposers, such as bacteria and fungi, break down organic matter, releasing nutrients back into the ecosystem.

Biosphere: The biosphere encompasses all the interconnected ecosystems on Earth, including the atmosphere, hydrosphere (water bodies), and lithosphere (solid Earth). It represents the global sum of all ecosystems and supports the existence and sustainability of life.

10.3 ENERGY FLOW AND MATERIAL OR NUTRIENT CYCLING (THE BIOGEOCHEMICAL CYCLES)

Energy flow and material or nutrient cycling are two fundamental processes that occur simultaneously in ecosystems.

Energy flow in ecosystems is a unidirectional and non-cyclic process. It involves the transfer of energy from the sun to the primary producers (plants) through the process of photosynthesis. The energy then moves through the food chain as organisms consume and are consumed by other organisms. At each trophic level, some energy is lost as heat or through metabolic processes. This unidirectional flow of energy ensures that energy is continuously supplied to sustain life in the ecosystem.

On the other hand, material or nutrient cycling involves the cyclic movement of essential elements through biotic and abiotic components of the ecosystem. Nutrients are taken up by primary producers from the environment and incorporated into their tissues. When organisms die or produce waste, decomposers such as bacteria and fungi break down the organic matter, releasing nutrients back into the environment in a process called mineralization. These nutrients can then be taken up by primary producers again, completing the nutrient cycle.

The activity of decomposers is in maintaining the cyclic movement of nutrients within ecosystems. Their role in breaking down organic matter and releasing nutrients ensures that essential elements are recycled and made available for uptake by plants and other organisms. Without decomposers, nutrients would become locked in dead organic matter and unavailable for use, leading to nutrient depletion and limiting the productivity of the ecosystem.

10.4 FOOD CHAINS, FOOD WEBS AND TROPHIC LEVELS

FOOD CHAINS

Food chains represent the transfer of food energy from one organism to another within an ecosystem. They illustrate the flow of energy as it moves through different trophic levels.

A food chain typically starts with a primary producer, which is usually a plant or photosynthetic organism capable of converting sunlight into chemical energy through photosynthesis. The primary producer is then consumed by a primary consumer, which is an herbivorous organism feeding directly on the primary producer. The primary consumer may, in turn, be consumed by a secondary consumer, and this process can continue with tertiary consumers and higher-order consumers.

For example, in an aquatic ecosystem, a food chain could be: Diatoms (primary producers) are consumed by mosquito larvae (primary consumers), which are then eaten by Tilapia fish (secondary consumers), which may be preyed upon by a Kingfisher (tertiary consumer), and ultimately, the Kingfisher may serve as food for a bird at an even higher trophic level.

Similarly, in a terrestrial ecosystem, a food chain could be: Plant leaves (primary producers) are consumed by grasshoppers (primary consumers), which are then eaten by toads (secondary consumers), followed by snakes (tertiary consumers), ducks (quaternary consumers), and potentially even consumed by humans (higher-order consumers) in the food chain.

Organisms that obtain their food from plants through the same number of feeding steps are considered to belong to the same trophic level. The length of a food chain represents the number of steps it takes for energy to flow from the primary producer to the final consumer. Generally, shorter food chains have a higher available energy at each trophic level compared to longer food chains. This is because energy is lost at each transfer due to metabolic processes and heat production.

TROPHIC LEVELS

Trophic levels are a way of categorizing organisms based on their feeding relationships and position in the food chain. Each trophic level represents a step in the transfer of energy through the ecosystem.

The first trophic level, known as the producer level, is occupied by green plants and other photosynthetic organisms. These organisms have the unique ability to harness energy from sunlight and convert it into chemical energy through the process of photosynthesis. They form the foundation of the food chain by producing organic compounds that serve as a source of energy for other organisms.

The second trophic level consists of herbivores, which are plant eaters. These organisms consume the primary producers to obtain energy and nutrients. They play an important role in transferring energy from plants to higher trophic levels.

Carnivores, or secondary consumers, occupy the third trophic level. They feed on herbivores or other carnivores to acquire energy. These organisms are also an essential part of the food chain, as they regulate population sizes of herbivores and contribute to energy transfer.

In some cases, there may be a fourth trophic level, known as tertiary consumers. These are carnivores that feed on other carnivores, further transferring energy up the food chain. However, not all ecosystems have a distinct tertiary consumer level, and it depends on the complexity of the food web and the availability of suitable prey.

It's worth noting that many organisms can occupy multiple trophic levels. For instance, omnivores, such as pigs and humans, have a diverse diet that includes both plants and animals. As a result, they can belong to different trophic levels depending on their specific food sources.

While organisms within an ecosystem occupy different trophic levels, it's important to recognize that the ultimate source of energy for all life on Earth is the sun. Through photosynthesis, green plants capture solar energy and convert it into chemical energy stored in organic compounds. This energy is then passed on from one trophic level to another as organisms consume and are consumed by others.

FOOD WEB.

Food webs are intricate networks of feeding relationships that exist in nature. They provide a more comprehensive understanding of the interactions between organisms within an ecosystem compared to simple linear food chains. In a food web, each trophic level has multiple connections and organisms can have diverse food relationships.

At the base of the food web are primary producers, such as green plants and algae, which capture sunlight energy through photosynthesis and convert it into chemical energy. These primary producers form the foundation of the food web and serve as a source of food for herbivores.

Herbivores, also known as primary consumers, feed directly on plants and consume primary producers. However, it's important to note that different species of herbivores can feed on the same primary producers or different plant species. This creates multiple feeding relationships within the herbivore trophic level.

Carnivores, including secondary and tertiary consumers, are organisms that feed on other organisms. They can consume herbivores, other carnivores, or a combination of both. This interconnectivity among different trophic levels forms a complex web of feeding relationships.

In a food web, organisms are not limited to a single linear chain of energy transfer. Instead, they can have multiple prey or predator relationships, allowing for more flexibility and stability within the ecosystem. This interconnectedness reflects the complexity and resilience of natural communities.

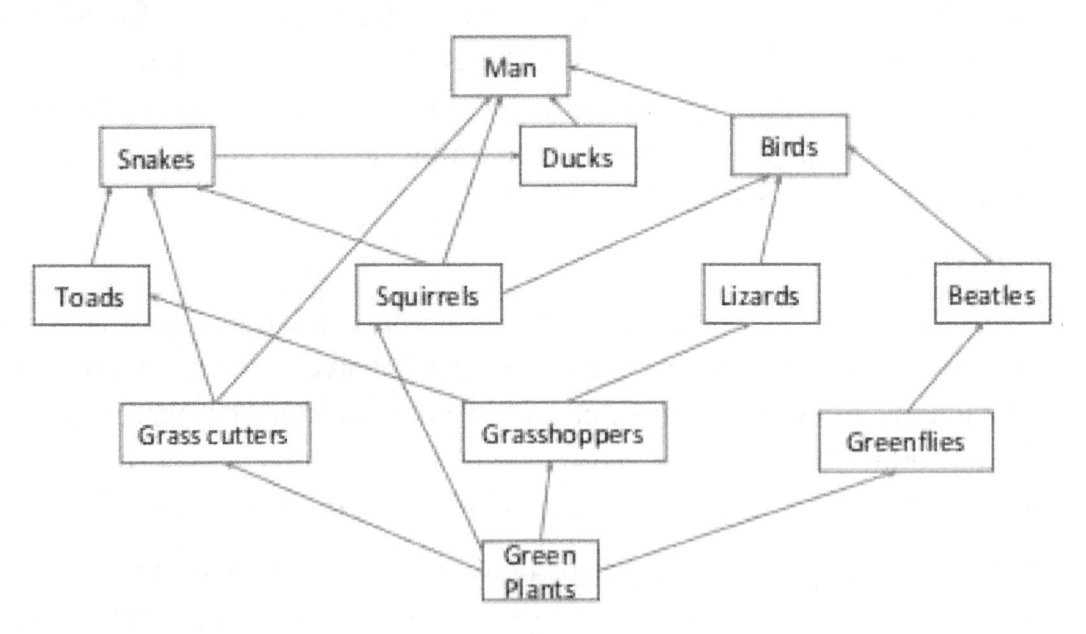

A simple Terrestrial Food Web

Here the green plant may provide the leaves also as food for squirrels, grass cutter and green flies apart from grasshoppers. The squirrels and grass cutter may be eaten by man, green flies by beetles and grasshoppers by lizards instead of toads. Next the beetles and lizards may be eaten by birds, then the birds by man.

The food web concept emphasizes the importance of understanding the broader ecological context in which organisms exist. It highlights the interdependence of species and the cascading effects that can occur if one population is affected. Changes in one part of the food web can have consequences throughout the entire ecosystem, impacting the abundance and distribution of other organisms.

Implicit in the relationship between autotrophs (producers) and heterotrophs (consumers) within a food web is the direction of energy flow through the ecosystem. This energy flow is unidirectional and non-cyclic, meaning it moves from the primary producers to the consumers and does not cycle back to the producers.

The non-cyclic, unidirectional flow of energy in ecosystems is crucial for understanding the dynamics of energy transfer. As energy moves through the different trophic levels of the food web, it undergoes losses at each transfer. These losses occur in the form of heat, metabolic processes, and other inefficiencies. Consequently, the amount of available energy decreases at higher trophic levels.

The explanation for this energy loss lies in the efficiency of energy utilization within each link of the food chain. As energy is transferred from one organism to another, a portion of it is used for growth, reproduction, and sustaining life processes. This utilization of energy within each organism limits the amount of energy available for transfer to the next trophic level.

The principle of one-way flow of energy is a fundamental concept in ecology. It signifies that energy enters the ecosystem from an external source, such as the sun, and gradually flows through the food web, sustaining life processes and supporting the various organisms within the ecosystem. This unidirectional energy flow ensures that energy is continuously supplied to the ecosystem and drives the ecological interactions and dynamics among organisms.

10.5 BIOGEOCHEMICAL CYCLE

Bio-geochemical cycles play a vital role in the circulation of chemical elements within the biosphere. These cycles involve the movement of essential elements, including those present in protoplasm, through characteristic pathways from the environment to organisms and back to the environment. They are often referred to as nutrient cycles.

Nutrient cycling is the process by which essential elements and inorganic compounds necessary for life circulate within an ecosystem. It helps conserve the nutrient supply and facilitates the repeated utilization of nutrients by organisms. Within each biogeochemical cycle, two compartments or pools can be identified: the reservoir pool, which represents a large, slow-moving component mainly comprised of non-biological factors, and the exchange or cycling pool, a smaller but more active portion that rapidly exchanges elements between organisms and their immediate environment.

In nutrient cycling, two simultaneous processes occur: mineralization and immobilization. Immobilization involves the uptake of inorganic elements (nutrients) from the soil, air, or water by organisms, which then convert these elements into microbial or plant tissues. These nutrients are utilized for growth and incorporated into organic matter. On the other hand, mineralization refers to the conversion of elements present in organic matter into mineral or ionic forms such as NH_3^+, Ca^{2+}, $H_2PO_4^-$, SO_4^{2-}, and K^+.

These ions then exist in the soil solution and become available for another cycle of immobilization and mineralization.

Mineralization, while essential for nutrient recycling, is a relatively inefficient process. During mineralization, a significant amount of carbon is lost as CO_2, and a substantial portion of energy dissipates as heat. Despite these losses, mineralization generates a surplus of nutrients that exceeds the needs of decomposers. This excess of released nutrients can be absorbed by plant roots, sustaining plant growth and productivity.

10.5.1   TYPES OF BIOGEOCHEMICAL CYCLES

From the perspective of the biosphere as a whole, biogeochemical cycles can be classified into two primary groups: gaseous cycles and sedimentary cycles.

GASEOUS CYCLES.

Gaseous cycles are characterized by the presence of a gaseous phase in their cycle. The atmosphere serves as a significant reservoir for the elements involved, existing in a gaseous form. These cycles typically exhibit little to no permanent alteration in the distribution and abundance of the elements. They are often considered relatively perfect due to the presence of self-regulating feedback mechanisms. When there is an increase in movement along one part of the cycle, adjustments occur along other parts to maintain balance.

Carbon and nitrogen are prime examples of biogeochemical cycles that possess a prominent gaseous phase. The cycling of these elements involves various processes occurring in the atmosphere, such as carbon dioxide (CO_2) exchange between plants and the atmosphere during photosynthesis and respiration. Additionally, hydrogen and oxygen cycles also exhibit gaseous characteristics.

Gaseous cycles are global in nature, transcending specific ecosystems and occurring on a large scale. The movements and transformations of these elements extend beyond local environments, impacting the entire biosphere. Understanding the dynamics of gaseous cycles is crucial for comprehending the global distribution and cycling of essential elements, and their influence on the overall functioning of ecosystems and the Earth's atmosphere.

SEDIMENTARY CYCLES

Sedimentary cycles, in contrast to gaseous cycles, involve the major reservoir of elements in the lithosphere, which is the Earth's crust. These cycles rely on the release of elements through weathering processes. The elements are gradually lost from biological systems due to erosion, eventually depositing in bodies of water, particularly the sea.

The replenishment or return of elements in sedimentary cycles to terrestrial ecosystems depends on various processes. These include the weathering of rocks, the addition of elements from volcanic gases, and the movement of elements from the sea to the land through biological mechanisms. Unlike gaseous cycles, sedimentary cycles are less perfect and more susceptible to disruption caused by human activities.

Two prominent examples of sedimentary cycles are the phosphorus and sulfur cycles. Phosphorus, an essential nutrient for organisms, primarily cycles through sedimentary processes. It is released from rocks through weathering and transported to aquatic systems, where it becomes available for uptake by organisms. Sulfur, although it has a gaseous phase, does not possess a significant gaseous reservoir, and its cycling is primarily sedimentary in nature.

Sedimentary cycles are more vulnerable to human-induced disturbances compared to gaseous cycles. Activities such as mining, deforestation, and excessive fertilizer use can impact the natural flow and balance of sedimentary cycles. Understanding and managing these cycles are essential for maintaining the availability of crucial elements in terrestrial ecosystems and minimizing the negative impacts of human activities on these cycles.

10.6 INPUTS AND LOSSES OF ELEMENTS

In ecosystems, elements are continually added and lost through various processes, influencing the availability and distribution of essential nutrients. The inputs of elements into an ecosystem can occur through several mechanisms:

Precipitation: Elements can be delivered to the ecosystem through rain or snowfall. Precipitation carries dissolved minerals from the atmosphere and deposits them onto the land surface.

Dust: Windblown dust particles can contain trace amounts of elements that are then deposited onto the land or water bodies. Dust deposition can contribute to nutrient enrichment in ecosystems.

Biological fixation: Certain biological processes, such as nitrogen fixation performed by certain bacteria, convert atmospheric gases into usable forms of elements. For example, nitrogen-fixing bacteria convert atmospheric nitrogen gas into ammonia, which can be utilized by plants.

Weathering of parent material: Elements present in rocks and minerals can be released through weathering processes. Over time, the breakdown of rocks and minerals releases nutrients into the soil, making them available for uptake by plants and other organisms.

Fertilizer application: Human activities often involve the application of fertilizers to enhance plant growth. Fertilizers contain essential elements such as nitrogen, phosphorus, and potassium, which are added to the ecosystem to supplement nutrient availability.

Conversely, elements are lost from ecosystems through various pathways:

Drainage waters: Water runoff or leaching can carry dissolved elements out of the ecosystem. When excess water flows through the soil, it can transport nutrients away from the system, leading to nutrient losses.

Plant and animal harvests: When plants or animals are harvested or consumed, elements present in their tissues are removed from the ecosystem. This occurs, for example, when crops are harvested for human consumption or when animals feed on plants.

Soil erosion: Erosion processes, such as wind or water erosion, can result in the physical removal of soil particles. These particles may contain elements, including organic matter and nutrients, which are then lost from the ecosystem.

Fires: During wildfires or controlled burns, elements stored in plant biomass can be released into the atmosphere as gases or particulate matter. This can result in the loss of elements from the ecosystem.

10.7 HUMAN IMPACT ON BIOGEOCHEMICAL CYCLES

Human activities have had a significant impact on biogeochemical cycles, disrupting the natural balance and functioning of ecosystems. These impacts have far-reaching consequences for plants, animals, and human well-being. Some of the key ways in which human activities have affected biogeochemical cycles include:

Land-use changes: The conversion of natural ecosystems for agriculture, urbanization, and other human activities alters the cycling of nutrients. Deforestation, for example, reduces the input of organic matter into the soil and increases the risk of nutrient runoff and erosion.

Burning of fossil fuels: The combustion of fossil fuels, such as coal, oil, and gas, releases large quantities of carbon dioxide (CO_2) into the atmosphere. This contributes to the accumulation of greenhouse gases, leading to global warming and climate change. The increased CO_2 levels also affect the carbon cycle, influencing the rate of photosynthesis and the balance of carbon storage in ecosystems.

Industrial activities and pollution: Industrial processes release various pollutants into the environment, including nitrogen compounds, sulfur compounds, heavy metals, and synthetic chemicals. These pollutants can disrupt biogeochemical cycles by altering nutrient availability, impairing soil and water quality, and affecting the health of organisms.

Excessive fertilizer use: Intensive agricultural practices often involve the excessive application of fertilizers to maximize crop yields. This can result in the runoff of excess nutrients, such as nitrogen and phosphorus, into water bodies, leading to eutrophication and harmful algal blooms. These changes in nutrient availability can disrupt aquatic ecosystems and negatively impact water quality.

Waste disposal: Improper disposal of waste, including sewage, industrial waste, and plastic pollution, can introduce harmful substances into ecosystems. These substances can contaminate soils, water bodies, and food chains, affecting the health of organisms and disrupting nutrient cycling processes.

Depletion of ozone layer: Human activities, particularly the release of chlorofluorocarbons (CFCs) and other ozone-depleting substances, have led to the depletion of the ozone layer in the stratosphere. This has increased the penetration of harmful ultraviolet (UV) radiation into the atmosphere, posing risks to human health, ecosystems, and agricultural productivity.

The cumulative effect of these human-induced changes in bio-geochemical cycles has resulted in environmental degradation, loss of biodiversity, and the alteration of ecosystems' ability to provide essential services. Recognizing and addressing these impacts is crucial for sustainable resource management, conservation efforts, and mitigating further disruption to bio-geochemical cycles.

10.8 HABITAT, MICROHABITAT, ECOLOGICAL NICHE.

Populations of organisms occupy specific places within a community, and their specific location along with the surrounding environment, which includes both living and non-living components, is referred to as their habitat. A habitat provides the necessary resources and environmental conditions for a population to survive, grow, and reproduce.

Within a community, different populations may have specific habitat requirements based on their ecological niche, which includes factors such as temperature, moisture, light availability, nutrient levels, and interactions with other species. As a result, the distribution of organisms within a community can vary, and certain species may be localized in specific areas due to micro differences in environmental conditions.

These localized areas within a habitat that provide unique micro-environmental conditions are known as microhabitats. Microhabitats can be small-scale variations within a larger habitat, such as a shaded area within a forest or a rocky crevice within a rocky shoreline. These microhabitats may have distinct ecological characteristics, such as differences in temperature, moisture levels, or availability of specific resources, that make them suitable for certain organisms but not others.

The concept of microhabitats highlights the intricate and dynamic nature of ecosystems, where even small variations in environmental factors can create diverse niches for different organisms to inhabit.

ECOLOGICAL NICHE

The ecological niche of a species refers to more than just its physical space in the community; it encompasses the role and function that the species performs within its ecosystem. It can be thought of as the organism's "occupation" within the community, describing its interactions with other species and its utilization of resources.

The ecological niche of a species includes various aspects such as its feeding habits, reproductive behavior, habitat preferences, and interactions with other organisms. It represents the specific set of ecological conditions and resources that a species requires to survive and reproduce successfully.

Some species have a broad ecological niche, meaning they have a wide range of food sources and can adapt to various environmental conditions. They may be generalists, feeding on different types of plants and animals, or have a diverse diet. On the other hand, some species have highly specialized niches, with very specific dietary requirements or habitat preferences. These specialists are adapted to thrive in specific ecological niches and may have evolved unique morphological, physiological, or behavioral traits to exploit those niches effectively.

The evolution of different ecological niches among species has occurred over long periods of time through natural selection and adaptation to specific environmental conditions. Each species has its own niche, and this diversity of niches is important for maintaining balance and stability within an ecosystem. By occupying different niches, species can minimize competition for resources and reduce direct conflict with one another, allowing for more efficient resource utilization and coexistence.

10.8.1 HABITATS OR ECOSYSTEMS OF THE WORLD

Habitats or ecosystems of the world can be broadly categorized into terrestrial and aquatic habitats. Terrestrial habitats refer to the environments found on land, while aquatic habitats are those found in water bodies.

Within terrestrial habitats, there are various natural habitats characterized by specific vegetation types and environmental conditions. Forests are complex habitats dominated by trees and can be further classified into tropical rainforests, temperate forests, and boreal forests, among others. Grasslands are habitats dominated by grasses and are found in both tropical and temperate regions. Deserts are habitats with low precipitation and sparse vegetation, characterized by extreme temperature fluctuations.

Aquatic habitats encompass freshwater and marine ecosystems. Freshwater habitats include lotic systems, which refer to running water environments such as rivers, springs, and streams. These habitats are dynamic and constantly flowing, offering a unique set of conditions for aquatic organisms. Lentic systems, on the other hand, are standing water environments such as lakes, ponds, and swamps. They tend to have slower water movement and often support a diverse array of plants and animals.

Marine habitats encompass the vast oceans and seas, which cover a significant portion of the Earth's surface. They are characterized by high salt content and various zones such as the intertidal zone, coastal zone, and open ocean. Marine habitats support a wide range of organisms, including coral reefs, kelp forests, and deep-sea ecosystems.

Each habitat or ecosystem has its own set of physical and biological characteristics, which determine the types of organisms that can thrive within them. These habitats provide specific resources and environmental conditions that shape the adaptations and interactions of the species that inhabit them.

10.9 TERRESTIAL/LAND HABITATS

Terrestrial or land habitats encompass a variety of biomes, which are large-scale community units characterized by distinct climax vegetation. The dominant vegetation in a particular biome plays a key role in identifying and defining it. For example, grass is the dominant climax vegetation in grassland biomes, although the specific species of grasses may vary across different regions where grasslands are found. However, it's important to note that biomes can also include other types of vegetation, such as early successional stages, subclimaxes in forests influenced by local soil and water conditions, and vegetation introduced by human activities like crops.

Biomes are recognized based on their unique combination of climate, soil conditions, and geographic factors. The characteristics of the dominant vegetation within a biome are shaped by these environmental factors. For instance, the presence of grasses in grassland biomes is often associated with moderate precipitation, seasonal variations in temperature, and well-drained soils. Other biomes include tropical rainforests, temperate forests, taiga (boreal forests), tundra, deserts, and Mediterranean shrublands, among others.

Each biome supports a specific array of plant and animal species that are adapted to its particular environmental conditions. These organisms have evolved and specialized to thrive within their respective biomes. The interactions between species and their environment in these habitats contribute to the overall functioning and biodiversity of terrestrial ecosystems.

Terrestrial biomes include:

(1) deserts

(2) tundra

(3) grasslands

(4) forests.

DESERTS.

Deserts are unique terrestrial habitats characterized by extreme aridity and limited precipitation. They can be caused by persistent cold temperatures in arctic, Antarctic, and alpine regions, or by dryness, as seen in the Sahara Desert. Deserts typically receive less than 255 mm (10 inches) of rainfall per year, which is often unevenly distributed throughout the annual cycle.

The defining feature of deserts is their dryness, which persists for most or all of the year. This aridity poses significant challenges for plant and animal life. Deserts are characterized by extremes of temperature and low humidity, which can have adverse effects on the survival and adaptation of organisms. Additionally, strong winds and sandstorms are common in desert climates.

The unique conditions of deserts have led to the evolution of specialized adaptations in the plant and animal species that inhabit these environments. Desert plants have developed various strategies to cope with arid conditions There are four notable plant life forms found in deserts:

Annuals: These plants complete their life cycle within a year, taking advantage of the limited periods of adequate moisture to grow and reproduce before the onset of drought.

Desert shrubs: These plants have multiple branches arising from a short basal trunk and possess small, thick leaves that can be shed during dry periods to reduce water loss through transpiration.

Succulents: Succulent plants, such as cacti, have adapted to desert conditions by storing water in their tissues. This enables them to survive extended periods of drought.

Microflora: Microscopic organisms like mosses, lichens, and blue-green algae make up the microflora of deserts. These organisms remain dormant in the soil but can quickly respond to cooler or wetter periods, allowing them to thrive in the desert ecosystem.

Desert plants face the ultimate stress of dehydration, as their protoplasm can easily become dehydrated in the arid conditions. To reduce competition for scarce water resources, desert vegetation tends to be spaced apart. This spacing allows each plant to have better access to water and other necessary resources.

Desert animals encounter several challenges related to their survival in harsh desert environments. They must breathe air, conserve water, and cope with extreme temperature fluctuations. Many desert animals employ strategies to evade these challenges, such as aestivation. Aestivation is a state of suspended animation or dormancy where animals reduce their metabolic activity, growth, and reproduction, thereby conserving energy and water. This dormant state enhances their resistance to heat, drought, and other climatic conditions.

Certain desert animals, including reptiles and insects, are naturally pre-adapted to desert life. Their impermeable integuments (outer coverings) and ability to produce dry excretions enable them to survive with minimal water intake. However, as a whole, mammals are not well-adapted to desert conditions. Nonetheless, some mammal species have undergone secondary adaptations to desert life. For example, camels can withstand periods of tissue dehydration and have physiological adaptations that allow them to tolerate water scarcity for extended periods.

Due to the dominant limiting factor of water, productivity in desert regions is primarily determined by rainfall. Desert ecosystems generally have low productivity due to the constraints imposed by drought. However, with suitable soils and irrigation, deserts can be converted into highly productive agricultural lands.

Compared to other ecosystems, deserts have been relatively unaffected by human activities. Humans are physiologically poorly adapted to desert environments, which has limited extensive human alteration of these ecosystems. However, it's important to note that human activities, such as land-use changes and water extraction, can still have localized impacts on desert regions.

TUNDRA

Tundra is a unique biome characterized by its treeless landscape. It experiences long and bitterly cold winters, with short summers where temperatures rise above freezing. The ground is typically free of snow during the summer months, allowing for the growth of tundra vegetation. Unlike deserts, the major limiting factor in the tundra is not water but rather heat, which is in short supply for biological processes.

Although precipitation in the tundra is generally low, water availability is not limiting due to the low evaporation rate. The tundra can be described as a wet arctic grassland or a cold marsh that freezes partially during the year. This ecosystem forms a ring around the land masses of the northern hemisphere.

Tundra vegetation is adapted to survive the extreme cold. It mainly consists of lichens, grasses, and sedges that have developed remarkable adaptations to withstand the harsh conditions. These plants have low growth forms, allowing them to conserve heat and resist strong winds.

The animals inhabiting the tundra have adapted to survive the seasonal changes from cold and darkness in winter to the warmth and light in summer. Some animals seek shelter underground during winter, while others endure the harsh weather in the open but take cover during severe storms. Many bird species migrate to warmer climates before winter arrives.

The tundra is home to various large animals, including musk oxen, caribou, reindeer, polar bears, wolves, and marine animals. Smaller creatures like lemmings are also present, known for their tunnelling activities within the vegetation mantle.

FORESTS.

Forests are ecosystems dominated by woody plants, typically with trees reaching a minimum height of 5 meters. They have a canopy that can be open or closed, and the presence of grass is generally limited. Most trees in forests are not fire-tolerant, meaning they are susceptible to damage or destruction by wildfires.

Forests are found in areas with relatively high rainfall and occur in both temperate and tropical regions. In temperate forests, the climate is characterized by distinct seasons, including cold winters and warm summers. Tropical forests, on the other hand, exist in regions with high and relatively constant rainfall throughout the year.

Tropical forests encompass a range of forest types. Broad-leaved evergreen rainforests are found in areas with abundant and evenly distributed rainfall year-round. These rainforests are known for their high species diversity and dense vegetation. Tropical deciduous forests, on the other hand, experience a dry season where some trees shed their leaves to conserve water.

The main plant components of tropical forests include:

(a) Forest trees: They form the dominant vertical structure of the forest and provide the canopy.

(b) Herbs: These are smaller non-woody plants that grow in the understory or forest floor.

(c) Climbers (vines and lianas): These are plants that use other plants for support and can climb towards the canopy to access sunlight.

(d) Stranglers: Certain species of plants start as epiphytes (growing on other trees) and eventually grow around and strangle the host tree.

(e) Epiphytes: These are plants that grow on the surface of other plants, such as trees, without parasitizing them.

(f) Saprophytes: These plants obtain nutrients by decomposing dead organic matter.

(g) Parasites: Some plants in the forest are parasites, deriving nutrients from other living plants.

Animals in forest ecosystems can be categorized into various ecological groups based on their adaptations and ways of life. Here are some examples:

Arboreal mammals: These mammals have adapted to live and move in trees. They have specialized features such as grasping hands or feet, prehensile tails, and sharp claws for climbing. Examples include primates like monkeys and lemurs, as well as tree-dwelling species like squirrels and koalas.

Terrestrial mammals: These mammals primarily inhabit the forest floor and need adaptations to navigate through dense vegetation. They may have strong limbs, keen senses, and camouflage to move stealthily in the understory. Examples include large herbivores like deer, ground-dwelling rodents, and predators like big cats.

Subterranean animals: While relatively scarce in forests, some animals have adapted to live underground. These include burrowing mammals like moles, certain rodents, and reptiles like burrowing snakes.

Arboreal birds: Forests are home to a variety of bird species. While cursorial (running) birds are less common in forested areas, many species have adapted for life in trees. They have strong feet and claws for perching, as well as specialized beaks for feeding on fruits, insects, or nectar. Examples include woodpeckers, parrots, and toucans.

Climbing reptiles and amphibians: Some reptiles and amphibians have evolved adaptations for climbing in forested habitats. This includes features like prehensile tails, adhesive pads on their feet, or muscular bodies for maneuvering through branches. Tree-dwelling snakes, tree frogs, and certain lizards are examples of climbing herpetofauna.

Shifting cultivation, a form of agriculture where land is cleared and cultivated for a few years before being abandoned and left to regenerate, has had significant impacts on forest ecosystems. This practice has led to the destruction of large areas of primary rainforests and has disrupted entire ecosystems, including the loss of habitat for many plant and animal species. Efforts to promote sustainable land use practices and conservation measures are crucial for preserving forest ecosystems and their biodiversity.

GRASSLANDS.

Grasslands are ecosystems characterized by a predominant vegetation of grasses, with forbs (non-grassy herbaceous plants) also present and scattered trees or shrubs.

SAVANNAH

Savannas, with their unique combination of grasses, trees, and shrubs, are remarkable ecosystems that thrive in tropical regions. The defining characteristic of savannas is the presence of a continuous grass stratum that is interspersed with scattered trees and shrubs. Unlike dense forests, savannas offer an open canopy that allows sunlight to reach the ground, promoting the growth of grasses and herbaceous plants.

One notable feature of savannas is the fire tolerance of the trees that inhabit them. These trees have adapted to survive and even benefit from periodic fires, which play a crucial role in shaping the structure and dynamics of the savanna ecosystem. The annual burning of the grassland helps control the encroachment of woody vegetation, stimulates the growth of new grass shoots, and maintains the open nature of the savanna.

Savannas are typically found in regions characterized by distinct wet and dry seasons. During the wet season, abundant rainfall nourishes the savanna, promoting the growth of grasses and supporting a diverse array of wildlife. However, as the dry season sets in, the availability of water decreases significantly, challenging the survival of many organisms. This cyclical pattern of rainfall and drought shapes the unique adaptations of plants and animals in the savanna.

Large herbivores, such as antelopes, zebras, and giraffes, are iconic residents of the savanna. These grazing animals have evolved to take advantage of the abundance of grasses during the wet season and are adapted to withstand the challenges posed by the dry season. Predators, including lions, cheetahs, and hyenas, thrive in the savanna by preying on the herbivores, forming intricate food webs and ecological relationships.

Unfortunately, human activities have had a significant impact on savanna ecosystems. Encroachment for agriculture, urbanization, and unsustainable land-use practices have led to the fragmentation and degradation of savanna habitats. These activities disrupt natural fire regimes, alter nutrient cycling processes, and threaten the survival of many plant and animal species that rely on the savanna ecosystem.

Conservation efforts are crucial for the protection and restoration of savanna habitats. Sustainable management practices, such as controlled burns and responsible grazing, can help maintain the ecological integrity of the savanna. Additionally, creating protected areas and promoting awareness about the importance of savannas can contribute to their long-term preservation and ensure the survival of their unique biodiversity.

Savannas are tropical grasslands characterized by a continuous grass stratum interspersed with scattered trees and shrubs. They are adapted to annual burning and thrive in regions with distinct wet and dry seasons. The rich diversity of plant and animal life in the savanna is intricately connected forming a complex and dynamic ecosystem. It is crucial to safeguard these unique ecosystems and mitigate human impacts to ensure the conservation of savannas for future generations.

TEMPERATE GRASSLANDS

Temperate grasslands are unique ecosystems that thrive in regions with distinct seasonal variations and moderate climates. These grasslands experience hot summers and cold winters, with precipitation levels that are relatively lower compared to other biomes. Within temperate grasslands, two main types are recognized: steppes and prairies.

Steppes are characterized by the presence of short grasses that are well-adapted to the climatic conditions of the region. They are often found in the

Eurasian steppes, which stretch across vast areas of Europe and Asia. These grasslands are known for their vast expanses of open grassy plains, where herbaceous plants dominate the landscape. Steppes are home to a diverse range of plant and animal species that have evolved to survive the harsh climate, including grazers such as gazelles and wild horses.

Prairies, on the other hand, are grasslands dominated by tall grasses that can reach impressive heights. Prairies are most commonly associated with North America, particularly the Great Plains, where they once covered extensive areas. These grasslands were historically home to iconic species such as bison, pronghorns, and prairie dogs. Prairies are known for their rich soil fertility, which has made them ideal for agriculture. As a result, much of the original prairie ecosystem has been converted for agricultural purposes, leading to the loss of native grassland species and habitat fragmentation.

In addition to steppes and prairies, temperate grasslands can also be found in other parts of the world. For example, the veldt in South Africa is a vast grassland characterized by a combination of grasses and scattered trees. The African savanna, with its distinctive mix of grasses, shrubs, and occasional trees, can also be classified as a type of temperate grassland. Similarly, the pampas in Argentina is a large expanse of fertile grassland known for its productive agricultural lands.

Temperate grasslands are ecologically significant as they support a unique assemblage of plants and animals adapted to the specific climatic conditions. They are important habitats for grazers and herbivores, which rely on the abundance of grasses for food. Additionally, grasslands provide nesting grounds for various bird species and support a diverse range of insect and invertebrate communities. Like other ecosystems, temperate grasslands face numerous threats due to human activities. Conversion of grasslands for agriculture, urbanization, overgrazing, and invasive species are among the main challenges to their conservation. The loss of native grassland species and the disruption of natural fire regimes have further impacted the ecological balance of these habitats.

Efforts are underway to preserve and restore temperate grasslands through sustainable land management practices, restoration projects, and the establishment of protected areas. Conserving the remaining grasslands and promoting their importance for biodiversity and ecosystem services is crucial for maintaining the ecological integrity of these valuable ecosystems.

Grasslands support a diverse array of large herbivores, which are a characteristic feature of these ecosystems. These herbivores are primarily represented by large mammals that have adapted to the grassland environment. Among them, two main life-forms can be distinguished: running types and burrowing types. Running types include ground antelopes and kangaroos, which are adapted for swift movement across the grassy terrain. On the other hand, burrowing types, such as ground squirrels and gophers, have developed the ability to excavate burrows to seek shelter and protection.

When humans utilize grasslands for livestock grazing, they often introduce their domesticated herbivores, such as cattle, sheep, and goats, in place of the native grazers. This human intervention can have a significant impact on the grassland ecosystem, altering the dynamics of herbivory and potentially affecting the composition and structure of the plant community.

Both savannas and temperate grasslands are prone to fires, which helps in shaping the structure and functioning of these ecosystems. Fire is a natural component of grassland ecosystems and can have both beneficial and detrimental effects. It helps control the encroachment of woody vegetation and stimulates the growth of grasses. However, frequent or intense fires can disrupt the natural balance and lead to the loss of certain plant and animal species.

Human activities have had a profound impact on grasslands worldwide. Large areas of grassland have been converted into agricultural land to meet the increasing demands for food production. This conversion involves the removal of native vegetation, alteration of soil composition, and changes in hydrological processes, leading to the loss of grassland biodiversity and ecosystem services. Conservation efforts are crucial to safeguard the remaining grassland areas and restore degraded grassland habitats, ensuring the preservation of these unique ecosystems and the species they support.

10.10 AQUATIC HABITATS

Aquatic habitats are divided into freshwater and marine ecosystems defined by salinity.

FRESHWATERS

Freshwater ecosystems, the study of which is known as **Limnology.** Freshwater habitats include bodies of water with low salinity, such as lakes, ponds, rivers, streams, and wetlands. These ecosystems are characterized by their relatively low salt content, which can vary depending on factors such as location, climate, and geology. Freshwater habitats support diverse communities of plants and animals that have adapted to survive in these unique conditions. They are essential for providing freshwater resources, supporting aquatic biodiversity, and regulating various ecological processes.

Freshwater habitats, including lakes, ponds, swamps, rivers, springs, and streams, form an essential component of the Earth's water cycle. These ecosystems are vital for supporting a wide range of plant and animal life, as well as providing important ecological services.

Lentic habitats, such as lakes, ponds, and swamps, are characterized by their still or slow-moving water. Lakes are large bodies of water, often surrounded by land, with diverse physical and chemical characteristics. They can vary in size, depth, and nutrient content, which influences the types of organisms that inhabit them. Ponds, on the other hand, are smaller and shallower than lakes, often with a more dynamic environment. Swamps are wetland areas characterized by the presence of trees and shrubs, with waterlogged soils that support unique plant and animal communities.

Lotic habitats encompass running water environments, including rivers, springs, and streams. These habitats are shaped by the continuous flow of water, which affects their physical characteristics, such as water velocity, depth, and substrate composition. Rivers are large, interconnected channels of flowing water, often originating from mountains and flowing towards the sea. Springs are natural outlets of groundwater that emerge from the Earth's surface, while streams are smaller tributaries that join and contribute to the flow of rivers. Lotic habitats exhibit high levels of oxygenation and are home to a diverse array of aquatic plants and animals adapted to the dynamic nature of running water.

The quality and biodiversity of freshwater ecosystems are influenced by various factors, including water chemistry, nutrient availability, temperature light levels, and the presence of pollutants. These habitats are sensitive to human activities, such as pollution, deforestation, water extraction, and climate change, which can have significant impacts on their ecological integrity and the species they support.

STREAMS AND RIVERS

Streams and rivers are dynamic freshwater habitats that play a vital role in shaping the landscape and supporting a wide range of organisms. They are characterized by the continuous flow of water, ranging from small, narrow streams to large, meandering rivers.

These aquatic ecosystems are of immense importance to human civilizations, providing various ecosystem services such as freshwater supply, transportation, energy generation, and recreational activities. Throughout history, humans have extensively altered rivers and streams through the construction of dams, levees, and artificial channels. These modifications aim to control water flow, mitigate flooding, and enhance water availability for agriculture and urban development. As a result, it has become increasingly challenging to find truly pristine and undisturbed rivers in many parts of the world.

Human activities, including water extraction, pollution from industrial and agricultural sources, and habitat degradation, have significantly impacted the health and biodiversity of stream and river ecosystems. Altering natural flow regimes, blocking fish migration routes, and reducing the availability of suitable habitats have led to declines in native fish populations and other aquatic organisms.

Efforts are being made globally to restore and rehabilitate degraded rivers and streams. River restoration initiatives focus on enhancing natural flow patterns, reintroducing native species, improving water quality, and re-establishing riparian vegetation. These actions aim to recreate more natural and sustainable river systems that can support the ecological integrity and services provided by these habitats.

Studies in river ecology examine the complex interactions between physical, chemical, and biological processes in rivers and streams. Researchers investigate factors such as water velocity, sediment transport, nutrient cycling, and the diverse array of plants, invertebrates, and fish species that inhabit these habitats. Understanding these dynamics is crucial for effective river management and conservation.

LAKES AND PONDS

Lakes and ponds are freshwater ecosystems that provide a diverse range of habitats for various organisms. These bodies of water can vary in size, from small ponds to large lakes, and they are typically formed by geological processes such as tectonic activity, glaciation, or volcanic activity.

The lifespan of ponds can vary greatly, ranging from temporary ponds that form during the rainy season and dry up within a few weeks or months, to more permanent ponds that can exist for several years. Lakes, on the other hand, are generally long-lived features that persist for many decades or even centuries.

Freshwater communities in lakes and ponds exhibit relatively low species diversity compared to other ecosystems. Many species, genera, and even entire families of organisms are widely distributed across different lakes and ponds within a given region. This can be attributed to the interconnectedness of freshwater systems through rivers, streams, and groundwater, allowing for the dispersal of organisms across different water bodies.

These ecosystems have well-defined boundaries, including the shoreline, the sides of the basin, the water surface, and the bottom sediment. Each of these zones provides distinct ecological niches and habitats for different organisms. Light availability, oxygen levels, and temperature gradients play crucial roles in shaping the distribution and adaptations of organisms within lakes and ponds.

A characteristic feature of lakes and large ponds is the presence of zonation and stratification. The littoral zone, or shallow water zone, is the area near the shore where light can penetrate to the bottom. It supports rooted vegetation such as aquatic plants, algae, and emergent macrophytes. This zone is important for many organisms, including fish, amphibians, and various invertebrates.

The limnetic zone represents the open water region of the lake or pond, away from the shoreline. It is dominated by plankton, including phytoplankton (microscopic algae) and zooplankton (tiny animals). This zone also supports free-swimming organisms, known as nekton, which include fish and other larger aquatic animals.

The profundal zone is the deep water region of the lake or pond, where light penetration is limited or absent. This zone is characterized by the absence of photosynthetic organisms and is primarily inhabited by heterotrophic organisms that rely on organic matter as a food source.

FRESHWATER MARSHES.

Freshwater marshes are unique wetland habitats that are characterized by their constant or frequent flooding, supporting the growth of grasses and shrubs. They serve as transitional zones between aquatic and terrestrial environments, providing important ecological functions and supporting a diverse array of plant and animal species.

Marshes are commonly found in lowland areas near rivers, lakes, or lagoons, where water levels fluctuate and create an ideal habitat for marsh vegetation. The continual flooding of marshes plays a crucial role in nutrient cycling and the breakdown of organic matter. The decay of organic materials in marshes occurs on a large scale, leading to the release of nutrients and the enrichment of the surrounding water.

One significant impact of this organic decay is the reduction of oxygen content in the water. The decomposition process consumes oxygen, creating an environment with lower oxygen levels compared to open water ecosystems. Despite this, marshes are highly productive and support a wide variety of plant and animal life.

Freshwater marshes play a role in maintaining water tables in adjacent ecosystems. They act as natural sponges, absorbing and storing excess water during periods of heavy rainfall or flooding. This helps to mitigate the impacts of floods and regulate water flow, ultimately contributing to the overall health of the surrounding ecosystems.

Plant life in freshwater marshes is well adapted to the wet and nutrient-rich conditions. Common plants found in marshes include various types of algae, water lettuce, duckweeds (Lemna), and floating ferns (Salvinia). These plants play a crucial role in stabilizing the marsh substrate, filtering water, and providing habitat and food sources for numerous organisms.

Marshes are also home to a diverse range of animal species. Amphibians such as frogs and toads thrive in the moist environment, utilizing the marshes for breeding and shelter. Fish species that can tolerate fluctuating water levels often inhabit marshes, providing a food source for various predators. Additionally, many bird species, including wading birds, rely on marshes as foraging grounds where they feed on fish, amphibians, and other small animals.

MARINE HABITATS

Marine habitats encompass vast bodies of saltwater, primarily represented by the oceans, but also including brackish and estuarine environments. These habitats exhibit distinct characteristics and support a diverse array of marine life.

One defining feature of marine habitats is their high salinity due to the presence of dissolved salts. Salinity is a measure of the concentration of salts in water and is typically expressed in parts per million (ppm). Seawater for instance, has an average salinity of approximately 35,000 ppm with common salt (sodium chloride - NaCl) accounting for about 30,000 ppm of that concentration.

The world's major oceans - the Atlantic, Pacific, Indian, and Antarctic - along with their interconnected seas and extensions, cover roughly 70% of the Earth's surface. These immense bodies of water provide a wide range of marine habitats and support a tremendous diversity of life.

In marine ecosystems, physical factors play a dominant role in shaping and influencing life. Waves, tides, currents, salinity levels, temperatures, pressures, and light intensities significantly impact the composition and distribution of biological communities. These physical factors, in turn, also have profound effects on the composition of bottom sediments and the concentrations of dissolved gases in the water.

The marine food chains begin with the smallest autotrophic organisms, such as phytoplankton. These microscopic plants convert sunlight and nutrients into organic matter through photosynthesis, forming the foundation of marine food webs. Phytoplankton are consumed by zooplankton, which serve as food for small fish and invertebrates. The energy transfer continues through successive trophic levels, culminating in the largest marine animals, including giant fish, squid, and whales.

Marine habitats exhibit a remarkable diversity of organisms, from microscopic plankton to massive marine mammals. They support an incredible range of species, adaptations, and ecological interactions. Coral reefs, kelp forests, mangrove swamps, and deep-sea hydrothermal vents are just a few examples of the unique marine habitats that exist worldwide.

ESTUARIES AND SEASHORES.

Estuaries and seashores encompass unique and dynamic ecosystems that serve as important transitional zones between the land and sea. They exhibit distinct characteristics and support a diverse array of life, playing crucial ecological roles.

The term "estuary," derived from the Latin word "aestus" meaning tide, refers to semi-enclosed bodies of water such as river mouths or coastal bays. Estuaries experience the mixing of seawater and freshwater, resulting in brackish water with intermediate salinity levels. Tidal action plays a significant role in regulating the physical conditions and providing energy subsidies within estuaries.

Estuaries are part of a broader band of ecosystems that serve as transition zones between the seas and continents. Along with estuaries, rocky shores, sandy beaches, intertidal mudflats, and tidal estuaries comprise the four major types of marine inshore ecosystems. These habitats support a vast array of species that have adapted to the unique challenges and opportunities presented by their dynamic environment.

Estuaries and inshore marine waters are known for their natural fertility, making them highly productive ecosystems. Three major types of autotrophs contribute to the high gross production rate within estuaries: phytoplankton, benthic microflora (algae living in and on mud, sand, rocks, or other surfaces), and macroflora (large attached plants such as seaweeds, submerged eel grasses, emergent marsh grasses, and mangrove plants in tropical regions). The interplay between these autotrophs supports a rich food web and provides critical habitats for numerous organisms.

Estuaries often act as efficient nutrient traps, enhancing their ability to absorb and process nutrients, particularly when organic matter has been reduced through secondary treatment. These nutrient-rich environments serve as important nursery grounds for many coastal shellfish and fish species. The young stages of these organisms grow rapidly in estuaries before venturing offshore.

Organisms in estuaries have evolved various adaptations to cope with the tidal cycles, enabling them to exploit the advantages of living in such dynamic environments. For example, animals like fiddler crabs possess internal biological clocks that help time their feeding activities to coincide with the most favorable parts of the tidal cycle.

DELTAS

Deltas represent unique and dynamic landforms that form at the mouths of rivers where sediment deposition occurs due to the decrease in water velocity. As rivers approach the sea or a lake, their flow slows down, leading to the settling of sediment carried by the water. Over time, these sediments accumulate and form a delta, typically characterized by fine-grained deposits.

A prominent example of a delta is the large delta at the mouth of the River Niger. Deltas are known for their fertility and are often highly productive areas. They provide favorable conditions for agriculture, as they possess fertile soils and abundant water supplies that can be utilized for irrigation. In natural river systems that are not controlled upstream, deltas are frequently subjected to flooding. These periodic floods bring in nutrient-rich sediment, enhancing the fertility of the delta and providing a regular influx of nutrients and fertile silt.

The agricultural value of deltas is significant, as they support productive crop lands and provide essential resources for local communities. The nutrient-rich soils and water availability make deltas favorable for cultivating a variety of crops. Deltas can play a crucial role in food production and contribute to the economic livelihood of the surrounding regions.

However, deltas are not without challenges. They are highly sensitive to changes in water flow, sediment supply, and human activities. Dams and upstream water management practices can significantly alter the natural flow patterns of rivers, impacting the sediment transport and the formation of deltas. The increasing pressure from human activities, such as land reclamation and pollution, can degrade the ecological integrity of deltas and compromise their agricultural productivity.

CHAPTER 11

ENVIRONMENTAL POLLUTION

Environmental pollution refers to the detrimental changes that occur in our surroundings, either wholly or largely as a result of human activities. These changes can manifest in various forms, including alterations in energy patterns, radiation levels, chemical composition, physical attributes, and the abundance of organisms within an ecosystem. Pollution can have direct or indirect effects, leading to adverse impacts on the environment and its inhabitants.

In the field of biology, pollution is defined more broadly as the introduction of any material into the environment that has a detrimental effect on the ecosystem. This definition encompasses both human-made pollutants and naturally occurring substances that can disrupt the balance and functioning of ecosystems.

While it is true that pollution can be caused by natural processes, such as volcanic eruptions or forest fires releasing harmful gases and particulate matter into the atmosphere, the term "environmental pollution" typically refers to the negative impacts resulting from human activities. These activities include industrial processes, transportation emissions, improper waste disposal, deforestation, agricultural practices, and the release of pollutants into air, water, and soil.

11.1 NATURAL SOURCES OF POLLUTION.

Natural sources of pollution can contribute to the release of various toxic elements and substances into the environment. These sources are inherent to natural processes and can have both local and global impacts. Some of the major natural sources of pollution include:

Volcanic eruptions: Volcanoes release gases such as sulfur dioxide (SO_2) and particles into the atmosphere. These emissions can contribute to air pollution and the formation of volcanic ash clouds.

Sea spray: The crashing of waves and movement of water in the oceans generates sea spray, which contains salt particles. These particles can be transported by wind and contribute to air pollution in coastal areas.

Vapour-phase outgassing: Certain elements, including arsenic (As), mercury (Hg), selenium (Se), and methane, can be released into the environment through natural processes such as outgassing from natural gas and coal deposits. These emissions can occur from geothermal areas, geological formations, and underwater methane seeps.

Re-entrainment of dust by wind: Dry weather conditions can lead to the re-entrainment of dust particles from soils and other surfaces. Wind can carry these particles over long distances, contributing to air pollution and reducing air quality.

Forest fires: Forest fires release a range of pollutants, including soot particles, nitrogen oxides (NO_x), sulfur dioxide (SO_2), and other gases. These emissions can have significant local and regional impacts on air quality and contribute to the formation of haze and smog.

Biological processes: Biological activities, both in terrestrial and aquatic ecosystems, can contribute to pollution. For example, carbon dioxide (CO_2) is released through respiration by plants and animals. Microbial processes in anaerobic wetlands produce methane, and incomplete oxidation of organic matter during forest fires can release methane as well as carbon monoxide (CO) and nitrous oxide (N_2O).

Radionuclides: Cosmic rays, as well as naturally occurring radioactive elements present in soils and rocks, can contribute to the release of radionuclides into the environment. These radioactive materials can have long-term impacts on human health and the environment.

11.2 ANTHROPOGENIC/ MADE-MADE SOURCES.

Anthropogenic or human-made sources of pollution significantly contribute to environmental degradation. These sources are directly linked to human activities and can have widespread impacts. Some major anthropogenic sources of pollution include:

Burning of fossil fuels, wood, and agricultural materials: The combustion of fossil fuels such as coal, oil, and gas releases carbon dioxide (CO_2), carbon monoxide (CO), nitrogen oxides (NOx), sulfur dioxide (SO_2), and particulate matter into the atmosphere. Similarly, the burning of wood and agricultural materials can release similar pollutants.

Metal mining and processing industries: Activities such as mining, smelting and processing of metals, including lead, cadmium, mercury, and other heavy metals, can result in the release of these toxic substances into the environment. These metals can contaminate water bodies, soil, and air, posing risks to ecosystems and human health.

Agricultural practices: Agricultural activities involve the use of pesticides, herbicides, fertilizers, and other chemicals to enhance crop yields. However, improper use and runoff of these substances can lead to water pollution, affecting aquatic ecosystems and human populations.

Chemical plants and waste burning: Industrial activities, including chemical plants and the burning of waste materials, can release various pollutants into the environment. Industrial compounds like polychlorinated biphenyls (PCBs) can be persistent and bioaccumulative, posing risks to ecosystems and human health.

Nuclear reactors and bomb tests: Nuclear activities, such as the operation of nuclear reactors and nuclear bomb tests, release radioactive substances and radionuclides into the environment. These radioactive materials can have long-term effects on ecosystems and human health.

While some pollutants from forest fires can be natural, the majority of pollution that we face today is anthropogenic in nature. Anthropogenic pollution tends to be more widespread and can have global impacts, unlike natural pollution, which is usually more localized.

11.3 POLLUTANTS

Pollutants are substances that, when introduced into the biosphere in significant quantities, can disrupt the functioning of ecosystems and have adverse effects on plants, animals, and humans. There are two main types of pollutants: degradable and non-degradable.

Degradable pollutants can be broken down or reduced to acceptable levels by natural processes. Some degradable pollutants are broken down by living organisms, such as specialized bacteria, and are referred to as biodegradable pollutants. For example, human sewage can be considered a biodegradable pollutant. Other degradable pollutants can be broken down by non-biological processes, such as the decay of radioactive substances. Degradable pollutants become problematic when they are introduced into the environment at a rate faster than natural processes can break them down.

Non-degradable pollutants, on the other hand, are not broken down by natural processes. Examples of non-degradable pollutants include certain types of plastics, mercury, and lead. Mercury and lead are non-degradable because they are elements, and breaking down an element would require specialized nuclear processes. Plastics tend to be non-degradable because they are synthetic materials that have not been present in the environment for an extended period, and organisms that can break them down have not yet evolved.

The challenge with non-degradable pollutants is that once released into the environment, they persist and accumulate over time. Therefore, the most effective approaches to addressing non-degradable pollutants are prevention, by not releasing them into the environment in the first place, and removal, through costly and often challenging cleanup efforts. These solutions require proactive measures, such as implementing regulations and waste management practices, to minimize the release of non-degradable pollutants and develop sustainable alternatives.

The severity of the effects of a pollutant is determined by three main factors:

Chemical nature: This refers to how active and harmful a pollutant is to specific types of organisms. Different pollutants have varying degrees of toxicity and can affect different species or ecosystems in different ways. Understanding the chemical properties of pollutants helps assess their potential risks and impacts.

Concentration: The concentration of a pollutant refers to the amount of the pollutant present per unit volume of air, water, soil, or body weight. Higher concentrations of pollutants generally increase the likelihood and intensity of adverse effects on living organisms. Monitoring and controlling pollutant concentrations are crucial in managing and mitigating their impacts.

One approach to reducing the concentration of pollutants is dilution, by adding them to a large volume of air or water. In the past, when pollution inputs were relatively low, dilution was a viable solution. However, with the increasing scale of pollution, dilution alone is no longer sufficient to address the problem effectively.

Persistence: Persistence refers to how long a pollutant remains in the air, water, soil, and organisms. Degradable or non-persistent pollutants can be broken down or reduced to acceptable levels through natural, physical, chemical, and biological processes over time. However, many substances and products introduced into the environment by humans take decades or even longer to degrade, making them persistent pollutants. Examples of slowly degradable or persistent pollutants include certain insecticides, DDT, most plastics, aluminum cans, and chlorofluorocarbons (CFCs).

The persistence of pollutants presents significant challenges, as they can persist in the environment for extended periods, leading to long-term accumulation and potential ecological and health risks. Managing and reducing the release of persistent pollutants is essential to minimize their impacts on ecosystems and human health.

11.4 TYPES OF POLLUTION

11.4.1 AIR POLLUTION

Air pollution is a pressing environmental concern characterized by the introduction of harmful substances into the atmosphere, predominantly originating from human activities. These pollutants, released as gases and particulates, have the potential to degrade and adversely impact both physical and biological systems. The addition of unwanted airborne matter alters the composition of the Earth's atmosphere, posing risks to life and materials.

Air pollutants encompass a wide range of substances, ranging from visible particles such as smoke and dust to invisible and odourless gases like carbon monoxide. Within the atmosphere, pollutants can undergo transitions from a dry gaseous phase to a liquid phase, eventually descending to the Earth's surface.

Among the most significant gaseous pollutants are sulphur dioxide (SO_2), hydrogen sulphide (H_2S), nitrogen oxides (NO, N_2O, NO_2), ammonia (NH_3), carbon monoxide (CO), carbon dioxide (CO_2), methane (CH_4), ozone (O_3), and peroxyacetyl nitrate (PAN). Additionally, there are vapour pollutants such as hydrocarbons and elemental mercury, as well as fine particulates with small diameters ($<\mu m$) that behave aerodynamically like gases and remain suspended in the atmosphere for extended periods.

Gaseous air pollutants can be categorized into primary and secondary forms. Primary air pollutants, including sulfur dioxide (SO_2), nitrogen oxides (NOx), carbon monoxide (CO), and unburnt hydrocarbons (HCs), are emitted directly into the atmosphere, exerting immediate effects upon release.

On the other hand, secondary air pollutants, such as ozone (O3) and peroxyalkyl nitrites (PANs), are formed through subsequent atmospheric reactions involving primary air pollutants and intense sunlight.

Once introduced into the atmosphere, air pollutants become integrated into the biogeochemical cycles through various pathways. Fine particulate matter and gaseous substances can be transported by atmospheric currents to distant locations, leading to their deposition as dry fallout. This deposition can subsequently enter nutrient cycles and food chains through water and soil interactions. Certain pollutants undergo chemical and photochemical reactions, resulting in the formation of secondary pollutants like sulfuric acid and ozone. Aerosols and other forms of fine particulate matter serve as condensation nuclei and eventually return to the Earth's surface as rainfall or other forms of precipitation.

Automobile Exhaust and Photochemical smog.

Automobile exhaust represents a persistent and challenging form of air pollution that is difficult to mitigate. It contains various pollutants, including carbon monoxide (CO), nitrogen oxides (NOx), hydrocarbons, and, in the case of leaded gasoline, particulate lead. Carbon monoxide, in particular, poses a significant health risk as it combines with haemoglobin in human blood, forming carboxyhaemoglobin, which impairs the transport of oxygen.

Nitrogen oxides (NOx) and hydrocarbons, such as methane, ethane, and toluene, are primary constituents of photochemical smog. When exposed to sunlight, these compounds react with oxygen, leading to the formation of photooxidative reactions. The complex reactions involve nitrogen oxides (NO), water, hydrocarbons, oxygen, nitrogen dioxide (NO2), peroxyacetyl nitrate (PAN), and ozone.

PAN and ozone, which are products of photochemical smog, are reported as total oxidants in air pollution data. These compounds are highly reactive and can be detrimental to human health when present in excessive concentrations. They also have adverse ecological effects, including damage to vegetation and agricultural crops.

Lead, another pollutant emitted by automobiles, has long-term environmental pollution effects. Significant concentrations of lead have been detected in rainfall, further highlighting the extent of its environmental impact.

AIR POLLUTION'S EFFECT ON WEATHER, CLIMATE AND ATMOSPHERIC PROCESS:

Air pollution has notable impacts on weather, climate, and atmospheric processes, giving rise to two significant global concerns: contamination of the upper atmosphere and the alteration of weather and climate patterns by air pollutants.

Local Weather Patterns: Pollution and population concentrations have a discernible influence on local weather patterns. The distribution and abundance of particulate nuclei in the lower atmosphere can modify local rainfall patterns. As a result, there is a notable increase in precipitation in and around cities attributed to pollution.

Continental and Global Effects: Air pollution also has the potential to affect weather on a continental or even global scale. Gaseous pollutants and fine aerosols released into the atmosphere can have fundamental effects on the penetration and absorption of sunlight. The increasing concentration or loading of particulate matter in the atmosphere can significantly impact the amount of sunlight energy reaching the Earth's surface.

Some scientists speculate that the escalating levels of particulate pollution may be diminishing the amount of solar radiation reaching the Earth's surface. This reduction in solar radiation could potentially lead to a cooling effect on global climate, which may ultimately trigger another ice age.

Note that the relationship between air pollution, climate change, and weather patterns is complex and still subject to ongoing research and investigation. While there are indications of pollution's influence on weather and climate, further scientific study is necessary to fully understand the extent and long-term implications of these interactions.

11.4.2 GLOBAL WARMING: 'GREENHOUSE EFFECT'

Global warming, attributed to the increase in concentrations of radiatively active trace gases such as carbon dioxide (CO_2), methane (CH_4), nitrous oxide (N_2O), chlorofluorocarbons (CFCs), and tropospheric ozone (O_3), is a significant consequence of the greenhouse effect.

The greenhouse effect arises from the interference of these gases with the Earth's natural process of dissipating absorbed solar radiation. While greenhouse gases allow incoming short-wave radiation to pass through, they absorb and trap the long-wave radiation emitted by the Earth's surface.

As a result, the presence of greenhouse gases raises the equilibrium temperature of the Earth, and further increases in atmospheric concentrations of these gases, largely due to human activities, contribute to global warming. When high-energy solar radiation enters the atmosphere, a portion is reflected back into space by clouds and other factors. Approximately half of the radiation reaches the Earth's surface, but most of it is reflected back as lower-energy long-wave infrared (IR) radiation.

However, the presence of radiatively active greenhouse gases in the atmosphere impedes the outward journey of this re-radiated long-wave IR radiation to space. Only a small proportion (around 10%) of this re-radiated energy escapes directly into space, while the majority is absorbed by the greenhouse gases in the atmosphere.

Subsequently, the absorbed infrared energy is re-radiated in various directions, including some back towards the Earth's surface (both directly from solar radiation and indirectly from re-radiated IR). This process contributes to the overall warming of the Earth's climate system, commonly referred to as global warming.

The greenhouse effect refers to the warming of the Earth's surface and atmosphere caused by the retention of heat by greenhouse gases in the atmosphere. The mechanism behind this effect can be likened to the process observed in a greenhouse, where the transparent glass and the humid atmosphere inside allow incoming radiation to pass through but absorb a significant portion of the outgoing long-wave infrared radiation, thus slowing down the radiative cooling of the interior.

One significant consequence of global warming resulting from the greenhouse effect is the partial melting of the world's polar ice-caps, leading to a rise in sea levels. This rise in sea level poses a threat to coastal cities worldwide, as they may become submerged under water.

The rising sea levels will cause flooding and the disappearance of vital coastal wetlands, such as marshes and mangrove swamps. These wetlands are home to numerous bird species and play a crucial role in the survival of seafood species like oysters, shrimps, crabs, and various fish species. These wetlands serve as important natural filters, breaking down pollutants that flow downstream from inland areas. Without their presence, we would face a loss of seafood, significant biodiversity in plants and animals, and a decline in our ability to protect the oceans from pollution.

11.4.3 OZONE LAYER DEPLETION

Ozone plays a critical role as an ultraviolet shield, protecting organisms on Earth's surface from the harmful effects of UV radiation. It acts as a barrier that filters out most of the incoming UV radiation, preventing it from reaching the surface at high intensities. This protective function of ozone is crucial for maintaining the health and well-being of various life forms.

In the upper atmosphere, a concerning issue arises with the accumulation of fluorocarbons, specifically chlorofluorocarbons (CFCs) or chlorofluoromethanes (CFMs). These compounds, commonly known as "freon," have widespread usage as aerosol propellants, refrigerants, foam blowing agents, insulation materials, solvents, and even in medical applications such as anaesthetics. Due to their chemical stability, CFCs have a long atmospheric lifetime, and there are no known natural processes that effectively remove them from the lower atmosphere.

As these inert gases disperse into the upper atmosphere, primarily the stratosphere (a layer located approximately 10-60 km above the Earth's surface), they accumulate at significant levels. Under the influence of intense shortwave ultraviolet radiation, CFCs release chlorine atoms through a photochemical reaction. These chlorine atoms then catalytically react with over 100,000 molecules of ozone, resulting in the conversion of ozone (O_3) to oxygen (O_2). This process, known as ozone layer depletion, leads to a reduction in stratospheric ozone concentration, allowing greater penetration of UV light to the Earth's surface.

The intensified UV radiation reaching the surface as a consequence of ozone layer depletion is a matter of concern. Scientists have raised alarm about the potential increase in skin cancer cases and adverse effects on various organisms, including humans. The damaging effects of UV radiation extend beyond health impacts and can also have implications for ecosystems, agriculture, and overall environmental balance.

Other reactions can also contribute to the depletion of ozone in the stratosphere. These include the direct photodissociation of ozone by UV radiation and reactions with trace gases such as nitrogen oxides (NO_X), nitrous oxide (N_2O), and fluorine-containing compounds. The complex interactions among these species further contribute to the overall reduction in stratospheric ozone levels.

The ozone layer in the stratosphere faces an additional threat from supersonic jets (SSJs). Nitrogen oxides released from the jet engines of supersonic aircraft flying at high altitudes can act as catalysts that destroy ozone molecules. The impact of SSJs on the ozone layer adds to the already existing concerns regarding ozone depletion and calls for careful consideration in the development and operation of such aircraft.

Preserving the ozone layer is of utmost importance to safeguard the well-being of ecosystems, human health, and the balance of atmospheric processes. International efforts, such as the Montreal Protocol, have been instrumental in reducing the production and consumption of ozone-depleting substances, leading to gradual recovery of the ozone layer. Continued vigilance and adherence to regulations are necessary to mitigate the risks associated with ozone depletion and ensure a sustainable future for our planet.

11.4.4 WATER POLLUTION

Water pollution refers to the human-induced alteration of the chemical, physical, or biological characteristics of water, resulting in a significant reduction in its utility or environmental value (Hester, 1986). It encompasses a wide range of aquatic contaminations, ranging from the excessive enrichment of biotic communities due to nutrient runoff from sewage or fertilizers to the contamination of bodies of water with toxic chemicals that can eliminate organisms and even eradicate all forms of life.

Water pollution occurs through various pathways, each with its own detrimental effects on aquatic ecosystems:

Physical pollution: This type of pollution occurs when solid debris is introduced into streams, rivers, or lakes, leading to the smothering of the stream-bed and the suppression of life within the water. Additionally, the discharge of heated water from factories and power stations can raise the temperature of the receiving water body. This increase in temperature reduces the availability of dissolved oxygen, crucial for supporting life and the self-purification processes that naturally occur in the water.

Biological pollution: Biological pollution takes place when living organisms, such as disease-causing pathogens of fecal origin, are introduced into water systems. The discharge of sewage effluent containing human pathogens is a prime example of biological pollution. Moreover, the release of biodegradable organic chemicals can disrupt the natural balance of organisms in a stream, leading to excessive growth and depletion of oxygen levels to critically low levels.

Chemical pollution: Chemical pollution refers to the introduction of chemical contaminants into water bodies. The primary source of chemical pollution is the discharge of wastewater from urban areas, which often contains a wide range of pollutants. Other significant sources include oil and industrial chemical spills, improper disposal of sludge into the sea, the deposition of solid wastes and landfill sites, and the use of fertilizers and pesticides in agricultural practices. These pollutants can have detrimental effects on water quality, aquatic organisms, and overall ecosystem health.

11.4.5 EUTROPHICATION

Eutrophication, while a natural process, can have negative effects when excessive nutrients are present in aquatic systems. Excessive nutrient enrichment can occur due to human activities, such as the runoff of agricultural fertilizers or the discharge of untreated wastewater. These excessive nutrients, primarily nitrogen and phosphorus, accelerate the natural eutrophication process, leading to a series of ecological changes and challenges.

When nutrient concentrations exceed natural levels, the effects of eutrophication become more pronounced:

Algal Blooms: High nutrient levels promote the rapid growth of algae, resulting in the formation of algal blooms. These dense accumulations of algae can reduce water clarity, hinder the penetration of sunlight, and disrupt the balance of the aquatic ecosystem. Some algal species may produce harmful toxins, posing risks to human health and the health of aquatic organisms.

Oxygen Depletion: As algae blooms die and decompose, bacteria break down the organic matter, consuming dissolved oxygen in the process. This leads to the depletion of oxygen levels in the water, creating oxygen-deficient zones known as hypoxic or anoxic conditions. Such conditions can be detrimental to fish and other aquatic organisms that require oxygen to survive, leading to fish kills and the decline of biodiversity.

Changes in Aquatic Life: Eutrophication can alter the composition and structure of the aquatic community. Some species of algae may dominate, while others, such as submerged aquatic plants, may struggle to compete for resources. This shift in the ecosystem can negatively impact fish populations, as well as other organisms that depend on specific habitats and food sources.

Nutrient Imbalances: Excessive nutrient levels can disrupt the natural balance of nutrients in the water. This can lead to the overgrowth of certain species, such as cyanobacteria (blue-green algae), which can further contribute to oxygen depletion and produce harmful toxins. The imbalance of nutrients can also impact the availability of essential elements for aquatic organisms, affecting their growth and reproduction.

Oligotrophic water bodies are characterized by a low concentration of nutrients and are considered in a natural state. In contrast, when excessive amounts of nutrients, such as nitrogen and phosphorus, enter streams and lakes from various sources like sewage, fertilizer, animal wastes, and detergents, eutrophication or cultural eutrophication occurs. This phenomenon leads to the overgrowth of microorganisms and aquatic vegetation due to the increased availability of nutrients.

Cultural eutrophication can have detrimental effects on water quality, significantly impacting the usability of water resources for human purposes. Unlike natural eutrophication, which is part of the aging process of lakes and ponds, cultural eutrophication is caused by human activities. The term "cultural eutrophication" distinguishes this human-induced process from the natural eutrophication process.

An intermediate condition between oligotrophic and eutrophic water bodies is referred to as "mesotrophic." Mesotrophic water bodies exhibit moderate nutrient concentrations and may have a balanced aquatic ecosystem with a mix of species.

One of the most noticeable symptoms of eutrophication is the occurrence of algal blooms, characterized by a substantial increase in the standing crop of phytoplankton. These algal blooms often result in a shift in the composition of algal species. In shallow water bodies, there may also be vigorous growth of vascular plants.

Excessive nutrient loading, particularly nitrogen and phosphorus, is widely recognized as the primary cause of cultural eutrophication. Nutrients such as ammonia nitrogen, nitrites, nitrates, and phosphates act as fertilizers, stimulating algal growth and leading to the formation of plankton blooms.

To address the issue of cultural eutrophication, efforts should focus on reducing nutrient inputs into water bodies. Implementing proper wastewater treatment, adopting responsible agricultural practices, managing animal waste effectively, and promoting the use of environmentally friendly detergents can all contribute to mitigating nutrient loading and preventing the adverse effects of eutrophication. By managing nutrient inputs, we can strive to maintain the balance and ecological health of water bodies, ensuring their sustainable use for both humans and the environment.

11.4.6 EFFECTS OF EUTROPHICATION IN WATER BODY

1. Production of odours and tastes and toxic metabolic products in water.

Plankton blooms, particularly those of blue Green algae produce obnoxious odours and tastes in water. Others such as dinoflagellate blooms or 'red tide' produce toxic metabolic products which can result in major fish kills.

2. Depletion of dissolved oxygen in water

Plankton blooms of green algae do not always produce undesirable odours or toxic products but can still create problems of oxygen supply in water. While these blooms exist under abundant sunlight, they contribute oxygen to the water through photosynthesis but under prolonged cloudiness they begin to die and decay and consume more oxygen than they produce, leading to oxygen depletion in the water causing kills of fish and other biota. This occurs when dissolved oxygen declines rapidly from favourable levels of 10 to 12 ppm to levels of 2 or 3 ppm at which point fish begin to die. Excessive levels of iron, manganese and fowl-smelling hydrogen sulphide in bottom waters may also occur as a result of oxygen depletion.

3. Impairment of fishing, bathing, fish spawning and navigation.

Excessive nutrient levels in aquatic systems can cause other kind of ecology consequence. They may lead to extensive growth of aquatic water weeds such as water hyacinth and many others which have become a worldwide problem. Excessive growth of these weeds can impair fishing, bathing, fish spawning and navigation and thus represent a major economic problem as well as a complete disruption of aquatic ecology.

4. Negative aesthetic impacts:

The excessive growth of algae and aquatic plants can have a visible impact on the aesthetic quality of a water body. With dense quantities of algae in a lake or reservoir the transparency of the water is greatly reduced and the water body can acquire undesirable 'pea-soup' green colour. Excessive algal densities can also interfere significantly with swimming and other recreational activities. Large 'mats' of dead algae can accumulate on beaches and with negative aesthetic impacts.

5. Potential negative health impacts: Cultural eutrophication, particularly in tropical and subtropical regions, can increase the prevalence of parasitic diseases such as schistosomiasis, onchocerciasis, and malaria. The excessive nutrients create favorable habitats for the organisms responsible for transmitting these diseases to humans.

These effects of eutrophication can significantly impact the multiple uses of water bodies, including drinking water, fisheries, recreation, and aesthetics. In some cases, controlled eutrophication is intentionally employed in certain countries to enhance fish production and aquaculture, aiming to maximize productivity while minimizing costs and efforts.

11.5 PESTICIDE CONTAMINATION

Pesticides (biocides) are substances that used to protect humans against the insect vectors of disease causing pathogens, to protect crop, plants and livestock from disease and degradation by fungi, insects, mites and rodents. The most widely used types of pesticides are insecticides (insect killers) herbicides (weed killers) fungicides (fungus killers and rodenticides (rodent killers). They have proved beneficial to human populations, reducing, eliminating insects and other animals which transmit diseases, destroy agricultural crops, damage homes and stored products and directly or indirectly affect human health and welfare.

Ecologically, however, pesticides have created major problems which were not anticipated.

1. Persistence and accumulation in the Environment.

Pesticides not reaching target pests end up in the soil, surface water, bottom sediments, food and non-target organisms. Concentrations of fat soluble, slowly degradable insecticides such as DDT, PCBs and other chlorinated hydrocarbons can be biologically amplified thousands to millions of times in food chain and webs. There is also disruption of such ecological processes as productivity and nutrient cycling.

2. Affected human health

Many of them have directly or indirectly affected human health. The World Health Organization estimates that each year about three million people are poisoned by pesticides, and 5000 to 20000 of them die. At least half of those poisoned and 75% of those killed are farm workers from Less Developed Countries, where educational levels are low, warnings are few and pesticides regulation and control methods are lax or non-existent. Accidents and unsafe practices in pesticide plants can expose workers, their families and sometimes the general public to harmful levels of pesticides or chemicals used in their manufacture. A percentage of the food bought in the markets has levels of residues of one or more of the active ingredients used in pesticides that are above level limit.

3. Development of genetic resistance

Many pest species especially insects can develop genetic resistance to a chemical poison through natural election. Most pest species especially insects and disease organisms can produce a large number of similarly resistant offspring in a short time. When an area is sprayed with a pesticide, a few organisms in a large population of a particular species usually survive because they have genes that make them resistant or immune to a specific pesticide. With repeated spraying each succeeding generation contains a higher percentage of resistant organisms.

Thus, eventually, widely used pesticides (especially insecticides) fail because of genetic resistance, in fact, widespread use usually leads to even larger populations of pest species , especially insects with large numbers of offspring and short generation time.

In temperate regions, most insects develop genetic resistance to a chemical poison within 5 to 10 years, it happens much sooner in tropical areas. Weeds and plant –disease organisms also develop genetic resistance but not as quickly as most insects. Because of genetic resistance, most widely used insecticides no longer protect people from insect transmitted diseases in many parts of the world leading to even more serious out breaks of diseases.

There is now abundant evidence that many chlorinated hydrocarbons used in agricultural and public health programme as pesticide are lethal to a variety aquatic and terrestrial organisms in small doses. Our present systems of agriculture and disease control require pesticides but it is now imperative that ecological safer control systems be developed and utilized. A sudden withdrawal of their use now would certainly be damaging to agricultural production and public health. Yet their continued indiscriminate use is ecologically damaging to the total environment.

We must therefore find new methods of insect and disease control which are less dangerous to the total environment and more selective in their effects.

11.5.1 ALTERNATIVE METHODS TO USE OF PESTICIDES

1. **Biological control**: Use of natural predators, parasites, competitors and other biological control to regulate pest populations.

2. **Genetic control**:

i. Autocidal control: One method of genetic sterile organisms (autocidal control). This is the release of great numbers of the pest species that have been sterilized by gamma rays or chemicals so that they mate with no progeny and exhaust the reproductive capacity of their fertile wild mates. It is used for insect pests. However, difficulties associated with large scale deployment of partially sterile males renders autocidal control impractical on a large scale.

ii Resistant crop breeding. Genetic control is achieved by breeding crop plants and animals that are more resistant to pests which can be produced by breeding techniques and also using modern, high tech genetic engineering techniques.

iii. Integrated pest Management. The overall aim of Integrated Pest Management is not eradication but keeping pest populations just below the size at which they cause economic loss. Within the context of Integrated Pest Management (IPM), an acceptable pest control is achieved by employing an array of complementary approaches. If successfully implemented, an IPM programme can greatly reduce but not necessarily eliminate the reliance on pesticides. It minimizes the hazard, human health, wildlife and the environment from the widespread use of chemical pesticides. An important component of IPM is the use of procedures that are as pest specific as possible, so that non-target damage can be avoided or reduced.

CHAPTER 12

PLANT FORMS AND EVOLUTION

12.1 FORMS OF PLANTS

Plants are known for the presence of chlorophylls which distinguish them from animals. It can be noted that movement is no longer exclusive to animals alone. Some lower plants can also move making use of appendages like flagella.

Plants exhibit a remarkable range of forms, which can be categorized into lower plants and higher plants based on their characteristics and reproductive strategies.

Lower plants, also known as cryptogams, encompass a diverse group of organisms that do not produce seeds. They lack well-developed vascular bundles, which are responsible for the transport of water, nutrients, and sugars within the plant. In addition, lower plants are not typically organized into distinct structures like leaves, stems, and roots.

Among the lower plants, viruses, algae, lichens, bryophytes, and pteridophytes can be identified. Viruses, although not classified as living organisms, can infect and affect plant physiology. Algae are a varied group of photosynthetic organisms that can be found in aquatic environments, ranging from single-celled species to larger multicellular forms. Lichens are unique symbiotic associations between fungi and photosynthetic organisms, often algae or cyanobacteria. Bryophytes, including mosses and liverworts, are non-vascular plants that rely on water absorption through their surfaces. Pteridophytes, such as ferns and horsetails, have true vascular tissues and reproduce through spores.

Higher plants, also referred to as phanerogams, are characterized by their ability to produce seeds. They possess a well-developed vascular system, allowing for efficient transport of fluids and nutrients throughout the plant. Higher plants also exhibit a more complex organization with distinct roots, stems, and leaves.

The two major groups of higher plants are gymnosperms and angiosperms. Gymnosperms are plants with exposed seeds, often housed in cones or similar structures, and they do not produce flowers. Examples of gymnosperms include conifers, cycads, and ginkgos. Many gymnosperms are exotic or foreign to specific regions.

Angiosperms, on the other hand, are the most common and diverse plants found in our environment. They are characterized by the production of flowers, which are specialized structures involved in sexual reproduction. Angiosperms have enclosed seeds that develop within fruits, contributing to seed dispersal and protection. They possess an extensive and efficient root system that anchors the plant and absorbs water and nutrients from the soil. Angiosperms dominate terrestrial ecosystems and include a wide range of plants, from trees and shrubs to grasses and flowers.

The study of evolution is crucial in understanding the diversity and complexity of life on Earth. Since the origin of life more than 3.5 billion years ago, the organisms inhabiting our planet have undergone significant transformations. Evolution refers to the process of change in living organisms over time, and it is not limited to biological entities alone; it also applies to non-living aspects of the universe.

In the 19th century, Charles Darwin and Alfred Wallace independently developed similar ideas regarding evolution. They proposed that organisms evolve through a mechanism called natural selection, wherein certain individuals have a higher likelihood of reproductive success than others. This theory of evolution by natural selection rapidly became a central concept in biology.

Fossils play a crucial role in understanding the history of life on Earth. These preserved remains or traces of ancient organisms provide valuable insights into past environments and the organisms that lived during different periods. By comparing fossilized organisms with present-day organisms, scientists can make important observations and draw conclusions about the changes that have occurred over time.

Comparative analysis of fossils from different geological periods allows us to trace the evolutionary trajectories of various species and identify patterns of diversification and extinction. By examining the anatomical structures, genetic information, and ecological relationships of fossilized organisms, researchers can reconstruct the evolutionary history of different groups of organisms.

The field of paleontology, which focuses on the study of fossils, has contributed significantly to our understanding of evolution. Fossil records provide evidence of extinct species, transitional forms, and evolutionary trends. This information, combined with other lines of evidence such as genetic studies and observations of living organisms, allows scientists to construct a comprehensive understanding of the history and mechanisms of evolution.

Viruses are unique organisms that exhibit a dual existence, existing in two distinct forms: intracellular and extracellular. In their intracellular form, they are considered to be in a living stage, whereas in their extracellular form, they are non-living. Viruses are significantly smaller than bacteria, typically ranging from 2 to 20 μm in size. They are capable of passing through the pores of filters due to their small size.

Most viruses are pathogenic and cause diseases in various organisms, including humans and plants. Unlike beneficial symbiotic organisms, viruses generally do not provide any advantages to their hosts. They are often named after the diseases they cause, such as the Tobacco Mosaic Virus (TMV), which affects tobacco plants, and the Human

Immunodeficiency Virus (HIV), which affects humans. It is worth noting that viruses can also infect bacteria, and these viruses are known as bacteriophages.

The structure of a virus can be divided into three main parts: the head, tail, and cell fibers. Viruses consist of an outer protein coat, called a capsid, which encloses the inner core. The inner core contains the genetic material of the virus, which can either be DNA or RNA. Viruses are classified as either DNA viruses or RNA viruses, and they do not possess both types of genetic material simultaneously.

There are two primary types of relationships between viruses and bacteria: lytic and lysogenic.

In a lytic relationship, the virus is considered virulent or harmful to the host. It undergoes a cycle of infection, replication, and cell lysis, leading to the destruction of the host cells.

In a lysogenic relationship, the virus integrates its genetic material into the host genome without causing immediate harm or cell lysis. This relationship is considered safer since the host cells are not immediately affected, and the viral genetic material may be passed on to subsequent generations of cells.

12.2 VARIETY CLASSIFICATION IN THE PLANT KINGDOM

12.2.1 INTRODUCTION

The study of the plant kingdom reveals a vast array of plant types, each with its unique characteristics and associations. This diversity may initially appear chaotic to an untrained observer, but by understanding the patterns and classifications within the plant kingdom, we can unravel the order hidden within nature's complexity.

The Earth, with its intricate physical features and diverse ecosystems, existed long before humans appeared. Whether one believes in creation or evolution, it is evident that the Earth's physical and biological aspects were already in place, presenting a seemingly disordered tapestry of life. It was the task of humans, upon their arrival, to discern and comprehend the underlying patterns and order intended by the Earth's "designer."

In order to make sense of the countless plant types and their associations, scientists have developed classification systems to organize and categorize them. These systems help us identify similarities and differences between plants and understand their evolutionary relationships.

The primary classification system used for plants is known as taxonomy. Taxonomy is based on the principles established by Swedish botanist Carl Linnaeus in the 18th century. It involves grouping plants into hierarchical categories based on shared characteristics.

The highest level of classification in the plant kingdom is the division or phylum. Divisions are based on major differences in plant structures and reproductive strategies. Examples of divisions include Bryophyta (mosses), Pteridophyta (ferns), Coniferophyta (conifers), and Magnoliophyta (flowering plants).

Within each division, plants are further classified into classes, orders, families, genera, and species. These classifications are based on increasingly specific characteristics and evolutionary relationships. For example, within the division Magnoliophyta, plants are classified into various classes such as Magnoliopsida (dicotyledons) and Liliopsida (monocotyledons). Each class is further divided into orders, families, genera, and species, allowing for precise identification and classification.

The classification of plants not only aids in organization but also provides valuable insights into their biological traits, habitats, and evolutionary history. By studying the relationships between different plant groups, scientists can gain a deeper understanding of plant evolution and their adaptations to different environments.

The apparent disorder and diversity within the plant kingdom can be deciphered by applying classification systems such as taxonomy. Through these systems, we can identify and categorize plant types based on their shared characteristics and evolutionary relationships. By unraveling the order hidden within nature's complexity, we gain a greater understanding of the plant kingdom and its vital role in shaping our planet.

12.2.2 VARIATION AND CLASSIFICATION IN BIOLOGY

Classification is an essential aspect of studying the diverse organisms found in the natural world. It involves the grouping and categorization of organisms based on their characteristics, habitat, or utility to humans. The level of criteria used in classification can range from a few to many, resulting in artificial or natural classifications, respectively. As the theory of evolution has advanced, the classification of organisms has increasingly aimed to reveal their evolutionary relationships.

When encountering an unknown plant, the initial step is identification, which involves assigning it to an established group. The purpose of identification is to place the unknown entity within a known category. Taxonomy is the scientific discipline concerned with classification, while systematics encompasses a broader domain that includes naming (nomenclature) and the collection and assessment of classification data.

Taxonomy employs a hierarchical system to organize organisms into categories, ranging from broad to specific classifications. These categories include domains, kingdoms, phyla, classes, orders, families, genera, and species. Each level represents a progressively more specific classification, allowing for a systematic arrangement of organisms.

Systematics goes beyond classification, incorporating various tools and techniques to establish evolutionary relationships among organisms. It utilizes molecular biology, morphology, and comparative anatomy to determine similarities and differences among organisms and infer their evolutionary history. By examining DNA sequences, physical structures, and other biological attributes, scientists can discern the evolutionary relationships and common ancestry of different groups.

The study of evolutionary relationships involves constructing phylogenetic trees or cladograms, which depict the branching patterns of evolution and represent the shared ancestry of organisms. These trees incorporate information about shared derived characteristics and genetic data, enabling scientists to infer the relationships between organisms and their common ancestors.

12.3 NOMENCLATURE AND HIERARCHIC RELATIONSHIPS OF ORGANISMS

Nomenclature has already been defined as a system of naming. It is common knowledge that practitioners of trades, professions or even belief systems need to agree on standard systems of naming of the object of their trade or professions to facilitate communication and avoid confusion. In Botanical and Zoological sciences, problem of naming are usually addressed in international congresses and through publications. Each organism is assigned two Latin names -a generic name - and a specific name – and the names are accompanied by the name of the person who first described the organism (authority). This idea of two names (the Binomial

System) was popularized by a Swedish naturalist Carolius Linneaus (1707 – 1778) but finally settled by the international botanical congress held in Amsterdam in 1935. The two names are always underlined or italicized. Some examples of names in then binomial system are: *Oryza sativa* L., *Vigna unguiculata var. unguiculata, Laportea aestuans* L., *Ocimum basilicum var. basilicum.* Note that it is possible to include varietal names as has been done for *Vigna unguiculata.*

An ideal system of classification should not only indicate the actual genetic relationship but should be also reasonably convenient for practical purposes. Individuals within a species that shows marked degree of variation in the form of size, shape, colour and other major characteristics are called varieties. A genus is a collection of species which bear close resemblance to one another in terms of morphology of floral or reproductive parts are concerned.

12.4 BOTANICAL NAMES OF SOME PLANTS

Rice – *Oryza sativa* Linn.

Mango – *Mangifera indica* Linn.

Maize – *Zea mays*

Pawpaw – *Carica papaya*

The first two name have their authority included while the last two do not have their authority.

Artificial System of Classification: it makes use of only one or at most a few characters that are selected arbitrary and plants are arranged into groups according to such characteristics.

- As a result closely related plants are often placed in different groups while unrelated plants are often placed in the same group because of presence or absence of particular characteristics.
- The system enables us to determine readily the names of plants but does indicate the natural relationship that exists among individual forming a group.
- The artificial system makes us to identify an unknown plant without much difficulty.

Natural System: all important characteristics are considered and plants are classified according to related characteristics (resemblance and differences). This system gives us a true idea of natural relationship existing between different plants based on sequence of evolution and meets the need to identify unknown plants. According to the natural system the plant kingdom has been divided into two divisions, Cryptograms (flowerless plants) Phanerogams (flowering plants). The phanerogams have been sub-divided into two subdivisions. The Gymnosperms (naked seeded plants) and Angiosperms (closed seeded plants). Angiosperms have been subdivided into classes, dicotyledons and monocotyledons. The classes are divided into orders, the orders into families, the families into genera and species, and sometimes species into varieties.

Genera that have common attributes are placed in one family: for example the tobacco genus, pepper genus tomato genus and the garden egg genus (*Nicotiana, Capsicum, Lycopersicum and Solanum,* respectively) all belong to the family Solanaceae with *Solanum* as the type genus. Related families such as Solanaceae and Convolvulaceae belong to the same order Polemoniales; related group of orders belong to the same class; different groups of classes belong to different divisions which make the Plant Kingdom. At the higher levels (above the family level) there is considerable disagreement on how to put various groups together. The table shows the Eichler's grouping of 1883 which we shall adopt.

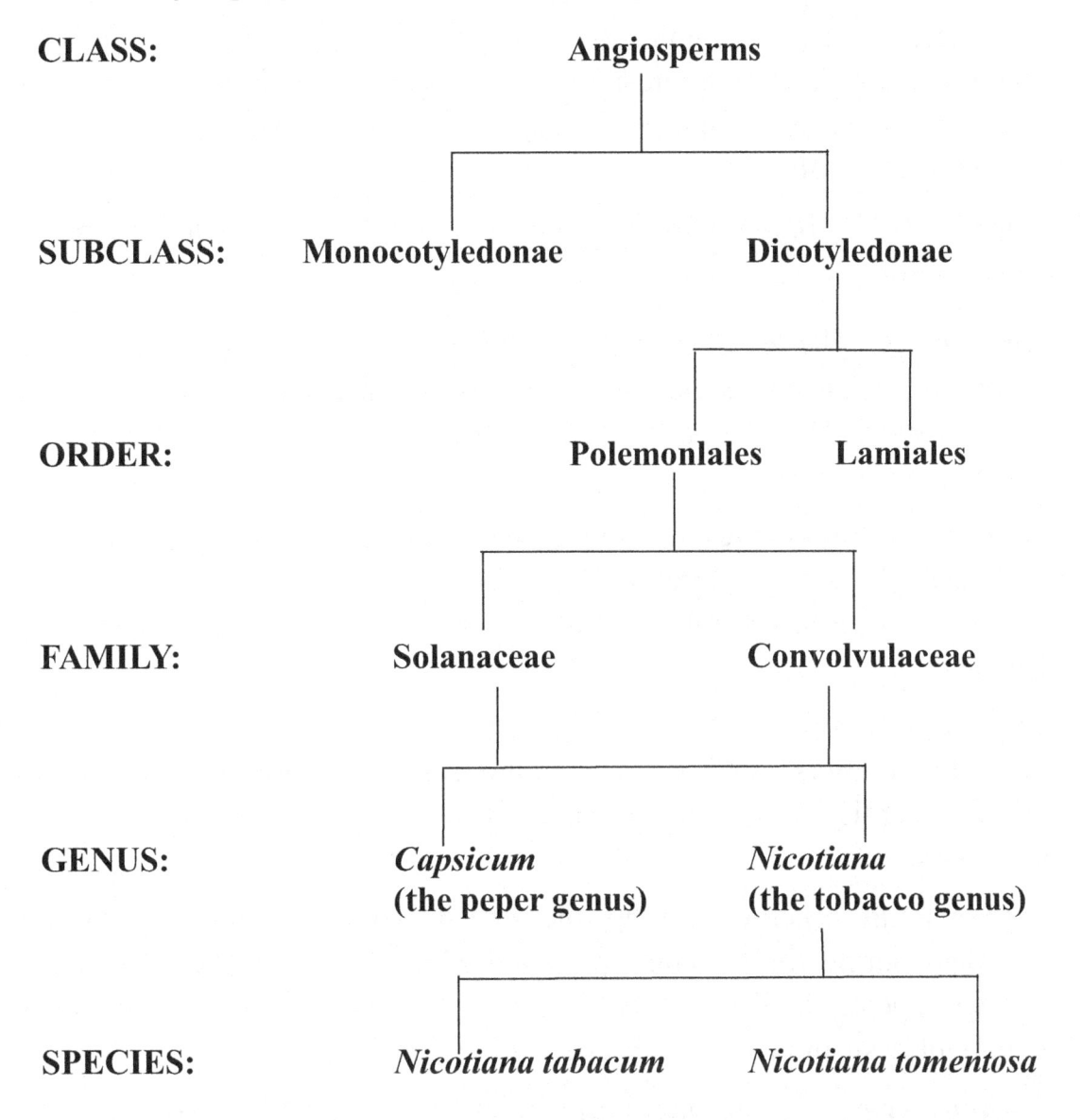

CLASS: Angiosperms

SUBCLASS: Monocotyledonae Dicotyledonae

ORDER: Polemonlales Lamiales

FAMILY: Solanaceae Convolvulaceae

GENUS: *Capsicum* (the peper genus) *Nicotiana* (the tobacco genus)

SPECIES: *Nicotiana tabacum* *Nicotiana tomentosa*

Hierarchic categories in plant classification using one of the classes of the Division Spermatophyta - the Angiospermae (the flowering seed plants) as an example.

12.5 EVOLUTION OF PLANTS

The evolution of plants has followed a remarkable path of development and diversification over millions of years. From simple algal mats to the complex and diverse angiosperms, plants have undergone significant evolutionary changes. The evolutionary sequence of plant groups can be summarized as follows:

Algal mats → bryophytes → lycopods → ferns → gymnosperms → angiosperms

Algal mats: The earliest forms of plant life were algal mats, which consisted of primitive photosynthetic organisms that lived in aquatic environments. These mats played a crucial role in the oxygenation of Earth's atmosphere.

Bryophytes: Bryophytes, including mosses and liverworts, were among the first land plants to emerge. They lacked vascular tissues and true roots, stems, and leaves. Bryophytes played a vital role in colonizing terrestrial habitats.

Lycopods: Lycopods, also known as clubmosses, were a diverse group of plants that dominated the landscape during the Carboniferous period. They had vascular tissues, enabling them to grow taller and exhibit more complex structures.

Ferns: Ferns emerged during the late Devonian period and became prevalent during the Carboniferous and Mesozoic eras. They reproduced through spores and had well-developed vascular tissues. Ferns were important in shaping ancient forests.

Gymnosperms: Gymnosperms, including conifers and cycads, appeared during the Paleozoic era and became widespread during the Mesozoic era. They were the first plants to produce seeds but lacked flowers. Gymnosperms played a significant role in establishing forests and adapting to diverse environments.

Angiosperms: Angiosperms, or flowering plants, emerged around 140 million years ago and quickly became the dominant plant group. They produce flowers and seeds enclosed in fruits. Angiosperms exhibit an extraordinary range of adaptations and have diversified into countless species, occupying almost every habitat on Earth.

12.6 CLASSIFICATION INTO THE PLANT AND ANIMAL KINGDOM

Living organisms are classified into two main groups based on the complexity of their cells: the Prokaryota and the Eukaryota. Prokaryota consists of organisms with primitive cells, while Eukaryota comprises organisms with true cells.

Eukaryotic cells are found in more evolutionarily advanced organisms such as green algae, fungi, bryophytes, and vascular plants. These cells exhibit a high degree of differentiation and adaptation. Eukaryotic cells possess a distinct nucleus, membrane-bound organelles such as mitochondria and plastids, and a mitotic apparatus involved in cell division.

Prokaryotic cells lack these membrane-bound organelles and exhibit simpler cellular structures. They do not possess a defined nucleus or mitochondria. However, it should be noted that some similarities exist between the membrane-bound organelles of eukaryotes and prokaryotic cells. These similarities include genetic material and ribosomes, which share common features with those found in prokaryotic cells.

The classification of organisms into the plant and animal kingdoms is based on these fundamental cellular differences. Plants, belonging to the Eukaryota, possess eukaryotic cells, while animals can also be found within the Eukaryota but exhibit different cellular characteristics.

Viruses are unique entities that possess characteristics of both living and non-living organisms. They consist of minute particles that contain genetic material, either DNA or RNA, enclosed within protein coats.

However, viruses lack the necessary cellular machinery, such as mitochondria, ribosomes, and other organelles, to carry out essential metabolic processes, energy production, and protein synthesis.

Due to their inability to replicate independently, viruses rely on host cells to complete their life cycle. They are obligate intracellular parasites, meaning they must invade and parasitize host cells to utilize their metabolic systems for energy production and other necessary metabolic activities. Once inside a host cell, viruses hijack the cellular machinery to replicate their genetic material and assemble new virus particles. This process often leads to the destruction of the host cell and the release of viral particles to infect new cells.

The reliance on host cells for replication and metabolic activities blurs the line between living and non-living organisms when it comes to viruses. While they possess genetic material and can undergo evolution, viruses lack the autonomous capacity for growth, metabolism, and reproduction seen in living organisms. Instead, they exhibit properties reminiscent of non-living entities, such as inert particles.

Viruses are often described as "obligate intracellular parasites" and are considered to occupy a unique position in the classification of organisms, straddling the boundaries between living and non-living.

We now have the following groups of living organisms: the prokaryota, the eukaryote, the protista, the monera; the last two were erected from the first two. The account below presents five system of classification into the plant and animal kingdom involving these groups.

1. ARISTOTLE'S SYSTEM

This system distinguished between plants and animals on the basis of **Movement, Feeding Mechanism and Growth Patterns.**

This system groups prokaryotes, algae and fungi as plants and protozoans as animals. With the increasing sophistication of laboratory methods and instrumentation the differences between prokaryotic and eukaryotic cells have been illuminated prompting a classification system that reflects them.

Plant Kingdom	**Animal Kingdom**
Prokaryotes, Algae, Fungi	Protozoans

2. CAROLIUS LINNEAUS (1735)

Linnaeus introduced the Binomial System of nomenclature and established hierarchical taxa such as kingdom, phylum, class, order, family, genus, and species. Single-celled organisms were observed but not categorized. In this system, plants and fungi were classified as the plant kingdom, while animals remained in the animal kingdom.

Plant Kingdom	**Animal Kingdom**
Plant + Fungi	Animals

3. GERMAN BIOLOGIST: ERNST HAECKEL (1866)

Haeckel proposed PROTISTA as the third Kingdom to include all single-celled organisms; some biologists also placed simple multicellular organisms in the **Protista** (e.g sea weeds)

PROTISTA	**PLANTAE**	**ANIMALIA**
Protozoans	Plants + Fungi	Animals
Diatoms		
Sea Weeds		

4. AMERICAN BIOLOGIST HERBERT COPELAND (1938)

Copeland introduced the kingdom Monera, which included only bacteria Prokaryotes (including bacteria) were classified as Monera, while eukaryotes were divided into Protista, Plantae, and Animalia. Protists such as amoebas and diatoms were included in Protista, plants were in Plantae, and animals were in Animalia. Fungi were placed separately.

PROKARYOTES	EUKARYOTES		
MONERA	**PROTISTA**	**PLANTAE**	**ANIMALIA**
	Amoeba, diatoms	Plants	Animals
Bacteria	Other single-celled eukaryotes; sometimes single multicellular organisms like sea weeds	Fungi	

5. AMERICAN BIOLOGIST CARL WOESE (1990)

Woese proposed a new category called the domain, based on molecular evidence. This system emphasized the genetic relationship between organisms. The three domains are Archaea, Bacteria, and Eukarya. Archaea and Bacteria encompass prokaryotes, while Eukarya includes eukaryotes such as protists, plants, fungi, and animals.

CHAPTER 13

CRYPTOGAMS

Cryptogam refers to a group of plants that do not produce seeds for reproduction. The term "cryptogam" comes from the Greek words "kryptos" meaning hidden and "gamos" meaning marriage or reproduction. Cryptogams reproduce through spores instead of seeds, and their reproductive structures are often inconspicuous or hidden.

The term "cryptogam" is primarily used in traditional botanical classification systems and has been replaced by more modern terms in contemporary taxonomy. However, it is still useful to understand the concept and historical context.

Cryptogams encompass a diverse range of plant groups, including algae, bryophytes (such as mosses and liverworts), and pteridophytes (such as ferns and horsetails). These plants play important ecological roles and can be found in various habitats, from aquatic environments to forests and even deserts.

13.1 ALGAE

Algae are diverse and predominantly aquatic organisms classified as cryptogamic plants. They inhabit a wide range of aquatic environments, including freshwater such as streams, rivers, lakes, and wells, as well as saltwater bodies like oceans and seas. One common characteristic of all algae is the presence of chlorophyll a, which is the primary pigment involved in photosynthesis.

Algae can be classified based on various parameters, including their morphology (physical form and structure), flagella (presence and arrangement of whip-like appendages for movement), habitat preferences, pigment or chlorophyll types, food reserve strategies, and cell types (prokaryotic or eukaryotic).

Unlike higher plants, algae lack true roots, stems, and leaves, making them cryptogamic in nature. Instead, they possess various structures for attachment and nutrient absorption, such as holdfasts and rhizoids. Algae exhibit a remarkable diversity, and they are classified into approximately 11 to 13 divisions, depending on the classification system used.

One notable division of algae is the blue-green algae, also known as cyanobacteria. Unlike other algae, blue-green algae are prokaryotic organisms, meaning they lack a membrane-bound nucleus. They contain chlorophyll a and their photosynthetic product consists of cyanophycean starch and oil. Blue-green algae are important in aquatic ecosystems, as they are often primary producers and can form large blooms in nutrient-rich environments.

13.1.1 BLUE-GREEN ALGAE (Cyanobacteria)

Blue-green algae, also known as cyanobacteria, are a group of algae that can be found in a variety of habitats, including freshwater, soil, and even on the surfaces of leaves. They exhibit remarkable adaptability and can thrive in environments with high temperatures, including hot springs and tropical regions. Blue-green algae are prokaryotic organisms, lacking a membrane-bound nucleus, and they have unique characteristics that set them apart from other types of algae.

Reproduction in blue-green algae primarily occurs through asexual methods such as binary fission or fragmentation. Unlike some other algae, blue-green algae do not exhibit sexual reproduction involving the fusion of gametes. Instead, they rely on rapid cell division to proliferate and colonize their environment.

Blue-green algae can exist as either unicellular or filamentous forms. Filamentous types, such as Oscillatoria and Anabaena, form long chains of cells that can be seen under a microscope. These filaments often have specialized cells known as heterocysts, which play a vital role in the nitrogen cycle. Heterocysts are enlarged cells that are responsible for fixing atmospheric nitrogen, converting it into a form that can be used by other organisms. This process, known as nitrogen fixation, is crucial for enriching the soil with nitrates, thereby enhancing soil fertility. Anabaena, for example, forms symbiotic relationships with certain plants, providing them with a source of nitrogen in exchange for nutrients and protection.

In addition to their ecological roles, blue-green algae have also garnered attention for their ability to produce a range of secondary metabolites with various biological activities. Some of these metabolites have shown potential as pharmaceuticals, including antibacterial, antiviral, and anticancer properties. Blue-green algae are being explored for their potential use in bioremediation, as they can help mitigate the effects of pollution by absorbing excess nutrients from water bodies.

13.1.2   CHLOROPHYTA (Green Algae)

Chlorophyta is a diverse division of algae that exhibits a cosmopolitan distribution, meaning they can be found in various habitats worldwide. These algae are commonly referred to as green algae due to their characteristic green color, which is a result of the presence of chlorophylls a and b, as well as additional pigments such as carotene and xanthophylls.

The members of Chlorophyta display a wide range of morphological forms. They include unicellular species, both flagellated and non-flagellated, as well as colonial forms. Among the colonial species, some are flagellated, such as Gonium, Volvox, and Pandorina, while others are non-flagellated, like Senedesmus and Ankistodesmus. Filamentous forms, such as Spirogyra and Chladophora, are also common in Chlorophyta. Additionally, there are unicellular non-colonial, non-flagellated species such as Chlorella, Cosmarium, and Closterium.

Reproduction in Chlorophyta can occur through both asexual and sexual means. Asexual reproduction is the more common form and often involves the fragmentation of body parts, allowing new individuals to develop from the broken fragments. Sexual reproduction, although less frequent, does occur in certain species. Chlamydomonas, for example, demonstrates sexual reproduction through three different modes: isogamy, anisogamy, and oogamy.

- In isogamy (*Chlamydomonas reinhardii*), two gametes fuse, and they are identical in shape and size.
- Anisogamy (*Chlamydomonas monoica*), involves the fusion of two gametes that are similar in shape but differ in size.
- Oogamy *Chlamydomonas coccifera*), involves the fusion of two gametes that are not identical in size.

Spirogyra, another member of Chlorophyta, displays a unique form of sexual reproduction. It involves the conjugation of two filaments where the contents of one filament are transferred to the other, resulting in the formation of zygospores.

The primary food reserve in Chlorophyta is starch, which is stored within the cells along with oils. These reserves serve as a source of energy for the organisms and contribute to their survival in various environments.

Apart from their ecological roles, green algae have also demonstrated the ability to produce vitamins and antibiotics. Some species of Chlorophyta have been found to synthesize these beneficial compounds, which can have implications for human health and pharmaceutical applications.

13.1.3 BACILLARIOPHYTA (Diatoms)

Diatoms, belonging to the division Bacillariophyta, are a group of algae known for their unique cell walls made primarily of silica (silicon dioxide) and pectin. These intricate cell walls give diatoms their characteristic intricate and geometric shapes. Diatoms can be found in diverse habitats, including both fresh water and marine environments.

The cell walls of diatoms are exceptionally durable and resistant to degradation. They can withstand harsh conditions and even resist dissolution in acids. However, when subjected to intense heat, the silica cell walls can be melted. This property becomes significant in the formation of diatomites, which are large clusters of diatoms buried under soil for extended periods. Over time, the heat generated during geological processes melts the diatomite, leading to the formation of crude oils beneath the ocean floor.

Diatomites have various applications due to their unique properties. They are used as reinforcing materials in building construction, as additives in toothpaste for their abrasive properties, and as additives to paints for their texture-enhancing effects. Additionally, diatomites have been employed in the prospecting of crude oil, as their presence can indicate the presence of oil deposits.

Photosynthetic pigments; diatoms possess chlorophylls a and c, which enable them to capture sunlight for photosynthesis. They play a crucial role in aquatic ecosystems as primary producers, contributing to the food web. Diatoms are a significant source of nutrition for many organisms, particularly in aquatic environments, serving as a vital food source for various organisms, including fish.

The primary food reserve or photosynthetic product in Bacillariophyta is chrysolaminarin, a type of carbohydrate, along with oils. These reserves serve as energy sources and contribute to the survival and growth of diatoms.

13.1.4   EUGLENOPHYTA (Euglenoids)

Euglenophyta, commonly known as Euglenoids, are a group of algae found in both fresh water and marine environments. They can also tolerate and thrive in polluted waters, making them useful indicators or monitors of water pollution levels. Euglenoids are characterized by their flexible cell structure, distinguishing them from the rigid cell walls of diatoms.

The cells of Euglena are elongated and possess a whip-like tail, called a flagellum, which enables them to move and navigate through the water. One distinctive feature of Euglena is the presence of an "eye spot" or stigma, which is a light-sensitive structure that allows the organism to detect and move towards light sources.

Euglenoids have the ability to undergo photosynthesis, utilizing sunlight to produce energy. In the presence of sunlight, they exhibit autotrophic nutrition, synthesizing their own organic compounds using chlorophylls a and b. However, when light is limited or absent, Euglenoids can switch to saprophytic nutrition, obtaining nutrients by absorbing organic matter from their environment.

The food reserve in Euglenoids is primarily starch, which serves as an energy storage molecule. They also accumulate oils as a supplementary energy source. These reserves support the organism's growth and survival during periods of limited or fluctuating nutrient availability.

Due to their unique characteristics and adaptability to various environmental conditions, Euglenoids contribute to the overall productivity of the aquatic ecosystem through their photosynthetic activity. Their presence in polluted waters can serve as an indicator of water quality and pollution levels, as they can thrive in environments that may be detrimental to other organisms.

13.1.5 RHODOPHYTA (Red Algae)

Rhodophyta, commonly known as Red Algae, is a division of algae that is primarily found in marine environments. They are characterized by their reddish color, which is due to the presence of the pigment called chlorophylls a and d, as well as additional pigments such as phycoerythrin.

Red Algae exhibit a wide range of morphological diversity, including filamentous forms, branching structures, and encrusting forms. They can be found in various marine habitats, from rocky shores to coral reefs and deep-sea environments. Some species of Red Algae can also inhabit freshwater ecosystems, although they are less common in such environments.

Reproduction in Red Algae is complex and involves both sexual and asexual modes. Sexual reproduction in Red Algae is highly developed, with distinct male gametes called spermatia and female gametes called carpogonia. Fertilization occurs when spermatia are transferred to carpogonia, leading to the formation of zygotes that develop into new individuals.

One group of Red Algae, known as Agarophytes, is particularly important in commercial applications. Agarophytes, including species like Gelidium and Gracilaria, are utilized in the synthesis of agar, a gelatinous substance widely used in food, pharmaceutical, and microbiological applications.

The food reserve in Red Algae is primarily Floridean starch, which is a unique type of storage carbohydrate found in these algae. Floridean starch, along with oils, serves as an energy source for growth and survival.

Red Algae are important primary producers, contributing to the overall productivity and biodiversity of coastal and marine habitats. Red Algae can form dense mats or provide shelter for a variety of marine organisms, serving as habitats and nurseries for other organisms.

13.1.6　　PHAEOPHYTA (Brown Algae)

Phaeophyta, also known as Brown Algae, is a group of algae primarily found in marine environments. They are characterized by their brown coloration, which is caused by the pigment called fucoxanthin, in addition to chlorophylls a and c.

Brown Algae are predominantly macroscopic, meaning they are visible to the naked eye and often form large, complex structures. They are commonly found along rocky coastlines and in intertidal zones, where they are exposed to both the sea and the air during tidal cycles.

One of the distinguishing features of Brown Algae is their advanced level of organization. They exhibit a plant-like form, with structures that resemble leaves, stems, and root-like attachments. For example, the genus Laminaria, commonly known as kelp, exhibits a blade-like structure (similar to leaves), a stipe (resembling a stem), and a holdfast (similar to roots), allowing them to anchor to the substrate.

The brown coloration of Phaeophyta is due to the presence of fucoxanthin, a pigment that absorbs light in the brown range of the spectrum. This adaptation allows Brown Algae to efficiently capture sunlight for photosynthesis in the underwater environment where other pigments, such as chlorophyll, may be less effective.

In terms of reproduction, Brown Algae exhibit a variety of reproductive strategies. Some species reproduce sexually through the release of male and female gametes, while others can also reproduce asexually through fragmentation or the formation of specialized reproductive structures.

The food reserve in Brown Algae is primarily stored as laminarin, a complex carbohydrate, along with oils. Laminarin serves as an energy storage molecule that can be utilized during periods of low light or nutrient availability.

13.2 LICHENS

Lichens are fascinating composite organisms composed of a symbiotic relationship between a fungus and an alga or cyanobacterium. The fungus, known as the mycobiont, does not possess the ability to photosynthesize, while the alga or cyanobacterium, known as the phycobiont or photobiont, carries out photosynthesis and provides nutrients for the partnership. The study of lichens, known as lichenology, is an interdisciplinary field that combines aspects of mycology, phycology, and ecology.

The structure of lichens is unique and differs from both the fungus and the alga individually. Lichens attach to their substrates, such as rocks, trees, or soil, using specialized structures called rhizines, which resemble roots but function differently. This attachment mechanism allows lichens to colonize diverse habitats, including harsh environments where other organisms may struggle to survive.

Lichens exhibit three main growth forms, each characterized by its distinctive morphology.

- Crustose lichens have a crust-like or powdery appearance, as seen in species like Lepraria.
- Fruticose lichens are hair-like in structure, exemplified by species like Cladonia.
- Foliose lichens have flat, leafy, and dorsiventral bodies, as observed in species like Parmelia.

These unique growth forms make lichens readily identifiable and provide clues about their ecological adaptations. They can thrive in a variety of habitats, including forests, deserts, tundra, and even urban environments. Lichens are known for their ability to colonize substrates with minimal nutrient availability and to tolerate extreme temperatures and pollution levels.

Lichens play a crucial role in environmental monitoring and are considered excellent indicators of atmospheric pollution. Their sensitivity to air quality makes them valuable bioindicators, as changes in their growth and composition can reflect changes in air pollution levels, including the presence of sulfur dioxide, heavy metals, and other pollutants. Lichens produce a variety of intracellular products. These include pigments like carotenoids, carbohydrates, free amino acids, and vitamins. Some lichens also contain unique compounds such as fumarprotocetraric acid, which imparts a bitter taste and makes them unpalatable to animals. This acid, along with other secondary metabolites produced by lichens, exhibits antibiotic properties, protecting the organisms from fungal and bacterial attacks.

13.3 BRYOPHYTES

Bryophytes are a group of small, non-vascular plants commonly known as mosses, liverworts, and hornworts. They are found in diverse habitats and play important ecological roles. Bryophytes lack certain characteristics found in higher plants, such as a cuticle to prevent water loss, roots for absorption, and a well-developed vascular system for transportation of water and nutrients.

The division Bryophyta can be classified into three main groups: mosses (Musci), liverworts (Hepaticae), and hornworts (Anthoceratae). Mosses are characterized by their leafy structures, while liverworts can exhibit either leafy or thalloid forms. Thalloid liverworts lack differentiation into distinct roots, stems, and leaves like higher plants.

Bryophytes exhibit a wide range of habitat preferences. They can be found growing in soil (terricolous), on the bark of trees (corticolous), on rock surfaces (saxicolous), on decaying wood (epixylic), and even on the bodies of animals (zoophyllous). This adaptability allows them to colonize various environments and fulfill important ecological roles.

One significant feature of bryophytes is their poikilohydric nature, meaning they can tolerate fluctuations in water availability. Bryophytes lack specialized tissues for water transportation and rely on the absorption of water through their leaf surfaces. This ability to adjust to water availability or non-availability is a key adaptation of bryophytes.

Bryophytes have several practical applications. They are used in pollution monitoring, as they can accumulate heavy metals and other pollutants from their surroundings. Some bryophytes, such as certain species of the genus Sphagnum, are effective in water retention and can be utilized to control erosion. Bryophytes also produce a variety of secondary metabolites, including antibiotics, which have potential pharmaceutical uses.

Structurally, bryophytes can exhibit two growth forms: erect (acrocarpous) and creeping (pleurocarpous). Acrocarpous mosses, such as Archidium, Hyophila, Philonotis, Bryum, and Barbula, grow upright. Pleurocarpous mosses, including species like Racopilum and Pelekium (formerly Thuidium), have a creeping growth habit.

Bryophytes reproduce through spores and are homosporous, meaning they produce only one kind of spore that is similar in size and shape. The spores germinate into tiny gametophytes, which produce gametes that fuse to form a sporophyte generation.

13.4 PTERIDOPHYTES

Pteridophytes are a group of plants that represent the most advanced members of non-seed bearing lower plants. Unlike algae and bryophytes, pteridophytes possess a vascular system, which includes specialized tissues for the transport of water, minerals, and organic compounds. The stem of pteridophytes is known as a "frond" and is typically composed of leaf-like structures called pinnae.

One notable characteristic of pteridophytes is the presence of a coiled and tender young shoot known as a **crozier**. This crozier is soft and often edible, making it a unique feature of pteridophyte development. The pinnae of pteridophytes may contain structures called sorus (singular) or sori (plural) on their undersides (abaxial surface). These sori house sporangia, which produce spores for reproduction. When a pteridophyte has sori, it is referred to as "fertile," while those lacking sori are considered "sterile."

Pteridophytes can exhibit different modes of reproduction. Some members, such as Cyclosorus, Thelypteris, and Nephrolepis, are homosporous, meaning they produce a single type of spore that is similar in size and gives rise to both male and female gametophytes. Other members, like Selaginella, are heterosporous and produce two types of spores: microspores, which are smaller and give rise to male gametophytes, and megaspores, which are larger and develop into female gametophytes.

These plants are commonly found in swampy areas, moist land, and can even grow epiphytically on palm trees and other trees. Pteridophytes have various economic uses. Some species are utilized for medicinal purposes, while others serve as camouflage for snakes and other reptiles due to their resemblance to the surrounding vegetation. Certain pteridophytes, such as Adiantum, are valued as ornamental plants. The crozier of ferns is consumed as food in some cultures. Pteridophytes also play a role in the remediation of polluted soils and can sometimes become nuisances in waterways.

Pteridophytes are a fascinating group of plants that bridge the evolutionary gap between non-vascular plants and seed-bearing plants. They display a range of structural and reproductive adaptations that have allowed them to thrive in diverse habitats and fulfill various ecological and human needs. Pteridophytes contribute to the ecological balance of their habitats. They play a role in soil stabilization and erosion control, particularly in swampy areas and moist land. The extensive root systems of some pteridophytes help bind soil particles together, preventing erosion and maintaining soil structure.

CHAPTER 14

GYMNOSPERMS

14.1 WHAT ARE PLANTS?

Plants have evolved unique adaptations to thrive on land, overcoming challenges such as desiccation and the need for gaseous exchange. To prevent water loss, the aerial parts of plants are coated with a waxy layer called cuticle, which acts as a barrier against dehydration. Another important adaptation is the development of stomata, specialized pores on the leaf surface that allow for the exchange of gases, such as carbon dioxide and oxygen, with the environment.

The origin of plants can be traced back to their aquatic ancestors, the green algae. Despite the transition to terrestrial habitats, plants still retain fundamental functions similar to those of their progenitors. One such function is photosynthesis, the process by which plants convert sunlight into energy-rich molecules. Within their cells, plants possess chloroplasts, specialized organelles that contain chlorophylls a and b, as well as various yellow and orange pigments known as carotenoids. These pigments capture light energy, which is then used to power the synthesis of organic compounds.

Plant cells are also characterized by the presence of a cell wall composed of cellulose, a complex carbohydrate. The cell wall provides structural support and protection to plant cells, giving them rigidity and shape. Additionally, plants store carbohydrates, such as glucose, in the form of starch. Starch is typically stored in plastids, including chloroplasts and other specialized plastids, and serves as a reserve energy source for the plant.

The reproductive apparatus of plants underwent significant changes during the transition to land. Unlike algae, which release their gametes directly into the aquatic environment, land plants have evolved specialized structures called gametangia. Gametangia are organs that enclose and protect the gametes, the reproductive cells involved in sexual reproduction. These protective jackets of sterile cells prevent the delicate gametes from drying out during their development, ensuring their viability.

In land plants, fertilization occurs within the female gametangium, where the egg is fertilized by a sperm cell. The resulting embryo is then retained and nourished within the parent plant for a period of time. This retention of the developing eggs within the parent plant is a distinct characteristic of land plants and provides the embryo with protection and nutrients for its early development.

14.2 USEFULNESS OF PLANTS TO MAN

Plants play a crucial role in meeting various needs and providing numerous benefits to humanity. One of the most significant ways in which plants are essential to humans is through their provision of food. Plants are the primary source of sustenance for humans and other animals. They produce a wide array of fruits, vegetables, grains, and nuts that form the basis of our diets, providing essential nutrients, vitamins, and minerals necessary for our growth and overall well-being.

Plants also contribute to the production of fibers used in textiles, paper, and cordage. Many plant species, such as cotton, flax, and hemp, provide fibers that are woven into fabrics, creating clothing, linens, and other textile products. Plants like bamboo and trees are used in the production of paper, which is an indispensable material for communication and record-keeping. Plant-based fibers also serve as the foundation for the creation of ropes, twines, and cords used for various practical purposes.

Plants have long been utilized as a source of fuel, serving as an energy resource for heating and cooking. Wood, derived from trees and other woody plants, has been a primary fuel source throughout human history.

Plants like peat, coal, and biomass crops are utilized for their energy content and serve as renewable or non-renewable sources of fuel.

The medicinal properties of plants have been recognized and utilized by humans for centuries. Many plant species contain chemical compounds with therapeutic effects, which are used in traditional medicine and modern pharmaceuticals. From aspirin derived from willow bark to potent anti-cancer drugs sourced from various plants, medicinal plants play a vital role in treating and preventing diseases, improving human health, and enhancing the quality of life.

Plants also contribute to the amelioration of the environment. They help to mitigate climate change by absorbing carbon dioxide from the atmosphere through photosynthesis and releasing oxygen. Trees and other vegetation help in stabilizing soil, preventing erosion, and promoting the health of ecosystems. Plants also enhance the aesthetic appeal of landscapes, parks, and gardens, providing recreational and psychological benefits to humans.

Plants are utilized in various industrial processes and materials. Plant-derived substances, such as oils, resins, and latex, serve as raw materials for the production of numerous industrial products. For example, vegetable oils are used in cooking, cosmetics, and biofuels, while natural rubber obtained from rubber trees is crucial in the manufacturing of tires and other rubber-based products.

Note that certain activities associated with plants can pose risks to human well-being. Plants can spoil and contaminate food, leading to foodborne illnesses. Some plant species, particularly saprophytes, can harbor pathogens that can cause diseases in humans. Weeds, invasive plant species, can compete with desired crops or disrupt natural ecosystems. Plants can contribute to allergies and respiratory problems in certain individuals.

14.3 ALTERNATION OF GENERATIONS

The life cycle of plants involves an alternation of generations, which refers to the alternating phases of a haploid gametophyte and a diploid sporophyte. The gametophyte generation is haploid, meaning it contains a single set of chromosomes (n), while the sporophyte generation is diploid, containing two sets of chromosomes (2n).

The life cycle begins with the production of spores by the sporophyte through a process called meiosis. These spores are released and develop into multicellular gametophytes. The gametophytes are independent organisms and can be either male or female. They produce gametes, such as sperm and eggs, through mitosis. When conditions are favorable, the male gametophyte produces sperm, while the female gametophyte produces eggs. The sperm is typically non-motile in land plants, and they are transported to the egg by various means such as wind, water, or insects. Once the sperm reaches the egg, fertilization occurs, resulting in the formation of a zygote.

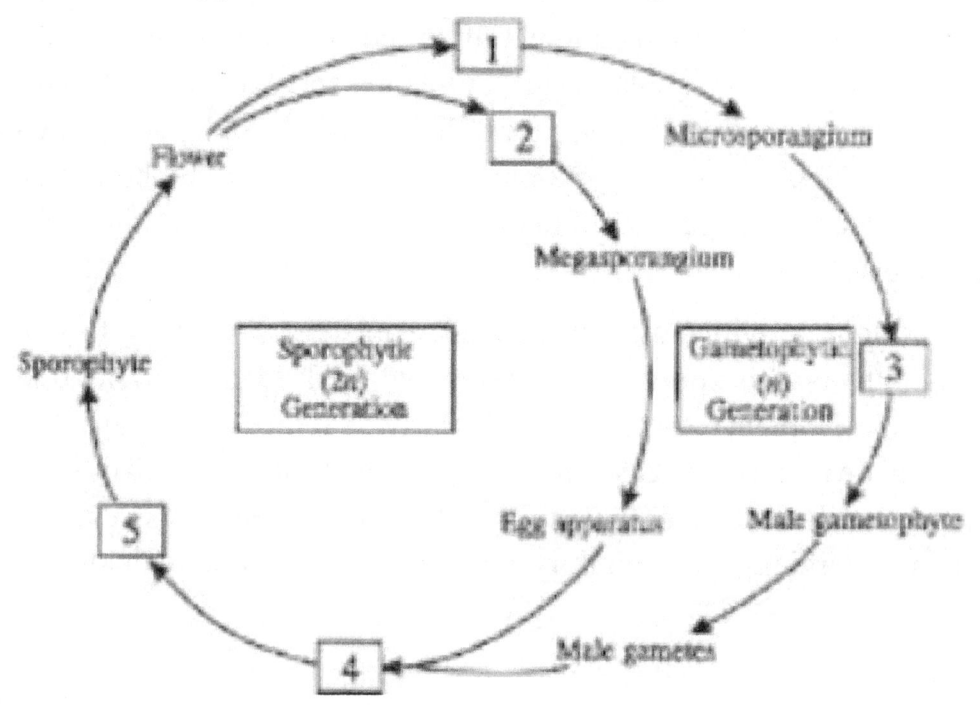

The zygote is the first cell of the new sporophyte generation. It undergoes cell division and development, eventually growing into a mature sporophyte plant. The sporophyte is the dominant phase in the life cycle of most land plants, and it is what we commonly recognize as the "plant" itself. The alternation between the haploid gametophyte and diploid sporophyte phases is a characteristic feature of plant life cycles and is known as alternation of generations. This phenomenon is distinct from the life cycles of animals, where the dominant phase is usually diploid.

During plant evolution, there has been a trend toward reducing the size and importance of the gametophyte generation compared to the sporophyte generation. This shift reflects the adaptation to the terrestrial environment, where the sporophyte is better equipped to deal with challenges such as water conservation and protection from environmental stressors. Another significant trend in plant evolution is the replacement of flagellated sperm with pollen. Pollen grains are structures that contain the male gametophyte, including the sperm cells. This adaptation eliminates the need for water-based fertilization and enables plants to reproduce successfully in diverse terrestrial habitats. The dispersal of pollen by wind, insects, or other means allows for efficient fertilization even in the absence of free-standing water.

14.4 HIGHLIGHTS OF PLANT PHYLOGENY

Plant phylogeny, or the evolutionary history of plants, can be traced through the fossil record and is marked by several key periods and events. Fossil records chronicle four major periods of plant evolution;

I. Late Ordovician Period: This period, occurring approximately 460 million years ago during the Paleozoic Era, witnessed the evolution of plants from their aquatic ancestors, green algae. One notable development during this period was the evolution of vascular tissue, which allowed for the transport of water, nutrients, and sugars within plants. It is important to note that most mosses, which belong to the group of non-vascular plants, lack vascular tissue.

II. Early Devonian Period: Around 360 million years ago, during the Early Devonian Period, the earliest vascular plants emerged. These plants did not yet possess seeds, a characteristic that still persists in modern ferns and a few other seedless vascular plants.

III. End of Devonian Period: The end of the Devonian Period, also around 360 million years ago, marked a significant milestone in plant evolution—the origin of seeds. Seeds are structures consisting of an embryo packaged with a food reserve and protected by a covering. The first vascular plants with seeds, known as gymnosperms, appeared during this time. These gymnosperms had naked seeds, meaning they were not encased in a protective fruit or ovary. Alongside gymnosperms, ferns and other seedless plants coexisted.

IV. Cretaceous Period: Occurring approximately 130 million years ago during the Mesozoic Era, the Cretaceous Period saw the emergence of flowering plants, known as angiosperms. Angiosperms are characterized by their complex reproductive structures called flowers. Flowers contain ovaries that protect and nurture the seeds. Unlike gymnosperms, the seeds of angiosperms are enclosed within fruits or ovaries, offering additional protection and aiding in seed dispersal. The rise of angiosperms was a significant event in plant evolution and led to the dominance of flowering plants in many terrestrial ecosystems.

14.5 THE CHALLENGES OF LAND COLONIZATION: THE ORIGIN OF VASCULAR TISSUES.

The colonization of land posed significant challenges for plants, and the development of vascular tissues was a crucial adaptation that allowed them to overcome these obstacles. Vascular tissues play a vital role in the transport of water, minerals, and nutrients throughout the plant body.

In vascular plants, the body is differentiated into two main components: the underground portion, consisting of roots, and the aboveground portion, consisting of stems and leaves. Roots absorb water and minerals from the soil, while the shoot system, composed of stems and leaves, photosynthesizes and produces food for the plant. However, in order for the shoot system to stand upright in the air, it requires support.

To address the need for support and conduction of water and minerals, plants developed vascular tissues that provide structural support and facilitate the transport of substances. One key adaptation was the accumulation of lignin in the cell walls of plant cells. Lignin is a complex polymer that provides rigidity and strength to cell walls. By incorporating lignin into their cell walls, plants were able to achieve structural support and upright growth.

The vascular system consists of two types of conducting tissues: xylem and phloem. Xylem is responsible for the transport of water and minerals from the roots to the aerial parts of the plant. It is composed of tube-shaped cells called tracheids and vessel elements, which are actually dead and lignified. The lignified cell walls of xylem cells provide additional support to the plant.

Phloem, on the other hand, is responsible for the translocation of sugars, amino acids, and organic nutrients throughout the plant. It is a living tissue composed of sieve tube elements and companion cells. The phloem allows for the distribution of photosynthetically produced sugars from the leaves to other parts of the plant, supporting growth and development.

Plants also evolved other adaptations to successfully colonize land. These include the evolution of seeds, which protect and nourish the developing embryo, as well as the replacement of flagellated sperm with pollen. Pollen grains, produced by the male reproductive organs of plants, enable efficient transfer of sperm to the female reproductive organs without the need for free-standing water. As plants adapted to land, there was a shift towards the increasing dominance of the diploid sporophyte generation in the alternation of generations. The sporophyte generation became the prominent and independent phase in the life cycle, while the gametophyte generation reduced in size and significance.

14.6 THE SEED PLANTS AND THE SIGNIFICANCE OF THE SEED HABIT

Seed plants, which include gymnosperms and angiosperms, are considered the most advanced and dominant group in the plant kingdom. They exhibit a wide range of life forms, from trees and shrubs to herbs, and can be found in various habitats, including autotrophs, parasites, semi-parasites, and insectivorous species. They are capable of adapting to different environmental conditions, ranging from extreme temperatures to varying soil-water and nutrient conditions.

Gymnosperms, such as cycads, pines, and palms, have naked seeds, meaning their seeds are not enclosed within an ovary. On the other hand, angiosperms, also known as flowering plants, have their seeds enclosed within an ovary, which develops into a fruit after fertilization. The evolution of angiosperms brought about significant modifications in their reproductive structures. The bracts, usually modified leaves, have transformed into flowers, often characterized by vibrant colors and intricate structures.

The development of flowers in angiosperms has provided several advantages. Flowers attract pollinators, such as insects, birds, or mammals, facilitating the transfer of pollen between plants and enhancing the efficiency of fertilization. The attractive colors and fragrances of flowers play a crucial role in attracting pollinators. Additionally, the fruits that develop from the ovary aid in seed dispersal, allowing plants to colonize new areas.

Seed plants have a tremendous impact on human life, as they fulfill various needs. They provide us with food, fruits, seasonings, fuelwoods, fiber, and ornamental plants. They have adapted to diverse environments, including brackish waters, freshwater ecosystems, arid regions, wetlands, and even harsh rocky terrains. Their ability to thrive in such diverse habitats has contributed to their global distribution and success.

14.7 THE GYMNOSPERMS

Gymnosperms are a group of seed plants that are characterized by their cone-bearing reproductive structures. Unlike angiosperms, the ovules of gymnosperms are not fully enclosed by tissues of the sporophyte at the time of pollination. This means that the seeds of gymnosperms are not protected within a fruit.

Some of the most well-known gymnosperms include pine trees, spruces, cedars, and firs. They are typically found in colder and sometimes drier regions of the world, such as boreal forests and mountainous areas. These plants have adapted to withstand harsh conditions, including low temperatures and limited water availability.

Gymnosperms serve as valuable sources of timber for construction and woodworking industries. The wood of gymnosperms is often used for producing furniture, flooring, and paper. Also certain species of gymnosperms, such as pines, produce resin, which can be used to make varnishes, adhesives, and turpentine.

Gymnosperms play a crucial role in the world's forest ecosystems. Conifers, a type of gymnosperm, make up a significant portion, up to 80%, of the world's forest trees. They contribute to the biodiversity of forests and provide habitats for numerous animal species. Moreover, the evergreen nature of many gymnosperms allows them to continue photosynthesising and providing oxygen throughout the year.

14.7.1   LIFE CYCLE OF GYMNOSPERMS

The life cycle of gymnosperms involves the production of male and female cones on separate branches of the trees. Female cones are larger and have woody scales. Within each scale of the female cone, two ovules develop at the base. Each ovule contains a megasporangium called the nucellus, surrounded by integument cells. The integument has a small opening called the micropyle.

Within the megasporangium, a single megaspore mother cell undergoes meiosis, resulting in the formation of four haploid cells. Three of these cells degenerate, while the remaining one develops into a megagametophyte over the course of one year. The megagametophyte can consist of thousands of cells, with 2-6 archegonia located at the micropylar end. Each archegonium contains a large egg cell.

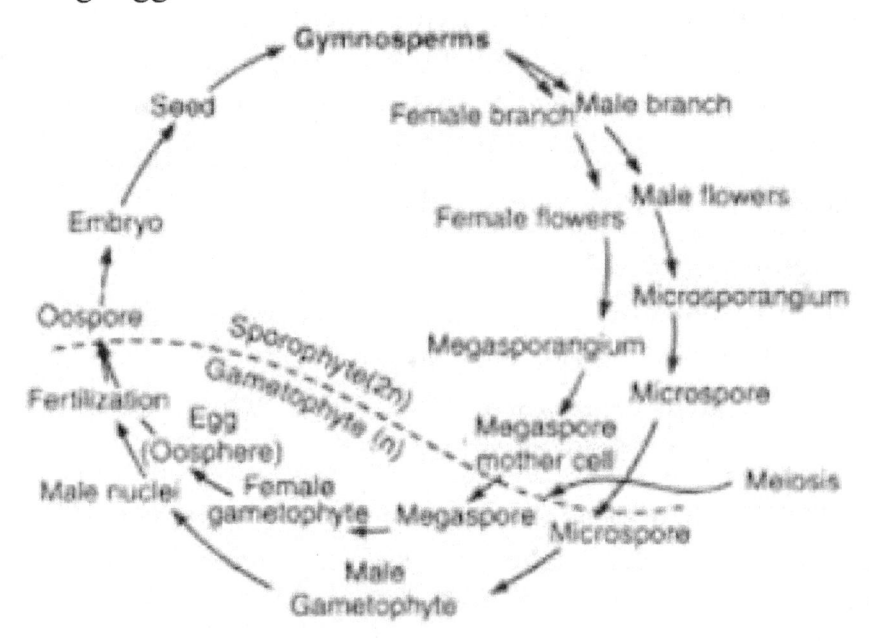

The scales of the female cones open, allowing pollen to enter the ovule through the sticky fluid in the micropyle. Pollen grains are produced in the male cones from microspores. Microsporangia, located on each scale of the male cone, contain numerous microspore mother cells. Meiosis occurs in these cells, leading to the formation of four haploid microspores. Each microspore develops into a four-celled pollen grain, which possesses sacs for buoyancy.

Pollen grains are released and carried by the wind. When the scales of the female cone are open, the pollen grains can land directly on the micropyle for fertilization. Unlike in flowering plants, gymnosperms lack a stigma, so the pollen can reach the micropyle without any intermediary structure. The pollen grain forms a pollen tube that grows through the nucellus, eventually reaching the archegonium. This process takes about fifteen months from pollination to fertilization.

Within the archegonium, one of the sperm nuclei fuses with the egg cell, resulting in the formation of a zygote. The other sperm nucleus and other cells degenerate. The zygote develops into a seed, which is covered by one layer of integument. The seed of gymnosperms is not enclosed within an ovary as it is in angiosperms. The mature seed can be dispersed and germinate under favorable conditions, giving rise to a new gymnosperm plant.

14.8 THE ANGIOSPERMS

Angiosperms, also known as flowering plants, represent the largest group of plants with approximately 250,000 known species. They are characterized by their unique reproductive structures, where the ovules are enclosed within diploid tissues at the time of pollination. This enclosing structure is called the carpel, which forms part of the flower.

Angiosperms belong to the phylum Anthophyta and are further divided into two major groups: Monocotyledonae (monocots) and Dicotyledonae (dicots). These divisions are based on differences in seed structure, leaf venation, floral characteristics, and other morphological features.

Monocots are characterized by having seeds with a single embryonic leaf, or cotyledon, within the seed. Examples of monocots include grasses, orchids, lilies, and palms. Monocot leaves typically have parallel veins, and their flower parts are usually in multiples of three (e.g., petals, sepals, and stamens).

Dicots, on the other hand, have seeds with two embryonic leaves or cotyledons within the seed. They exhibit a wider range of morphological diversity compared to monocots. Examples of dicots include roses, sunflowers, beans, and oak trees. Dicot leaves typically have net-like veins, and their flower parts are often in multiples of four or five.

14.8.1   THE ANGIOSPERM LIFE CYCLE

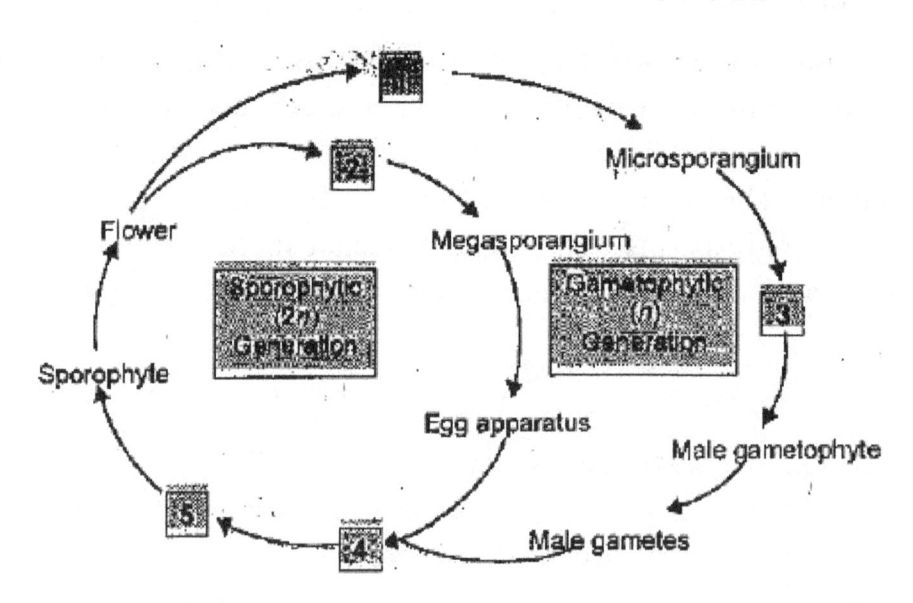

Meiosis takes place in the anther (microsporogenesis) and in the ovary (megasporogenesis). At the end of the meiotic division in the ovary, a linear tetrads of cells (four cells linearly arranged) result. Three of the cells degenerate leaving the megaspore to continue the meiotic process. The megaspore undergoes three mitotic divisions leading to 8 haploid nuclei three of which cluster at one end (the antipodal cells), two at the centre (the polar nuclei) and the other three clustering at the micropylar end of the embryo sac (the megagametophyte) which consist of the egg cell sandwiched between two synergids.

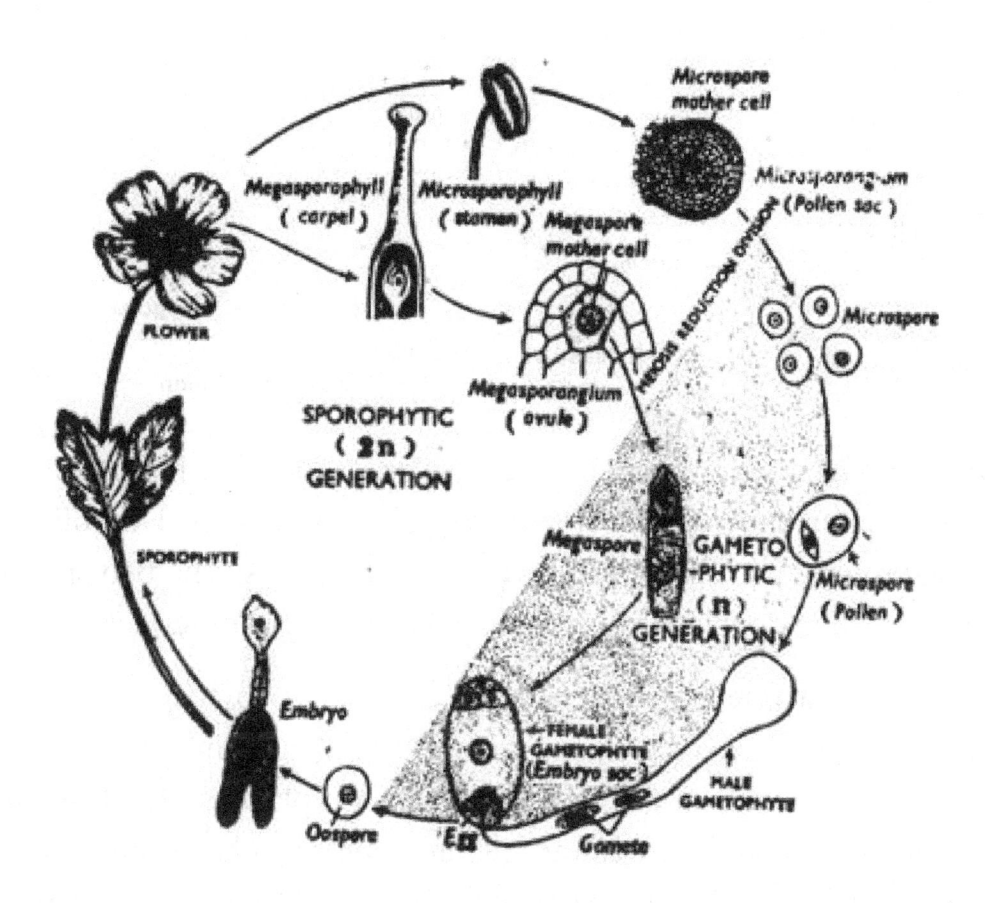

The purpose of meiosis is to produce viable egg cells and pollen grains, which can effect fertilization to produce a new zygote, the seed. During fertilization, a functional pollen grain lands on the stigmatic surface and germinates by extruding its protoplast as a pollen tube through a pore. This tube carries the two sperm nuclei to the embryo sac. One of the sperm nuclei fuses with the polar nuclei to form the endosperm (3n) while the other fuses with the egg cell to form the embryo (2n). This is the process of double fertilization. The integuments form the coat of the new seed.

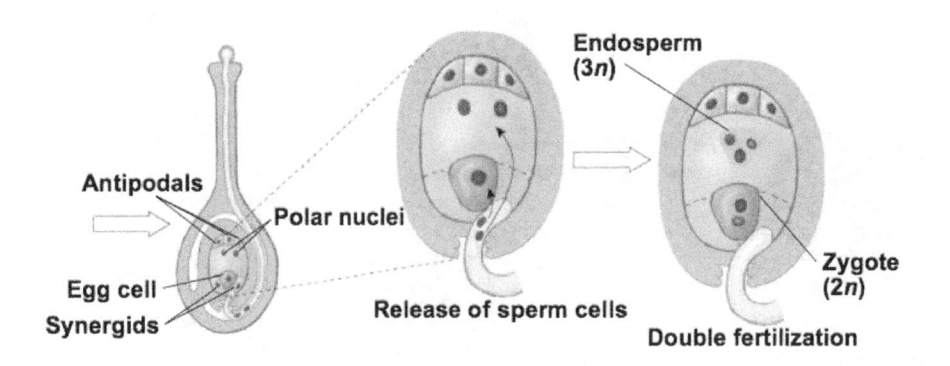

Microsporogenesis is simpler. The four products of meiosis (called tetrad) each develop into functional pollen grains through two mitotic division which produce two sperm nuclei and a tube nucleus. The two sperm nuclei are involved in double fertilization that has just been described.

A seed is define as a fertilized ovule. An ovule is define as a megasporangium covered by one, two or rarely more integuments. We have described the details of fertilization in the seed plants so far. It is important to know that unlike in lower vascular and non-vascular plants, the male sex is not motile in seed plants.They are transported to the female sex cells by wind or by animals.

The most important attribute of the seed habit is probably the growth of the embryo within a megasporangium that contains stored food. This situation makes it possible for the embryo to exist in a state of dormancy of various durations. This is a most important evolutionary achievement of the angiosperms. It is also important to note that gymnosperm seeds are produced from single fertilization while angiosperm seeds are produced from double fertilization.

CHAPTER 15
ANGIOSPERMS-MORPHOLOGY

Angiosperms, also known as flowering plants, exhibit a wide range of morphological variations in their plant organs. These variations include the roots, stems, leaves, inflorescences (flower clusters), and fruits. Angiosperms comprise a large group of plants that produce flowers and have seeds enclosed within a carpel. This diverse group includes herbaceous plants, shrubs, grasses, and most trees.

Roots in angiosperms serve to anchor the plant in the soil and absorb water and nutrients. They can vary in size, shape, and branching patterns, depending on the species and environmental conditions.

Stems in angiosperms provide support for the plant and transport water, minerals, and sugars between the roots and leaves. They can range from herbaceous stems in non-woody plants to woody stems in trees. The stems may also exhibit variations in thickness, texture, and growth habit.

Leaves are important for photosynthesis, as they contain chlorophyll and other pigments that capture sunlight. Angiosperms display a wide array of leaf shapes, sizes, arrangements, and venation patterns. These variations allow plants to adapt to different environmental conditions and optimize their exposure to sunlight.

Inflorescences refer to the arrangement and structure of flowers on a plant. Angiosperms exhibit diverse inflorescence types, including solitary flowers, clusters, spikes, umbels, and panicles. The arrangement and number of flowers can vary significantly, influencing the pollination and reproductive strategies of the plant.

Fruits are unique to angiosperms and develop from the ovary after successful fertilization. They serve to protect the developing seeds and aid in seed dispersal. Fruits exhibit a wide range of forms, including fleshy fruits like apples and berries, dry fruits like nuts and grains, and specialized structures like pods and capsules.

Angiosperms are incredibly diverse and can be found in almost every habitat, ranging from forests and grasslands to aquatic environments and deserts. They display a tremendous variety of life forms, including towering trees, small herbs, submerged aquatic plants, bulbous plants, and epiphytes that grow on other plants.

Within the angiosperms, certain plant families stand out in terms of their size and species diversity. Examples include Orchids (Orchidaceae), Compositae or daisies (Asteraceae), and Legumes or beans (Fabaceae). It is estimated that there are over 300 families, 8,000 genera, and 352,000 species of flowering plants or angiosperms worldwide.

Angiosperms, or flowering plants, exhibit a wide range of life spans and growth forms. They can be categorized into annuals, biennials, and perennials based on their life cycle and duration of growth.

Annual angiosperms complete their entire life cycle within a single growing season. They germinate from seeds, grow, produce flowers and fruits, and eventually die within a year. Examples of annuals include common garden plants like marigolds, sunflowers, and zinnias.

Biennial angiosperms have a two-year life cycle. During the first year, they primarily focus on vegetative growth, developing leaves and roots. In the second year, they produce flowers, fruits, and seeds before eventually dying. Beet (Beta vulgaris) and carrot (Daucus carota) are well-known examples of biennial plants.

Perennial angiosperms have a longer life span and can live for multiple years, often flowering annually. They may exhibit various growth forms, including herbs, shrubs, trees, climbers, twiners, and lianas. Perennial herbs such as mint and lavender regrow from the base each year, while perennial trees like oak and maple can live for decades or even centuries.

Angiosperms also vary greatly in size. Some species, like the minute Wolffia microscopica, have a size of less than 2 millimeters (0.08 inches), while others can reach impressive heights of over 100 meters (300 feet). The towering Eucalyptus trees are a notable example of tall angiosperms.

The longevity of angiosperms can also vary significantly. Some angiosperms have relatively short life spans, living only for a few weeks or months. Peas and gram are examples of plants with relatively short life spans. On the other hand, certain angiosperms, such as the Bodhi tree (Ficus religiosa), can live for many years. In fact, some individuals of the Bodhi tree are known to be over 2,500 years old, making them some of the longest-lived angiosperms.

The diverse life spans and growth forms of angiosperms allow them to adapt to different ecological niches and environmental conditions. Their ability to occupy a wide range of habitats and exhibit various growth strategies contributes to their success and abundance as a dominant group of plants on Earth.

15.1 PLANT MORPHOLOGY

Plant morphology is the study of the external features and forms of plants. It encompasses the examination and understanding of various plant organs, including roots, stems, leaves, flowers, seeds, and fruits. Through the study of plant morphology, scientists can analyze the structural and functional characteristics of plants and gain insights into their evolutionary adaptations.

The complexity of a multicellular plant body is the result of evolutionary specialization over time. Through this specialization, different plant organs have developed distinct morphological and physiological characteristics. This differentiation has led to the establishment of plant organs as conceptually separate entities with specific functions.

The fundamental parts of the plant body are the aerial portion, known as the stem, and the subterranean portion, known as the root. The stem provides support to the plant and serves as a pathway for the movement of water, nutrients, and photosynthetic products. It also houses the reproductive structures, such as flowers and fruits. The root, on the other hand, anchors the plant in the soil and absorbs water and minerals. It also stores nutrients and provides a means for vegetative propagation.

Leaves are another essential organ of plants. Leaves are responsible for photosynthesis, the process by which plants convert sunlight into chemical energy. They have specialized structures, such as chloroplasts, which contain chlorophyll and enable the absorption of light for photosynthesis.

Flowers are the reproductive structures of angiosperms and play a crucial role in the formation of fruits and seeds. They are composed of various parts, including sepals, petals, stamens, and carpels. These floral organs have specific functions in the process of pollination and fertilization.

Seeds and fruits are formed as a result of successful fertilization in flowering plants. Seeds are reproductive structures containing the embryonic plant and a food source, enclosed within a protective seed coat. Fruits, on the other hand, develop from the ovary after fertilization and serve to protect and disperse the seeds.

15.2 ROOT MORPHOLOGY

Roots are vital organs of plants that play essential roles in anchoring the plant to the ground and absorbing water and nutrients from the soil. They also store food reserves and participate in vegetative reproduction. Roots typically grow underground, although there are also aerial roots in some plant species.

Root morphology can vary among different plant species, but there are common characteristics and structures found in most roots. These include:

Primary Root: Also known as the main root, the primary root is the first root to emerge from the seed during germination. It grows vertically downward and forms the central axis of the root system. In some plants, the primary root persists throughout the plant's life, while in others, it may be replaced by adventitious roots.

Lateral Roots: Lateral roots branch off from the primary root and extend horizontally into the soil. They play a crucial role in absorbing water and nutrients from a larger soil volume. Lateral roots can further branch into smaller secondary and tertiary roots, forming a complex network.

Root Hairs: Root hairs are tiny, finger-like extensions that increase the surface area of the root for enhanced nutrient and water absorption. They are found near the tips of young roots and greatly increase the root's absorptive capacity.

Root Cap: The root cap is a protective structure found at the tip of the root. It covers and shields the delicate growing region behind it, called the root meristem. The root cap also secretes lubricating substances that facilitate the root's penetration through the soil.

Adventitious Roots: Adventitious roots develop from non-root tissues, such as stems, leaves, or other roots. They can arise in response to environmental stimuli or plant growth requirements. Examples of plants with prominent adventitious roots include corn, where the roots develop from the lower part of the stem.

Root morphology can also vary depending on the plant's habitat and specific adaptations. For example, some plants have specialized roots like prop roots, which provide additional support, or pneumatophores, which help with oxygen exchange in waterlogged environments.

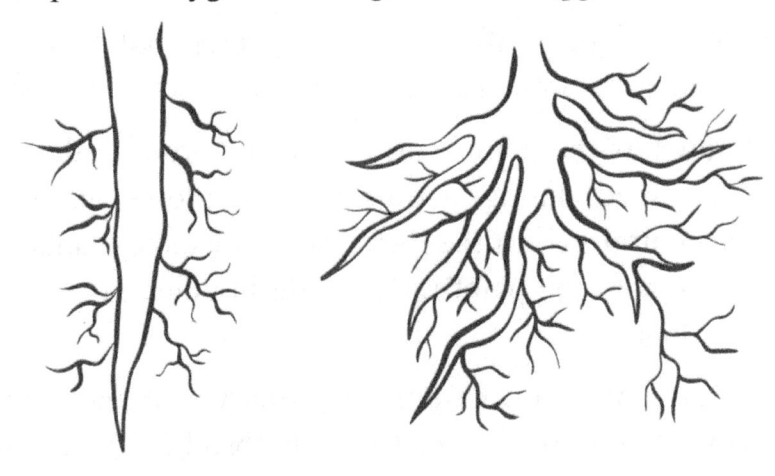

Tap Root   **Fibrous Root**

15.2.1   CHARACTERISTICS FEATURES OF ROOTS

Geotropism: Roots exhibit positive geotropism, meaning they grow in the direction of gravity. This helps the roots penetrate into the soil and anchor the plant securely.

Root Cap: The root tip is covered and protected by a cap-like structure called the root cap. The root cap aids in the penetration of the root through the soil and protects the delicate meristematic tissues behind it.

Absorptive Structures: Roots possess root hairs, which are tiny, finger-like projections that increase the surface area of the root for efficient absorption of water and nutrients from the soil.

Lack of Buds: Unlike stems, roots do not commonly bear buds. However, in some cases, vegetative buds may develop on the root for vegetative propagation.

Lateral Root Development: Lateral roots, which branch off from the primary root, develop endogenously from the pericycle, a layer of cells in the root. Lateral roots help increase the surface area for water and nutrient absorption.

Shape and Structure Variability: Roots exhibit a wide range of shapes and structures depending on their function and environmental conditions. They can be thick and fleshy, thin and fibrous, or modified for specific functions such as storage (as in carrots or sweet potatoes).

No Chlorophyll: Roots lack chlorophyll, the pigment responsible for photosynthesis. Instead, they rely on the leaves and stems to produce food through photosynthesis and transport it to the roots.

Root Systems: Plants can have different types of root systems. Some have taproots, where the primary root is dominant and gives rise to lateral roots. Others have fibrous root systems, where numerous thin roots develop from the base of the stem.

15.2.3   MODIFICATIONS OF ROOTS

Roots, the descending portion of the plant axis responsible for anchorage and absorption, can undergo various modifications to perform specific functions beyond their normal roles. These modifications occur in both tap root systems and adventitious root systems and are essential for adaptation to the plant's surrounding environment. Common modifications observed in roots

Tap Root Modifications: Tap roots, the primary root of the plant, can undergo various modifications for specific functions:

Storage Roots: Some tap roots, such as carrots and potatoes, are modified to store food reserves, often in the form of starch or sugars.

Pneumatophores: In certain plants like mangroves (e.g., Rhizophora, Avicennia), the tap roots grow vertically above the ground to obtain oxygen in oxygen-depleted soils. These specialized roots are called pneumatophores.

Adventitious Root Modifications: Adventitious roots are roots that develop from any part of the plant other than the primary root (radicle). They can undergo modifications for different purposes:

Storage Roots: Plants like ginger, cassava, and sweet potatoes develop adventitious roots that are modified to store carbohydrates and other nutrients.

Prop and Stilt Roots: Some plants, such as maize (corn), develop adventitious roots called prop roots or stilt roots, which grow out from the stem and provide additional support to the plant.

Climbing Roots: Plants like black pepper (Piper nigrum) have adventitious roots that function as climbing structures, allowing the plant to attach itself to other surfaces for support.

Sucking Roots: Parasitic plants like Orobanche develop adventitious roots that penetrate into the host plant to extract nutrients.

Respiratory Roots: Some plants, like Jussiaea, develop adventitious roots that grow above the water surface, enabling gas exchange in habitats with waterlogged or poorly oxygenated soils.

Epiphytic Roots: Epiphytic plants, such as orchids, develop adventitious roots that cling to tree branches or other surfaces, allowing them to grow as epiphytes.

15.2.4 FUNCTIONS OF THE ROOT

The root, a vital organ of the plant, performs various functions essential for the plant's growth and survival. These functions can be categorized as mechanical and physiological, including fixation, absorption, conduction, storage, and assimilation.

Fixation: The root, whether taproot or fibrous, plays a crucial role in anchoring the plant firmly in the soil. This ensures stability and prevents the plant from being uprooted by external forces such as wind or water movement.

Absorption: One of the primary functions of roots is to absorb water and essential minerals from the soil. Root hairs, located near the root tips, increase the surface area for efficient absorption. Water is absorbed through osmosis, while minerals are taken up through active transport mechanisms.

Conduction: Once absorbed, the roots transport water and minerals upwards through the plant via the xylem tissue. This process, known as conduction, ensures the distribution of water and nutrients to the stem, leaves, and other plant parts for growth and metabolic activities.

Storage: Certain plants store carbohydrates, proteins, and other organic compounds in specialized root structures. These storage roots, such as taproots in carrots or tuberous roots in sweet potatoes, serve as reservoirs of nutrients. The stored reserves can be utilized during periods of dormancy, adverse conditions, or reproductive stages.

Assimilation: While leaves are the primary sites for photosynthesis, certain plants have roots capable of performing limited photosynthetic activities. These specialized roots, known as photosynthetic roots, contain chlorophyll and can produce organic compounds through photosynthesis. This allows the plant to supplement its energy requirements and contribute to its overall metabolic activities.

15.3 STEM MORPHOLOGY

The stem, as a vital organ of a plant, exhibits distinct morphological characteristics that contribute to its overall structure and functionality. In botanical terms, the stem is defined as the ascending axis of a plant, which typically grows in an opposite direction to the root or descending axis. It serves as the principal body or stalk of a plant or shrub, with the capability of rising above the ground or, in certain instances, remaining subterranean.

Originating directly from the plumule, the stem plays a pivotal role in the development and sustenance of a plant. It serves as a central axis from which branches, leaves, flowers, and fruits emerge. The structure of the stem exhibits remarkable variations, as it is adapted to perform diverse functions based on the specific requirements of the plant.

Aerial stems are commonly observed, with their upward growth positioning them above the ground. They may possess erect and robust characteristics that enable them to maintain an upright posture. Alternatively, some aerial stems exhibit reduced strength, causing them to trail along the ground or rely on neighboring plants or objects for support. Furthermore, certain stems have a sub-aerial nature, partially extending above and below the ground surface. In contrast, underground stems remain permanently below the soil, periodically giving rise to aerial shoots.

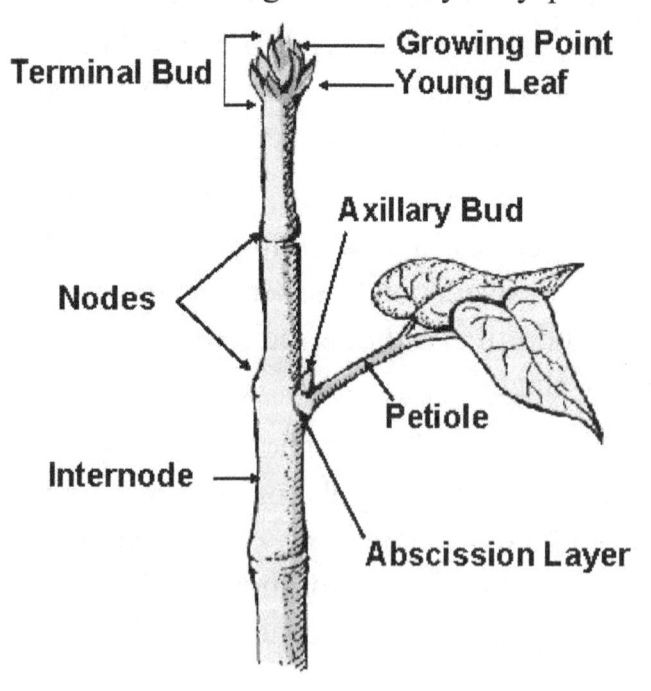

15.3.1 CHARACTERISTICS OF STEM

The stem, a crucial component of the plant axis, possesses distinct characteristics that contribute to its structure and functionality. Understanding these characteristics provides insights into the growth, development, and overall physiology of plants. The key features of the stem are as follows:

Development: The stem originates from the plumule and epicotyl of the embryo, serving as the primary axis of plant growth.

Aerial and Ascending: Typically, the stem is an above-ground structure, ascending in its growth pattern, opposite to the root's downward growth.

Terminal Bud: At the apex of the main stem and lateral branches, a terminal bud is present. It is responsible for the elongation and growth of the stem in length.

Nodes and Internodes: The stem is differentiated into nodes, the points of leaf attachment, and internodes, the segments between the nodes.

Leaf Emergence: Leaves emerge from the nodes of the stem and its branches, facilitating photosynthesis and other metabolic processes.

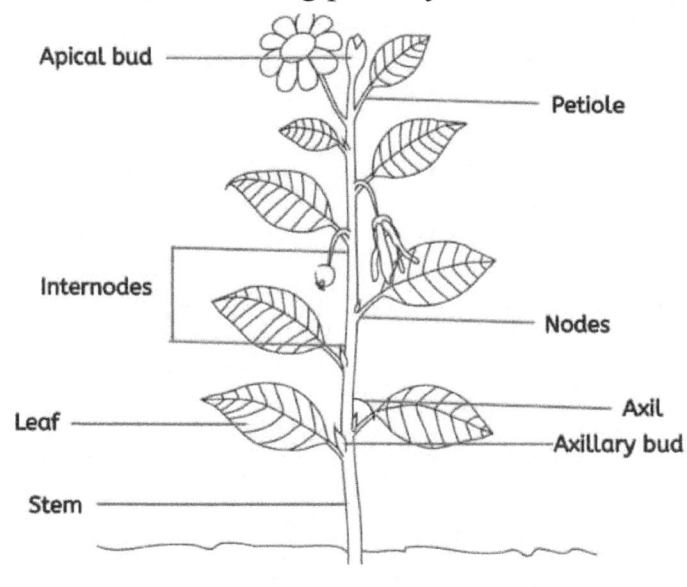

Photosynthesis: In its early stages of development, the young stem is green and can perform photosynthesis, contributing to the plant's energy production.

Multicellular Hairs: The stem may bear multicellular hairs, providing additional surface area for various functions, including water absorption or protection.

Exogenous Origin: Both the branches of the stem and its leaves originate externally from the stem's tissues.

Reproductive Structures: Flowers and fruits develop on the mature stem, facilitating sexual reproduction and seed production.

Geotropic and Phototropic Responses: Generally, the stem exhibits negative geotropism, growing against the force of gravity, and positive phototropism, orienting its growth towards a light source.

15.3.2   THE MODIFICATIONS OF THE STEMS

In most of the plants, the stem is aerial, vertically upwards and bears leaves, flowers and fruits. However, some stems are modified into various shapes to carry on special functions such as perennation, vegetative propagation, synthesis and storage of food etc. these modifications can be categorized into three;

- Underground modifications of Stem
- Sub-aerial modifications of Stem
- Aerial modifications of Stem.

```
                    Stem modification
                           │
        ┌──────────────────┼──────────────────┐
        ▼                  ▼                  ▼
      Aerial           Sub-Aerial         Underground
   modification       modification        modification
```

Sub-Aerial modification → Runner, Stolon, Sucker, Offset

Aerial modification → Creepers, Trailer, Climber

Trailer → Procumbent, Decumbent, Diffuse

Climber → Root Climber, Stem Climber (Twiner), Tendril Climber, Hook climber, Lianas

Underground modification → Bulb, Corm, Rhizome, Tuber

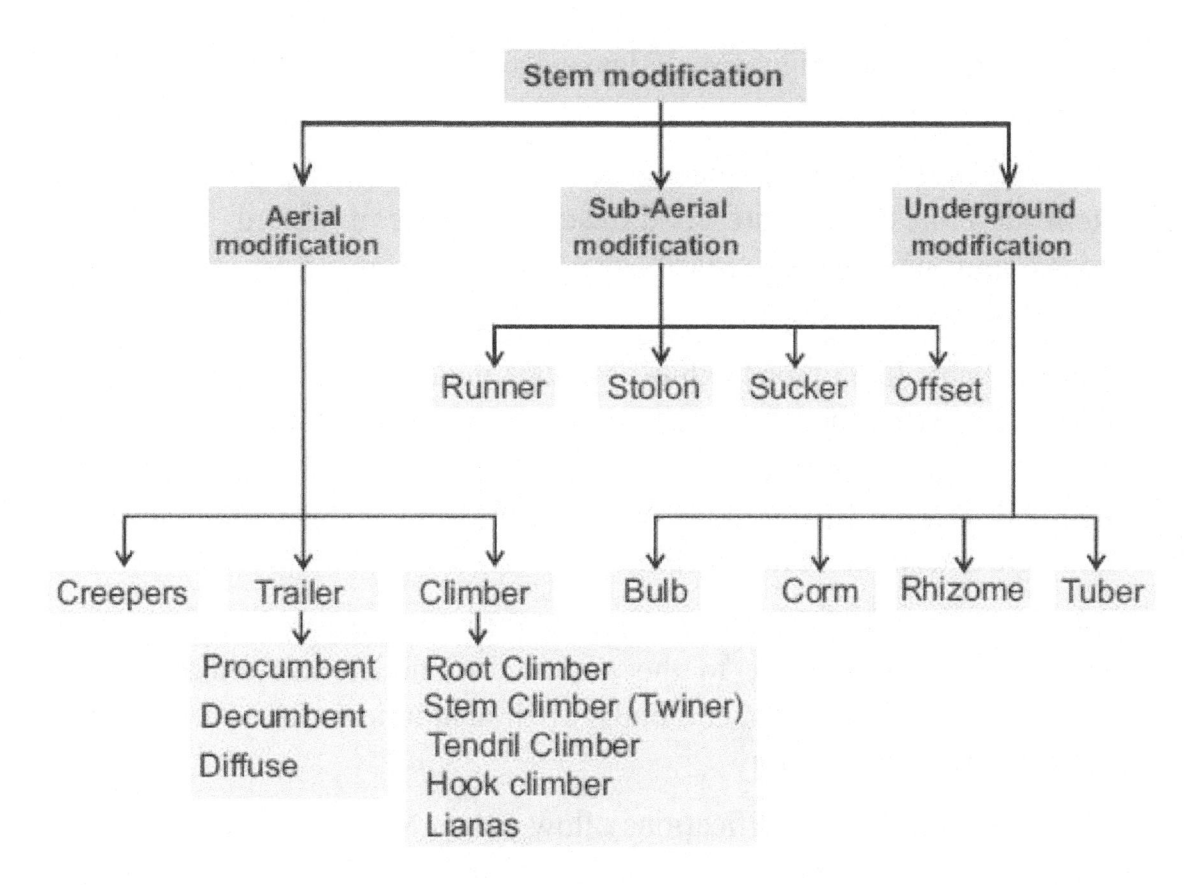

15.3.3 UNDERGROUND MODIFICATIONS OF STEM

Certain plants have evolved underground modifications of their stems, which serve various purposes such as perennation, vegetative propagation, and storage. These adaptations enable the plants to survive unfavorable seasons, ensure their long-term survival, and facilitate the production of new individuals. The underground modifications of stems exhibit unique characteristics and structures that distinguish them as stem structures. Examples of underground stem modifications:

Rhizomes: Plants like ginger, curcuma, turmeric, and water lily possess rhizomes, which are horizontal, underground stems. Rhizomes store food reserves and enable perennation and vegetative reproduction. They possess nodes and internodes, scale leaves, adventitious roots emerging from the nodes, buds at the axils of scale leaves, and anatomical features resembling stems.

Tubers: Tubers, found in plants such as potatoes, are swollen, underground stem structures that store food reserves. These modified stems possess nodes and internodes, scale leaves, adventitious roots, and buds that contribute to vegetative propagation and perennation.

Bulbs: Plants like onions exhibit bulbous underground stem modifications. Bulbs consist of modified stem tissue surrounded by fleshy scale leaves. They store food reserves and support perennation. Bulbs exhibit nodes and internodes, scale leaves, adventitious roots, and buds at the apices of the underground structure and its branches.

Despite lacking typical green coloration, these underground structures are recognized as stems due to the presence of nodes and internodes, scale leaves, adventitious roots, and buds. Their anatomical characteristics further confirm their stem identity.

These underground modifications allow plants to survive harsh conditions, such as winter, by storing food reserves in protected underground structures. As favorable conditions return, these plants produce new aerial shoots from the stored nutrients, ultimately flowering and completing their life cycle. The underground stem modifications enable vegetative reproduction, allowing for the formation of new individuals without the need for seed production.

15.3.4 SUBAERIAL MODIFICATIONS OF STEM

Some plants exhibit subaerial modifications of their stems, which play a crucial role in vegetative propagation and perennation. These adaptations enable the plants to rapidly propagate and ensure the long-term survival of the plant colony. Subaerial modifications occur in weak herbaceous plants and involve specialized branch stems that give rise to small daughter plants. The following are examples of subaerial stem modifications:

Runners: Plants like strawberries produce runners, also known as stolons. Runners are horizontal, above-ground stems that grow along the surface of the soil. They have nodes from which new plants, known as daughter plants, arise. These daughter plants develop roots at their nodes and eventually become independent individuals. This mode of propagation allows the plant colony to expand rapidly and establish new individuals in suitable locations.

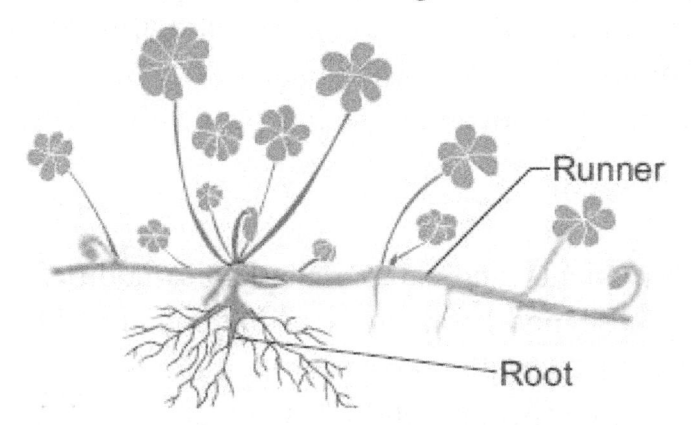

Stolons: Stolons are similar to runners and are found in plants such as mint and spider plants. They are above-ground, horizontal stems that grow along the soil surface. Stolons produce nodes from which new plants develop, allowing for vegetative propagation. These daughter plants can take root at the nodes and establish themselves as independent individuals.

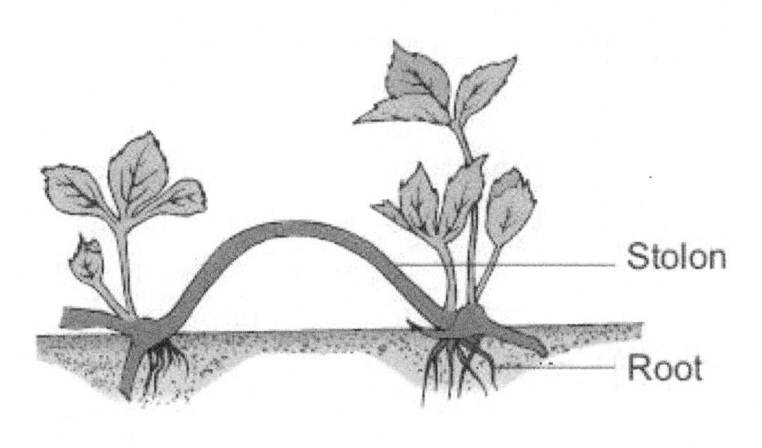

Suckers: Suckers are subaerial stem modifications commonly seen in plants like banana trees and roses. Suckers emerge from the base of the main stem or lateral branches and grow vertically. They develop adventitious roots, allowing them to establish themselves as new plants. Suckers contribute to vegetative propagation and aid in the formation of plant colonies.

Offsets: Plants such as hens and chicks (Sempervivum) produce offsets. Offsets are miniature plants that develop at the base of the main plant. These structures are modified stems that grow partially or wholly underground. Once the offsets have developed their own root system, they can be detached from the parent plant and grow independently.

15.3.5 AERIAL MODIFICATIONS OF STEM

The aerial stem of certain plants undergoes remarkable modifications to serve specific functions. These modifications are so extreme that it becomes challenging to recognize them as stems, except for their origin and position. These structures are considered metamorphosed stems due to the extent of their modifications. The following are examples of aerial stem modifications:

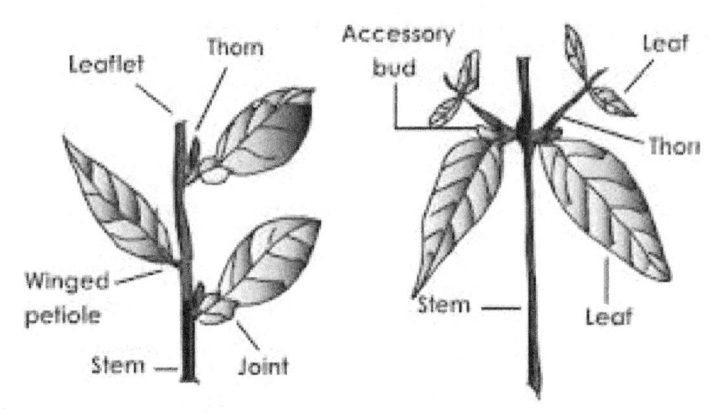

Thorns: Thorns are sharp, pointed structures found in plants such as roses and cacti. They are modified stems that arise from axillary buds. Thorns serve as a defense mechanism, deterring herbivores from feeding on the plant.

Despite their appearance, thorns are derived from stem tissues.

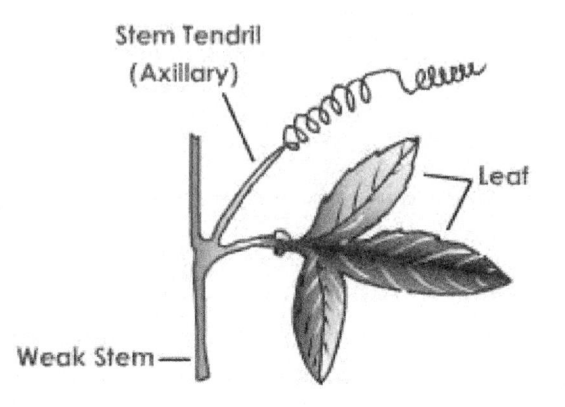

Stem Tendrils: Stem tendrils are slender, coiling structures that aid climbing and provide support to plants. They are commonly observed in plants like grapes and sweet peas. Stem tendrils are modified stems that arise from axillary buds. They possess a sensitive tip that can detect touch or contact with a support structure, allowing the tendril to coil around it and provide support for the plant.

Phylloclades: Phylloclades, also known as flattened stems or cladodes, are leaf-like structures found in plants like cacti and certain succulents. In these plants, the leaves are reduced or absent, and the flattened stem carries out photosynthesis. Phylloclades are modified stems that have taken on the role of leaves, allowing the plants to efficiently capture sunlight for photosynthesis.

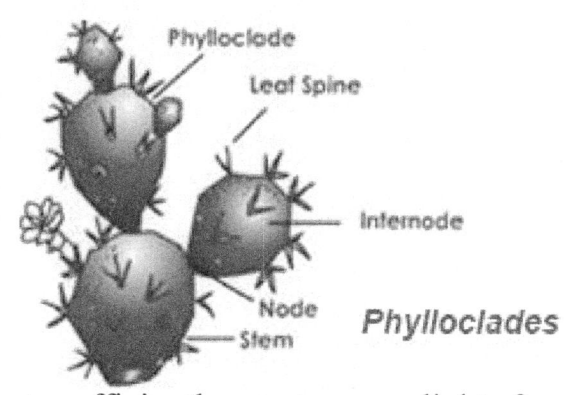

Bulbils: Bulbils are small, rounded structures found in plants like onions and lilies. They are modified stems that develop in the axils of leaves. Bulbils serve as a means of vegetative reproduction, capable of giving rise to new plants when they detach and fall to the ground These structures often contain stored nutrients enabling them to survive adverse conditions and produce new individuals.

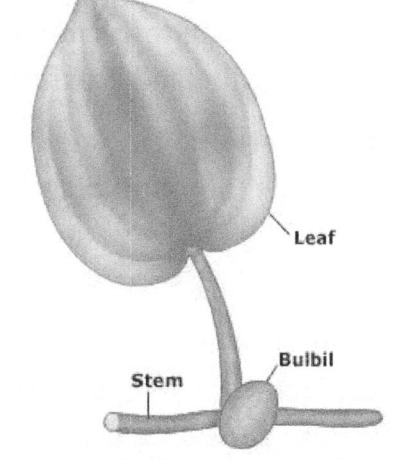

These aerial stem modifications demonstrate the remarkable adaptability of plants to their environment. By undergoing extreme transformations, the stems assume specialized functions such as defense (thorns), climbing (stem tendrils), photosynthesis (phylloclades), and vegetative propagation (bulbils). These modifications enhance the plant's ability to survive, reproduce, and occupy different ecological niches.

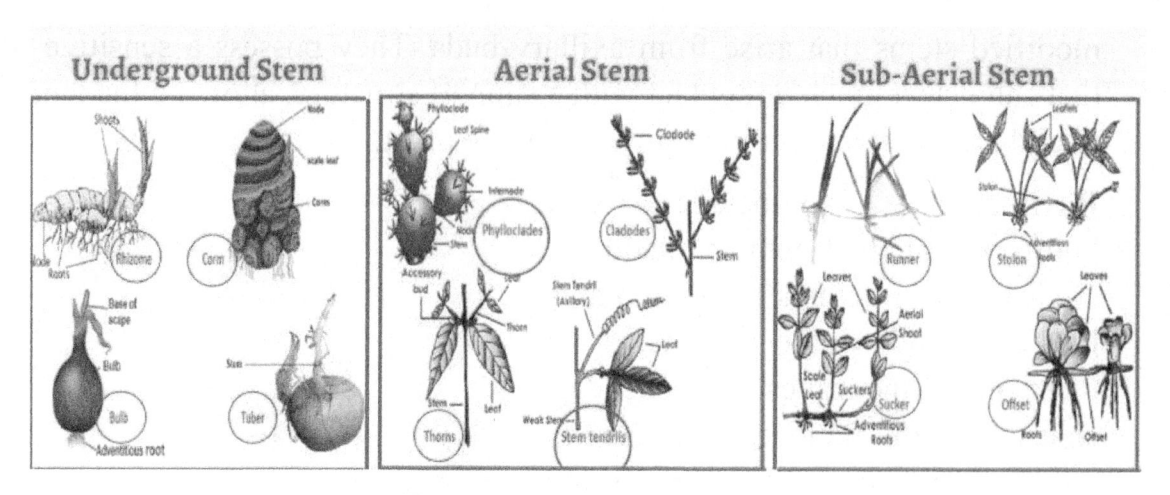

15.3.6 FUNCTIONS OF THE STEM

Mechanical Support: The main stem of a plant is typically thick, aerial, and sturdy, providing structural support to keep the plant in an upright position. This mechanical function helps plants withstand external forces such as wind and gravity.

Conduction: The stem serves as a conduit for the transport of water, minerals, and nutrients from the roots to the leaves and other parts of the plant. This upward movement of fluids and substances through the stem is facilitated by specialized tissues like xylem and phloem.

Food Storage: In certain plants, particularly those with underground stems, the stem acts as a storage organ for storing reserve food materials such as starch, sugars, and proteins. These stored nutrients can be utilized during periods of dormancy, adverse conditions, or for supporting new growth.

Photosynthesis: Young, green stems can contain chloroplasts, the site of photosynthesis in plants. These stems are capable of producing food through the process of photosynthesis, converting sunlight, water, and carbon dioxide into carbohydrates and other organic compounds.

Protection: Some stems undergo modifications, such as the development of thorns or spines, which serve as a protective adaptation against grazing animals and potential threats. These modified stems help deter herbivores and safeguard the plant's vulnerable parts.

Water Storage: Certain plants, like cacti and succulents, have fleshy stems capable of storing significant amounts of water. This adaptation enables the plant to survive in arid environments with limited water availability.

Perennation: Underground stems, such as rhizomes, tubers, and bulbs, function as perennating organs. These specialized stems allow plants to survive unfavorable conditions, such as harsh winters or dry seasons, by remaining dormant underground and resuming growth when conditions become favorable again.

Vegetative Propagation: Some modified stems aid in vegetative propagation, allowing plants to reproduce asexually. Examples include runners, stolons, and offsets, which give rise to new plants when they come into contact with the soil or detach from the parent plant.

15.4 LEAF MORPHOLOGY

Leaves, as lateral appendages of the stem, constitute the principal organs of vascular plants collectively referred to as foliage. They play a crucial role in the life of plants, particularly in their autotrophic nature. Through photosynthesis, plants utilize carbon dioxide, water, and light energy to synthesize organic matter, predominantly glucose and sucrose, which are further transformed into complex compounds like cellulose—fundamental components of plant cell walls. The leaf serves as the primary site for this vital process by integrating these three essential ingredients.

Typically, leaves possess a broad, flat, and thin structure, characterized by a dorsiventral (top-bottom) flattening, which serves to optimize the surface area exposed directly to light. This facilitates the penetration of light into leaf tissues, allowing for efficient photosynthesis. While leaves are generally arranged on the plant in a manner that minimizes mutual shading, variations and complexities arise due to factors such as climate, light availability, herbivory, nutrient access, and interspecific competition.

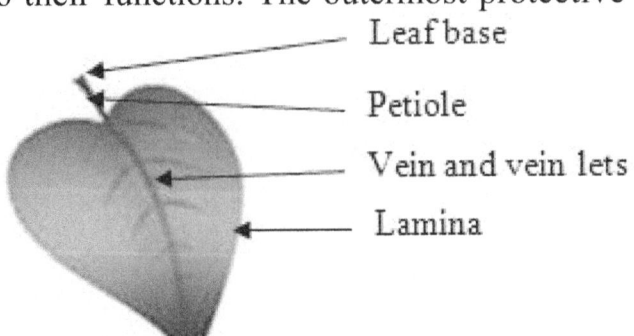

Leaf morphology exhibits substantial diversity in terms of shape and structure across plant species. These variations are primarily attributed to their adaptation to specific environmental conditions. Leaves may undergo modifications to enhance specific functions, such as reducing water loss in arid habitats or maximizing light absorption in shaded environments. Internally, leaves comprise specialized tissues and structures that contribute to their functions. The outermost protective layer, known as the epidermis, shields the leaf, while chloroplast-containing cells within the leaf, called chloroplasts, capture light energy through the pigment chlorophyll. Vascular tissue, consisting of xylem and phloem, forms leaf veins responsible for transporting water, nutrients, and sugars throughout the leaf.

Leaf arrangement on the stem follows distinct patterns referred to as phyllotaxy. Common arrangements include alternate, opposite, whorled, and rosette forms, each offering advantages in terms of light capture and resource allocation.

15.4.1   PARTS OF THE LEAF

The leaf, a vital organ of a plant, consists of several distinct parts that contribute to its structure and functions:

Lamina: The lamina, also known as the leaf blade, is the broad, thin, and flat structure of the leaf. It is the primary site for photosynthesis, where light energy is captured and converted into chemical energy.

Petiole: The petiole is the stalk-like structure that connects the leaf to the stem. It provides support and allows the leaf to be positioned optimally for light absorption. In some plants, the leaves are directly attached to the stem without a petiole, a condition known as sessile leaves.

Veins: The lamina of the leaf possesses a network of veins, which are vascular tissues responsible for the transport of water, minerals, and sugars throughout the leaf. Veins provide structural support and distribute essential resources to different parts of the leaf.

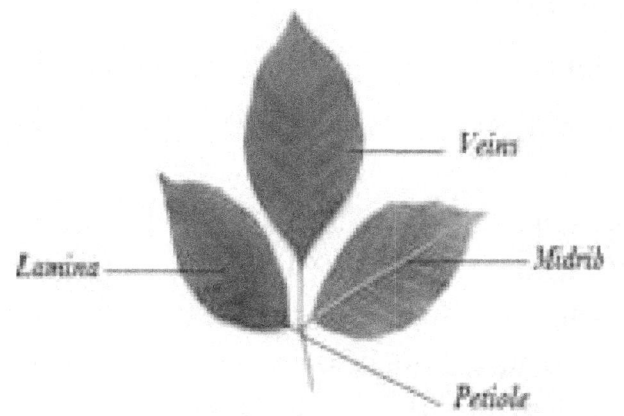

a. Midrib: The midrib is the largest and most prominent vein in the leaf. It runs centrally through the leaf blade from the base to the apex, providing structural support and serving as a pathway for vascular tissues.

b. Lateral Veins: Emerging from the midrib, lateral veins branch out and spread across the leaf lamina. They transport water, nutrients, and sugars to the leaf cells, ensuring efficient distribution of resources.

c. Veinlets: The lateral veins further divide into smaller veins called veinlets. These thinner veins extend throughout the lamina, forming an intricate network that supplies water and nutrients to individual leaf cells.

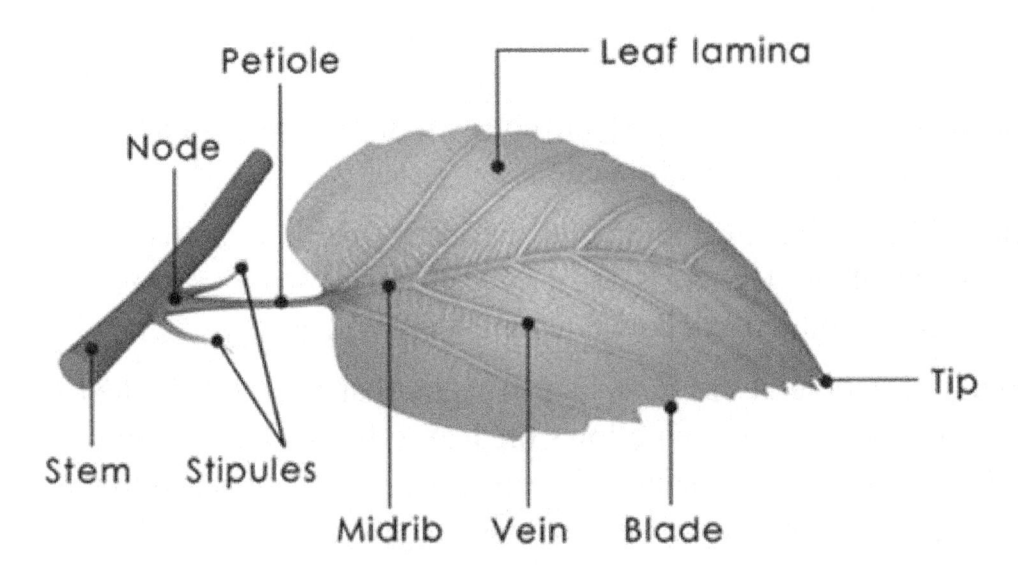

HETEROPHYLLY AND ANISOPHYLLY

In leaf morphology, certain plants exhibit interesting variations in the shapes and sizes of their leaves, giving rise to the terms heterophylly and anisophylly.

Heterophylly refers to the phenomenon where a plant species produces different types of foliage leaves. This is particularly observed in plants that grow partially submerged in water and partially in the air. Submerged leaves, which are underwater, tend to be finely dissected or divided into smaller segments, while aerial leaves, which are exposed to air, are generally more intact and undivided. This occurrence can be observed in aquatic plants such as water chestnut (Trapa), Ranunculus aquatilis,

Ranunculus multifidus, Myriophyllum indicum, Limnophila heterophylla, and Sagittaria. Heterophylly can also be found in certain terrestrial plants like coriander (Coriandrum sativum), Pterospermum, Artocarpus chaplasa, Ficus heterophylla, among others.

Anisophylly, on the other hand, is a distinct phenomenon observed in plants with opposite phyllotaxy, where leaves are arranged in opposite pairs along the stem. Normally, opposite leaves are of equal size, but in rare cases, such as Goldfussia glomerata of the Acanthaceae family and Pellionia daveanana of the Urticaceae family, they display unequal sizes. This condition is referred to as anisophylly.

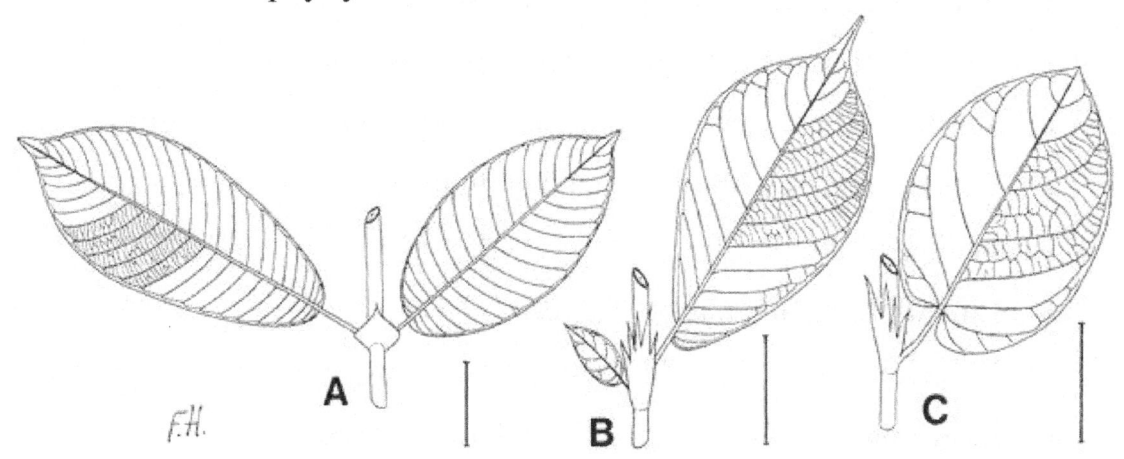

Anisophylly in Sabicea floribunda (A), Sabicea mildbraedii (B) and Pseudosabicea sthenula (C).

15.4.2 TYPES OF LEAF

Leaf can be differentiated on the following basis:

According to Petiole

- **Petiolated (Stalked):** Leaves that have a distinct stalk called a petiole.
- **Sessile (Unstalked):** Leaves that lack a petiole and are directly attached to the stem.

According to Blade

- **Simple Leaf:** Leaves with an undivided blade or, if divided, the divisions do not reach the midrib.
- **Compound Leaf:** Leaves with a fragmented blade where the divisions reach the midrib. Each fragmented division is called a leaflet and resembles a single leaf.

According to Edge

- **Entire:** Leaves with a smooth margin.
- **Sinuate:** Leaves with slightly wavy or curved edges.
- **Dentate:** Leaves with small teeth along the margin.
- **Serrate:** Leaves with small, saw-like teeth along the margin.
- **Lobed:** Leaves with divisions that do not reach the midrib.

According to Shape

- **Linear:** Long and narrow leaves.
- **Elliptic:** Leaves shaped like an ellipse, being 2-3 times longer than wide
- **Lanceolate:** Leaves shaped like a lance or spear.
- **Acicular:** Needle-shaped leaves.
- **Ovate:** Leaves shaped like an egg.
- **Cordate:** Leaves with a heart shape.
- **Hastate:** Leaves resembling a sagittate leaf with two lobes directed outward.

| Professor Carl William Matthias

needle shaped	linear	oblong	elliptic	lanceolate
oblanceolate	ovate	obovate	cordate	obcordat
deltoid	obdeltoid	cuneate	rhomboid	reniform
peltate	orbicular	spathulate	hastate	sagittate
lunate	pandurate	flabellate	fan-shaped	subulate
palmatifid	palmatisect	pinnatifid	pinnatisect	

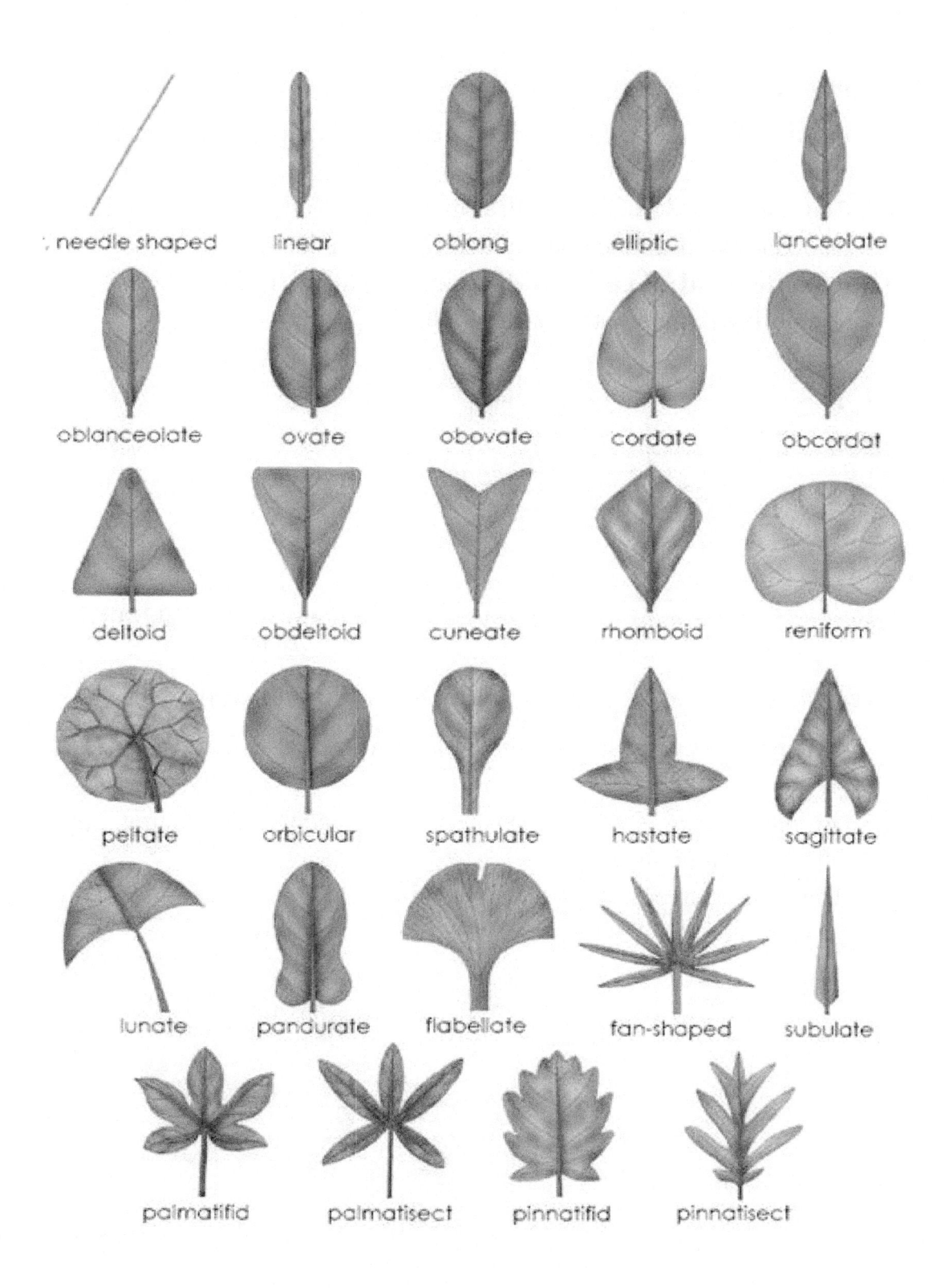

According to Veins

- **Parallel Vein Leaf:** Leaves with veins running parallel to each other at equal distances, commonly found in monocots.
- **Pinnate Leaves:** Leaves with a main nerve (midrib) from which other secondary veins branch out, typically seen in dicots.
- **Palmate Leaf:** Leaves with veins radiating from a central point, resembling fingers in the palm of a hand.

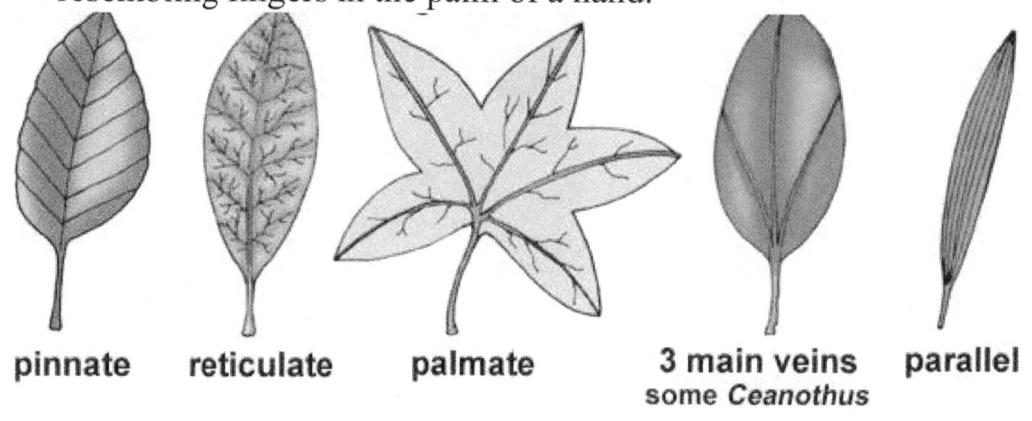

pinnate reticulate palmate 3 main veins some *Ceanothus* parallel

According to the arrangement on the Stem (Phyllotaxy)

- **Alternate:** Leaves arising singly from nodes at different levels along the stem.
- **Opposite:** Leaves occurring in pairs at each node, facing opposite directions along the stem.
- **Whorled:** Leaves arranged in several or many at the same level around the stem.

Alternate Spiral Opposite Whorled

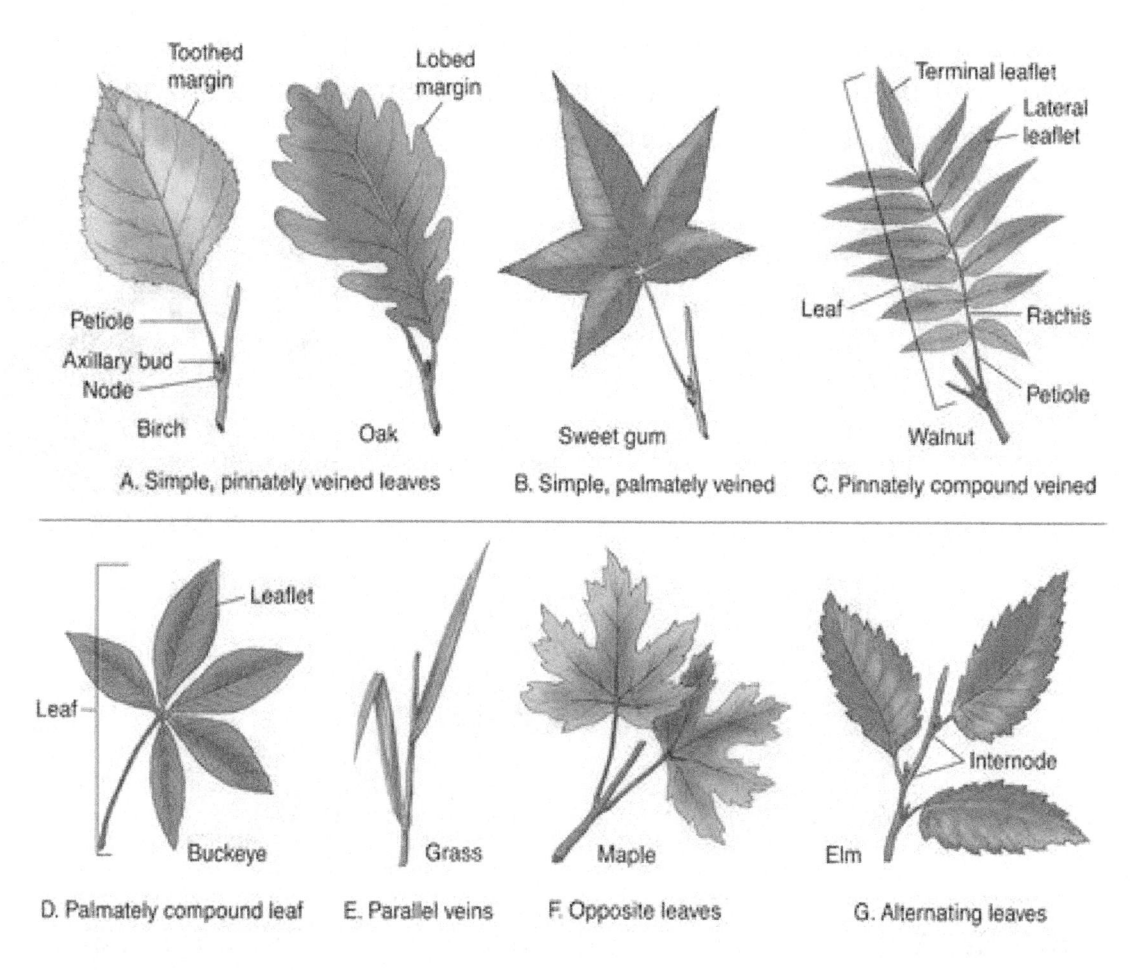

A. Simple, pinnately veined leaves — Birch, Oak
B. Simple, palmately veined — Sweet gum
C. Pinnately compound veined — Walnut
D. Palmately compound leaf — Buckeye
E. Parallel veins — Grass
F. Opposite leaves — Maple
G. Alternating leaves — Elm

15.4.3 MODIFICATIONS OF LEAVES

Leaves in various plant species undergo specialized modifications that enable them to perform specific functions

Scale Leaf: These are thin, dry, papery structures without a stalk. They are often brown in color but can also be thick and fleshy.

Leaf Spines: Some plants modify their leaves into spines as a defense mechanism against herbivory. These spines deter animals from feeding on the plant.

SCALE LEAVES
(Ex: *Casurina*)

Leaf Tendril: Certain plants develop slender, wire-like, coiled structures called tendrils. These modified leaves aid in climbing and provide support to the plant by attaching to nearby objects.

LEAF MODIFICATIONS AS TENDRILS

ENTIRE LEAF EXCEPT STIPULES AS TENDRIL
(Ex: *Lathyrus*)

TERMINAL LEAFLETS AS TENDRILS
(Ex: *Pisum*)

LEAF APEX AS TENDRILS
(Ex: *Gloriosa*)

PETIOLE AS TENDRIL
(Ex: *Clemantis*)

STIPULES AS TENDRIL
(Ex: *Smilax*)

Phyllode: A phyllode is a flattened petiole that resembles a leaf, especially when the leaf blade is absent or reduced. The phyllode takes on the role of photosynthesis and other leaf functions.

PHYLLODE
(Ex: *Acacia*)

PHYLLODE
(Ex: *Parkinsonia*)

Pitcher: The leaf blade of certain plants undergoes modification to form a pitcher-like structure. This specialized leaf adaptation helps in the capture and digestion of insects or other small prey, providing additional nutrients to the plant.

TYPES OF TRAP LEAVES

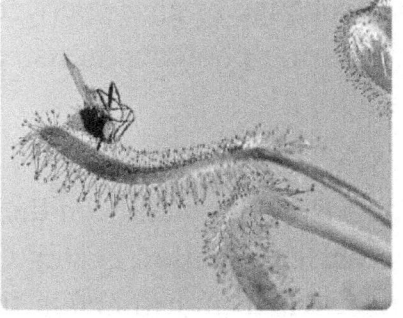

Dionaea

Drosera

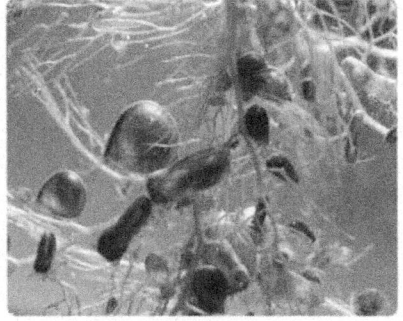

Nepenthes

Utricularia

15.4.4 FUNCTIONS OF THE LEAVES

The leaves perform three main functions such as manufacture of food, interchange of gases between the atmosphere and the plant body and evaporation of water. Some of the functions are de-scribed below. The functions of modified leaves are not dealt as they carry on specialized functions as discussed earlier.

Manufacture of Food: Green leaves are primarily responsible for photosynthesis, the process through which they produce food. Using sunlight and the green pigment chlorophyll, leaves convert water and carbon dioxide into glucose and oxygen. This energy-rich glucose serves as a nutrient source for the plant.

Gas Exchange: Leaves facilitate the exchange of gases, serving both respiration and photosynthesis. Stomata, small openings found on the leaf surface, enable the exchange of gases between the plant and the atmosphere Oxygen is released during photosynthesis, while carbon dioxide enters for use in the process.

Evaporation of Water: Transpiration is the process by which water absorbed by the roots is evaporated from the leaf surface during the daytime Stomatal openings on leaves allow water vapor to escape, facilitating the upward movement of water and minerals from the roots to the rest of the plant.

Storage of Food: Some leaves, such as the fleshy leaves of Indian aloe and Portulaca, as well as the scale leaves of onion, store food and water for future use. In xerophytic plants found in desert areas, succulent leaves store significant quantities of water, mucilage, and food reserves to cope with arid conditions.

Vegetative Propagation: Certain plants, including Bryophyllum, Begonia, and Kalanchoe, have leaves capable of producing buds. Through vegetative propagation, these buds develop into new plants when the leaf touches the ground, producing roots at the point of contact and forming a bud that grows into a new individual.

Protection: Leaves also serve a protective role, particularly when modified into thorns and spines. Examples include plants like Berberis and Aegle, where modified leaves deter grazing animals and help safeguard the plant's vulnerable parts, such as the axillary bud.

15.5 INFLORESCENCE

An inflorescence is a distinctive arrangement of flowers on a stem in seed plants. It represents a modified part of the shoot where flowers are formed and organized in a specific pattern. The study of inflorescences is crucial for understanding the reproductive strategies, pollination mechanisms, and diversity of plant species.

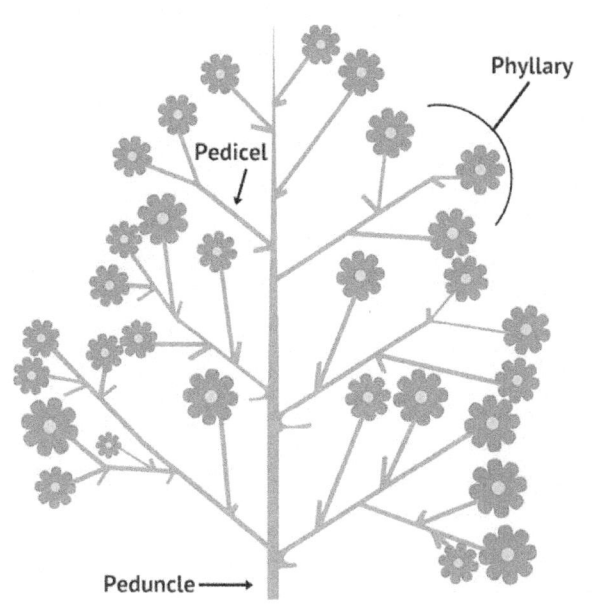

Morphologically, inflorescences exhibit a wide array of modifications involving the length and nature of the internodes, as well as the phyllotaxis. Additionally, variations in the proportions, compressions, and reduction of main and secondary axes contribute to the remarkable diversity observed in inflorescence structures.

The stem that holds the entire inflorescence is referred to as the peduncle, while the axis that bears the flowers or additional branches within the inflorescence is known as the rachis. Each individual flower is attached to the inflorescence by a stalk called a pedicel. However, when a flower is solitary and not part of an inflorescence, its stalk is referred to as a peduncle.

Within an inflorescence, the individual flowers may be called florets, especially when they are small and densely clustered, as seen in a pseudanthium. These compact clusters of flowers resemble a single flower, but they are composed of multiple florets arranged in a particular manner.

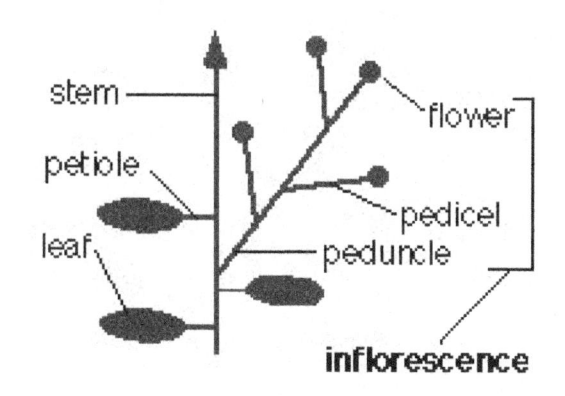

As the inflorescence matures and enters the fruiting stage, it is known as an infructescence. This stage represents the transformation of the flowers into fruits, further contributing to the reproductive success and propagation of the plant.

Inflorescences may be Terminal or Axillary, and may be branched in several ways. This way, depending on the type of branching, several kinds of inflorescences have been developed and the may be classified into two distinct groups:

- Racemose (indefinite)
- Cymose (definite)

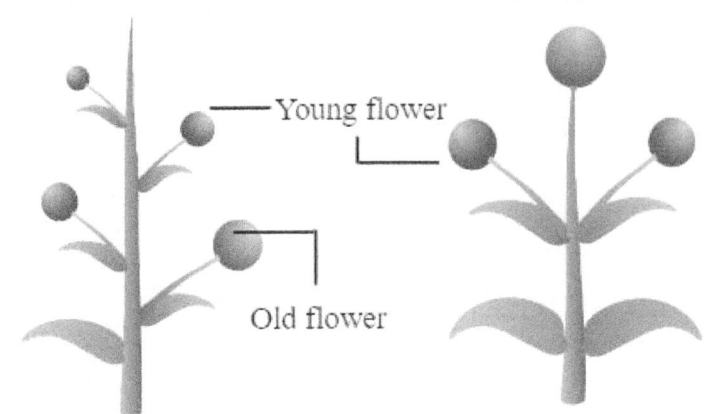

15.5.1 RACEMOSE INFLORESCENCE

Racemose inflorescence is characterized by the continuous growth of the main axis, which does not terminate in a flower but produces flowers laterally in acropetal succession. This means that the lower flowers are older than the upper ones along the elongated axis or at the center of a truncated axis.

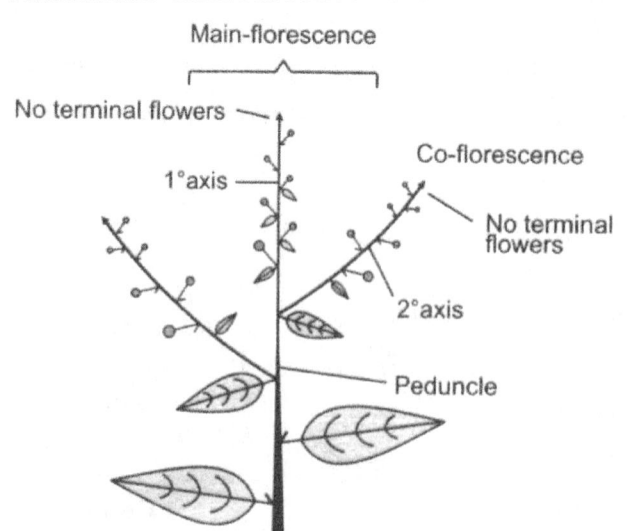

The arrangement of flowers in racemose inflorescence allows for the progressive development of new flowers towards the apex of the axis. As a result, the youngest flowers are found at the top of the elongated axis or in the central region of a truncated axis. This acropetal succession of flower development ensures that the older flowers are located towards the base of the inflorescence.

Indeterminate inflorescences are characterized by the continuous growth of the main axis, which results in the development of various types of racemose inflorescences. These include the raceme, panicle, spike, catkin, corymb, umbel, spadix, and head. Each type exhibits unique characteristics and patterns of flower arrangement along the axis.

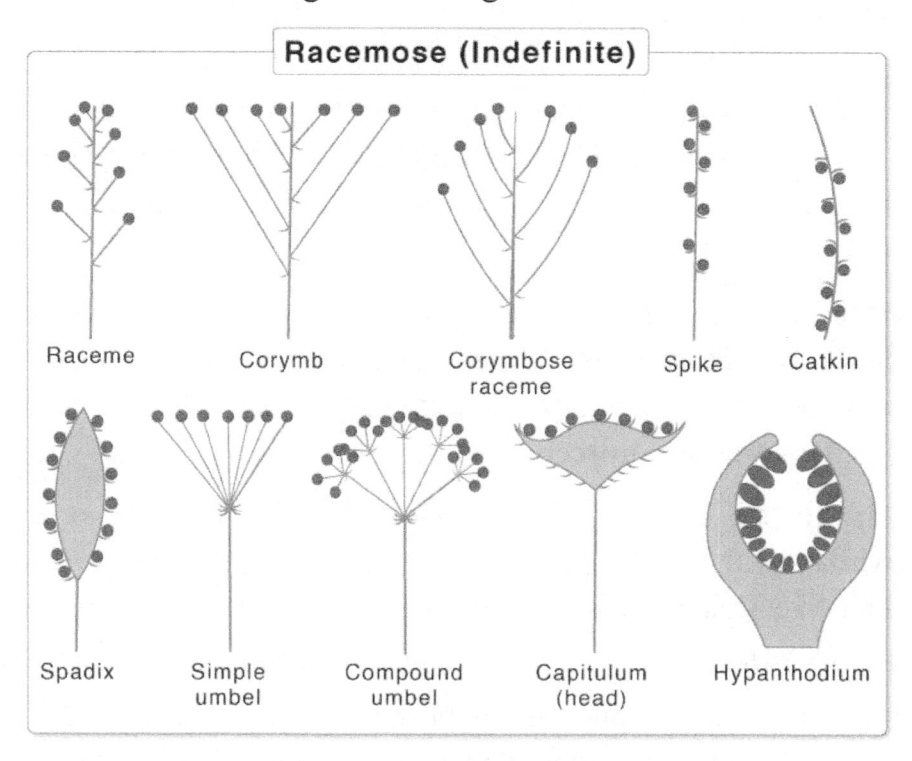

Racemose (Indefinite)

Raceme | Corymb | Corymbose raceme | Spike | Catkin

Spadix | Simple umbel | Compound umbel | Capitulum (head) | Hypanthodium

The racemose inflorescence type is commonly observed in various plant families and species. It provides a flexible and continuous method of flower production, allowing for prolonged flowering and increased reproductive success. The elongated axis of the racemose inflorescence offers ample space for the development and display of numerous flowers, contributing to the attractiveness and functionality of the inflorescence.

Raceme: In a raceme, flowers develop at the upper angle or axil between the stem and branch of each leaf along a long, unbranched axis. Each flower is borne on a short stalk called a pedicel. A classic example of a raceme is found in the snapdragon (Antirrhinum majus).

Panicle: A panicle is a branched raceme in which each branch bears more than one flower. This branching pattern can be observed in plants such as astilbe (Astilbe).

Spike: A spike is another type of raceme in which the flowers develop directly from the stem without the presence of pedicels. Barley (Hordeum) and Amaranthus are examples of plants that exhibit a spike inflorescence.

Catkins: also known as aments, are spikes in which the flowers are either male (staminate) or female (carpellate). They are usually pendulous and may have reduced or absent perianths. Catkins are commonly seen in oak trees (Quercus).

Corymb: A corymb is a raceme in which the pedicels of the lower flowers are longer than those of the upper flowers, giving the inflorescence a flat-topped appearance. Hawthorn (Crataegus) is an example of a plant that displays a corymb inflorescence.

Umbel: In an umbel, the pedicels of the flowers originate from approximately the same point at the tip of the peduncle, resembling an umbrella-like shape. This type of inflorescence can be observed in wax flowers (Hoya).

Spadix: A spadix is a spike that is borne on a fleshy stem and is commonly found in the family Araceae. Plants such as Philodendron exhibit a spadix inflorescence, with the subtending bract known as a spathe.

Head also referred to as a capitulum, is a short, dense spike in which the flowers are borne directly on a broad, flat peduncle. This arrangement gives the inflorescence the appearance of a single flower, as seen in the dandelion (Taraxacum).

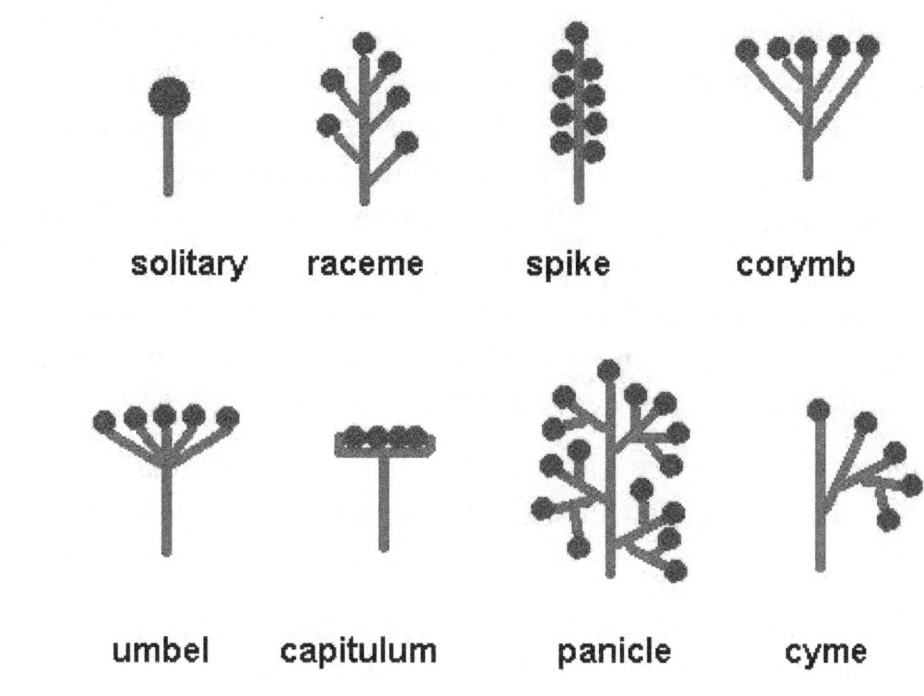

solitary   raceme   spike   corymb

umbel   capitulum   panicle   cyme

15.5.2 CYMOSE INFLORESCENCES

Cymose inflorescences are characterized by the cessation of growth in the main axis due to the development of a flower at its apex. The lateral axes, which develop the terminal flower, also culminate in flowers, resulting in a determinate pattern of growth. Several key features distinguish cymose inflorescences.

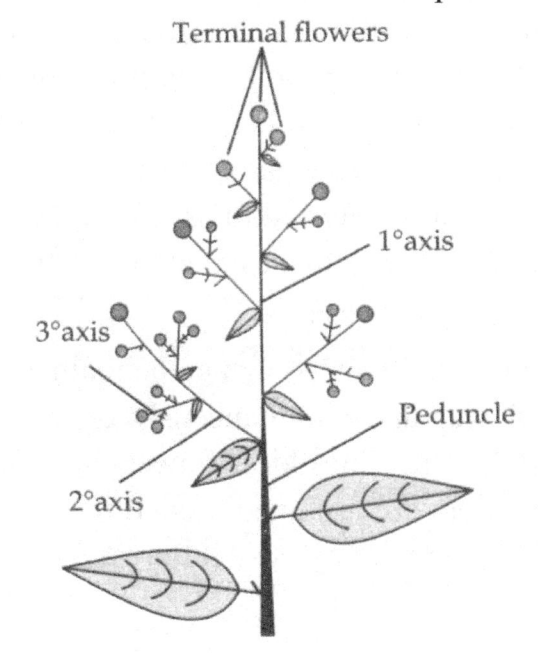

The flowers in a cymose inflorescence may be pedicellate, meaning they are attached to the main axis by individual stalks, or sessile, where they lack stalks and are directly attached to the axis. This distinction in flower attachment contributes to the overall structure and appearance of the inflorescence.

In terms of flower development, cymose inflorescences follow basipetal succession. This means that the terminal flower, which is the first to develop, is also the oldest, while the lateral flowers that subsequently form along the axis are younger. This developmental sequence gives rise to a characteristic arrangement of flowers within the inflorescence. Cymose inflorescences can exhibit various forms, including the dichasium, helicoid cyme, scorpioid cyme, and monochasium, among others. Each form has a specific pattern of branching and flower arrangement, contributing to the diversity and complexity of cymose inflorescences.

The cymose inflorescence can be classified into four main types, each exhibiting distinct characteristics in terms of growth and flower arrangement:

Uniparous or Monochasial Cyme: In this type, the main axis of the inflorescence terminates in a flower and produces only one lateral branch at a time, with each lateral branch ending in a flower. This results in a single-file arrangement of flowers along the axis. Examples of plants displaying this type of cyme include Gossypium (cotton) and Solanum (nightshade).

Biparous or Dichasial Cyme: The biparous cyme is characterized by the production of two lateral branches or flowers at each branching point along the main axis. The main axis terminates in a flower, and each lateral branch further branches into two lateral flowers or branches, creating a branched structure with pairs of flowers. This type is also referred to as a True Cyme or Compound Dichasium. Examples of plants with dichasial cymes include Ixora and Teak.

Multiparous or Polychasial Cyme: In the multiparous cyme, the main axis of the inflorescence culminates in a flower and simultaneously produces multiple lateral flowers or branches around it. The oldest flower occupies the central position, and the younger flowers are arranged around it. Plants such as Calotropis and Hamelia exhibit this type of cymose inflorescence.

Cymose Capitulum: In this type, the peduncle of the inflorescence is condensed or reduced to a circular disc-like structure. The oldest flowers develop at the center of the disc, while the youngest flowers are found towards the periphery. Mimosa and Albizzia are examples of plants that display a cymose capitulum.

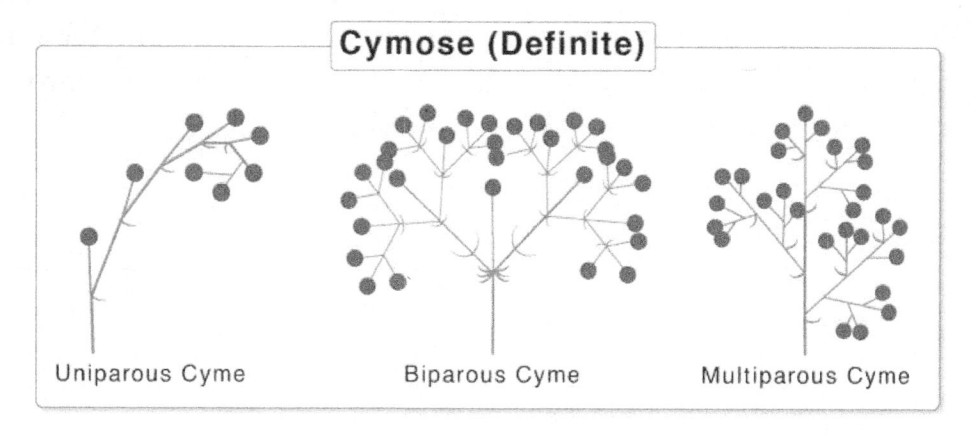

Cymose (Definite)

Uniparous Cyme | Biparous Cyme | Multiparous Cyme

15.5.3 COMPOUND INFLORESCENCE

Compound inflorescence refers to a type of inflorescence in which the main axis, known as the peduncle, undergoes repeated branching in a racemose or cymose manner. As a result, the inflorescence takes on a compound structure, either resembling a compound raceme or a compound cyme.

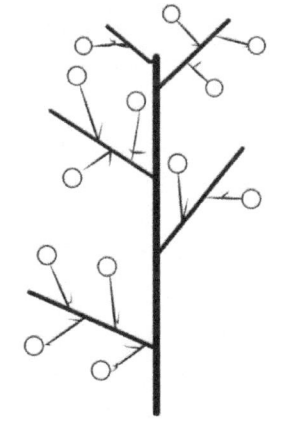

Compound Raceme: In a compound raceme, the main axis produces secondary branches, each of which bears multiple flowers along its length. These secondary branches may themselves produce tertiary branches with additional flowers. This branching pattern gives the inflorescence a multi-level, branched appearance. Examples of plants with compound racemose inflorescences include Lupinus (lupine) and Delphinium (larkspur).

Compound Cyme: In a compound cyme, the main axis branches repeatedly in a cymose manner. Each branch terminates in a flower, and these lateral flowers may further branch into additional flowers. This results in a complex, branching structure where multiple flowers are arranged in a cymose fashion. Umbelliferae family (e.g., carrot) and Allium (onion) are examples of plants exhibiting compound cymose inflorescences.

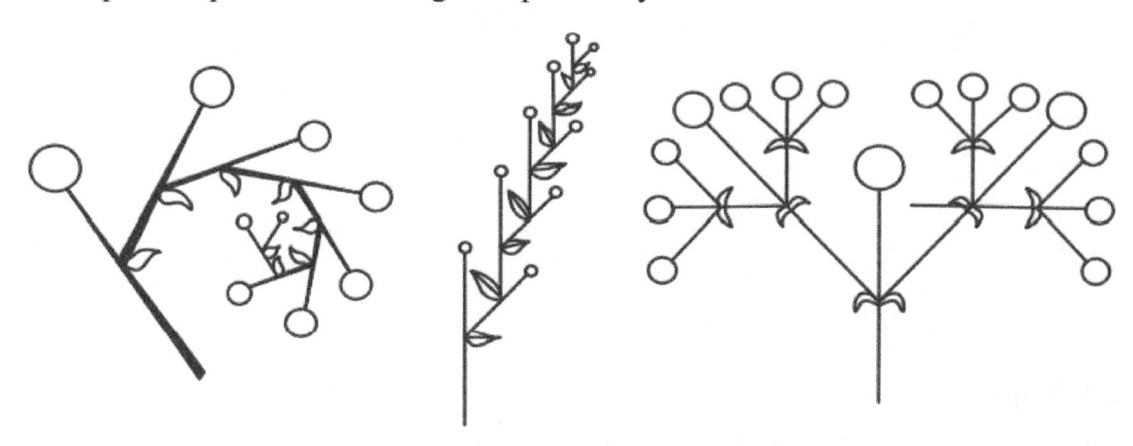

15.6 FRUITS

Fruits are the mature and ripened ovaries of flowering plants, which also contain the developed seeds. The ovary, located within the flower, serves as the reproductive structure that houses the ovules. After pollination and fertilization, the ovary undergoes changes and transforms into a fruit.

The formation of a fruit begins with the development of the seed within the ovary. The ovules, which contain the potential embryo, undergo fertilization and start to develop into seeds. Concurrently, the ovary itself undergoes significant changes, including enlargement, differentiation, and sometimes the development of specialized structures.

As the ovary matures, it often undergoes further transformations, such as changes in color, texture, and taste. The mature ovary, now referred to as the fruit, serves several important functions. Firstly, it protects the seeds within, providing them with a safe and favorable environment for development and dispersal. Additionally, fruits play a crucial role in seed dispersal, attracting animals through various means such as bright colors, enticing aromas, or nutritious pulp. Animals consume the fruits, effectively dispersing the enclosed seeds as they travel and eventually excrete them.

The edible portion of the fruit is derived from the enlarged and ripened ovary. In some instances, other floral parts may contribute to the structure commonly referred to as the fruit. These include receptacles, floral tubes, or other accessory structures that may fuse or become incorporated into the fruit.

Fruits exhibit remarkable diversity in size, shape, color, and flavor, reflecting the wide range of flowering plant species and their unique adaptations for seed dispersal. They provide an essential source of nutrition for both animals and humans and have significant ecological and economic importance.

15.6.1 STRUCTURE OF A FRUIT

The structure of a fruit consists of various parts that contribute to its overall form and function. The two main components of a fruit are the pericarp and the seed.

The pericarp is the wall of the fruit derived from the ovary, and it is composed of three distinct layers: the exocarp or epicarp, the mesocarp, and the endocarp.

The exocarp, also known as the epicarp, is the outermost layer of the pericarp. It forms the protective skin or rind of the fruit and can vary in texture, thickness, and color depending on the specific fruit type.

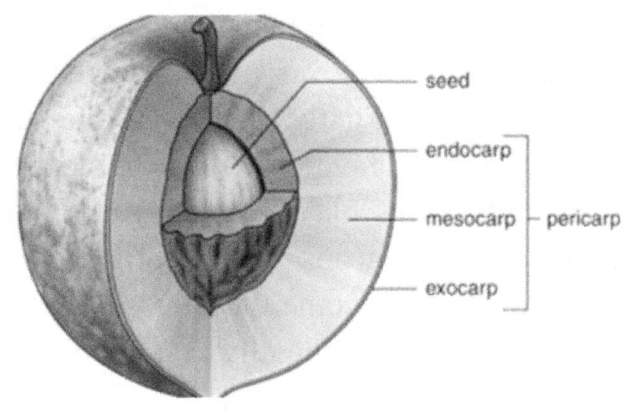

Beneath the exocarp lies **the mesocarp,** which is the middle layer of the pericarp. The mesocarp can exhibit different characteristics based on the type of fruit. In some fruits, such as berries and drupes, the mesocarp is fleshy, soft, and often juicy, serving as the edible part of the fruit. In other fruits, such as pomes (e.g., apples and pears) the mesocarp is more fibrous and less juicy.

The innermost layer of the pericarp is the endocarp. It surrounds the seed(s) and can have various structures and textures depending on the fruit type. In some fruits, the endocarp becomes hard and stony, forming a protective shell around the seed(s). Examples of fruits with stony endocarps include peaches and cherries. In contrast, other fruits have a thin and papery endocarp, as seen in tomatoes and peppers.

Within the pericarp, the seed(s) develop. The seed is the reproductive structure of the plant, containing the embryo and often enclosed within a protective seed coat. The seed plays a vital role in plant propagation and dispersal. Some fruits may have additional structures or modifications, such as wings, hairs, or specialized appendages, which aid in dispersal mechanisms.

15.6.2 CLASSIFICATION OF FRUITS

On basis of the number of ovaries and the number of flowers involved in their development, fruits are classified into three categories:

- Simple Fruits
- Aggregate Fruits
- Multiple Fruits

15.6.3 SIMPLE FRUITS

These fruits develop from a single ovary of several carpels. These fruits are further divided into Dry fruits and Fleshy fruits depending upon the pericarp. There are two classifications of simple fruits– dry and fleshy.

DRY FRUITS

Dry fruits are characterized by their non-fleshy nature, and their pericarp (fruit wall) is not differentiated into distinct layers. There are different types of dry fruits based on their mode of dehiscence (splitting open) or indehiscence (not splitting open) after ripening.

DEHISCENT FRUITS (CAPSULAR FRUITS)

Particular of these fruits is that their pericarp rupture after- ripening and the seeds are dispersed. Within the category of dry fruits, there are dehiscent fruits, also known as capsular fruits, which have a pericarp that ruptures or splits open after ripening, allowing the seeds to be dispersed. These fruits can be classified into five different types based on their mode of dehiscence:

Legume: Legumes are fruits formed from a monocarpellary pistil (a pistil with a single carpel) and dehisce along both the dorsal and ventral sutures.

They are characteristic fruits of the family Leguminosae (Fabaceae), which includes plants like peas and beans.

Follicle: Follicles are fruits formed from a monocarpellary pistil and dehisce along only one suture. An example of a follicle is the fruit of the larkspur (Delphinium).

Siliqua: Siliquas are long, cylindrical fruits formed from a bicarpellary, syncarpous pistil (pistil with two fused carpels) with an ovary having two parietal placentas. They split open along two sutures, and the seeds are attached to a central partition called the replum. Mustard plants (Brassicaceae) are examples of plants with siliqua fruits.

Silicula: Siliculas are similar to siliquas but are shorter and as long as they are broad. They contain fewer seeds compared to siliquas. An example of a plant with a silicula fruit is candytuft (Iberis).

Capsule: Capsules are fruits originating from a polycarpellary, syncarpous pistil with an exceptional ovary. They can also be formed from an inferior ovary, where the sepals, petals, and stamens are attached above the ovary. Examples of plants with capsule fruits include iris and willow.

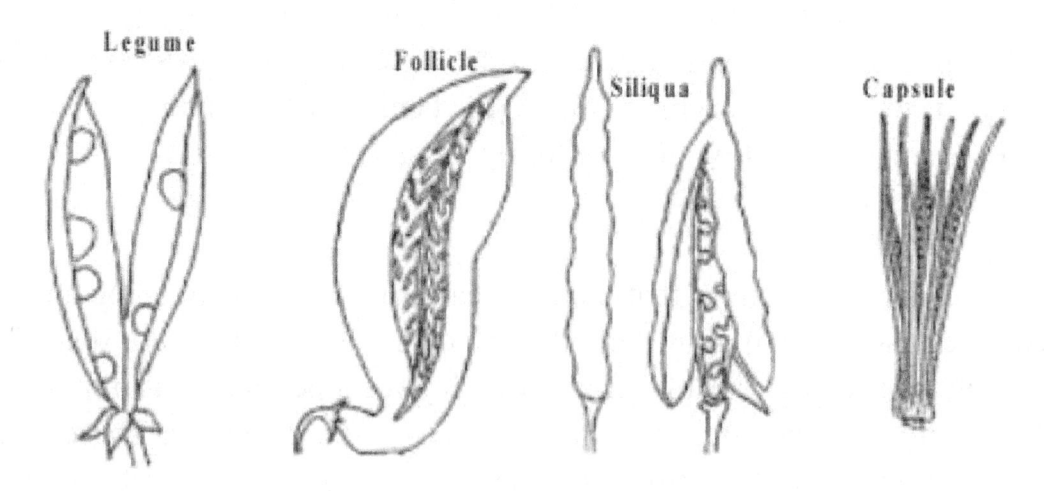

INDEHISCENT FRUITS (ACHENIAL FRUITS):

Indehiscent fruits, also known as achenial fruits, are characterized by their pericarp, which does not burst or open upon ripening, resulting in the seeds remaining enclosed within the fruit. These fruits have a single-seeded ovary, and the pericarp tightly surrounds and protects the seed. Indehiscent fruits can take on various forms and structures. Some common examples include:

Achene: An achene is a small, dry fruit with a single seed that is not fused to the pericarp. The pericarp of an achene is typically thin and tightly adheres to the seed. Examples of plants with achenial fruits include sunflowers (Helianthus) and dandelions (Taraxacum).

Nut: A nut is a hard-shelled fruit with a single seed enclosed within a stony or woody pericarp. Nuts are typically larger in size compared to achenes. Examples of plants with nut fruits include acorns from oak trees (Quercus) and chestnuts (Castanea).

Samara: A samara is a winged fruit in which the pericarp extends into a wing-like structure that aids in wind dispersal. The seed is located at the base of the wing. Examples of plants with samara fruits include maple trees (Acer) and ash trees (Fraxinus).

Caryopsis: A caryopsis, also known as a grain, is a type of fruit in which the pericarp is fused tightly to the seed, and the seed coat is not distinguishable. Caryopses are characteristic of grasses such as wheat, rice, and corn.

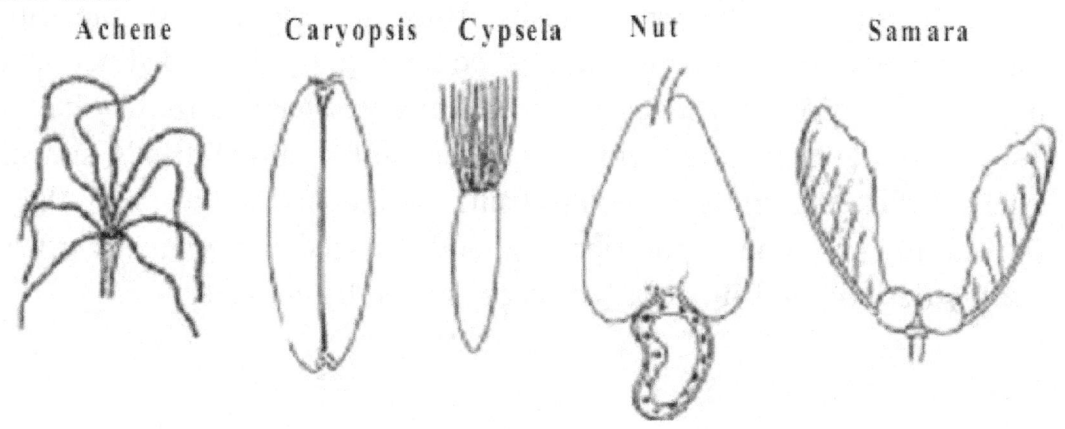

SCHIZOCARPIC FRUITS (SPLITTING FRUITS):

Schizocarpic fruits, also known as splitting fruits, are dry fruits that consist of multiple one-seeded parts that separate or split upon ripening. These fruits undergo a process called schizocarpy, where the fruit splits into individual segments or carpels, each containing a single seed. The two main types of schizocarpic fruits are mericarps and cocci, which differ in their dehiscence characteristics:

Mericarps: Mericarps are the one-seeded parts of a schizocarpic fruit that remain indehiscent, meaning they do not naturally split open to release the seed. Each mericarp is a distinct unit, often resembling a seed itself, and retains its individual identity after the fruit has split. Examples of plants with mericarp fruits include fennel (Foeniculum vulgare) and caraway (Carum carvi).

Cocci: Cocci are the one-seeded parts of a schizocarpic fruit that are dehiscent, meaning they naturally split open at maturity to release the seed. Unlike mericarps, which remain intact, cocci separate and disperse individually. Each cocci typically represents a carpel of the original fruit. Examples of plants with cocci fruits include plants from the Apiaceae family, such as coriander (Coriandrum sativum) and parsley (Petroselinum crispum).

SUCCULENT FRUITS (FLESHY FRUITS):

Succulent fruits, also known as fleshy fruits, are a type of simple fruit in which the pericarp, or fruit wall, is composed of three distinct layers: epicarp, mesocarp, and endocarp. These fruits are characterized by having a fleshy or juicy mesocarp, which is responsible for the fruit's succulent texture. Unlike dry fruits, succulent fruits are indehiscent, meaning they do not naturally split open upon ripening, and the seeds are typically released after the flesh of the fruit decays or is consumed by animals.

There Are 3 Types Of Succulent Fruits:

Drupes: Drupes exemplify succulent fruits wherein the mesocarp assumes the role of the edible portion, while the endocarp hardens into a shell or stone, encapsulating the seed within. Such fruits are recognized by the presence of a solitary seed enclosed by a tough, stony layer. Renowned examples of drupes encompass the illustrious mango (Mangifera indica), the versatile coconut (Cocos nucifera), the luscious peach (Prunus persica), and the nourishing almond (Prunus dulcis).

Pomes: Pomes represent a distinct group of succulent fruits that, despite their fleshy characteristics, are classified as false fruits. This distinction arises from the fact that pomes derive from the floral tube rather than the ovary. While the fleshy thalamus envelops the pome and serves as the edible portion, the true fruit resides within this structure. Prominent examples of pomes include the beloved apple (Malus domestica), the delectable pear (Pyrus communis), and the tantalizing quince (Cydonia oblonga).

Berries: Berries are succulent fruits endowed with a pericarp differentiated into three layers: the epicarp, mesocarp, and endocarp. Within the pulp formed by one or more of these layers, the seeds are nestled. True berries originate from a superior ovary, positioned above the attachment point of other floral parts. Notable instances of true berries encompass the esteemed grape (Vitis vinifera), the versatile tomato (Solanum lycopersicum), and the revered eggplant (Solanum melongena). Conversely, false berries arise from an inferior ovary, where the thalamus and pericarp unite. Illustrative examples of false berries include the iconic banana (Musa spp.) and the succulent guava (Psidium guajava).

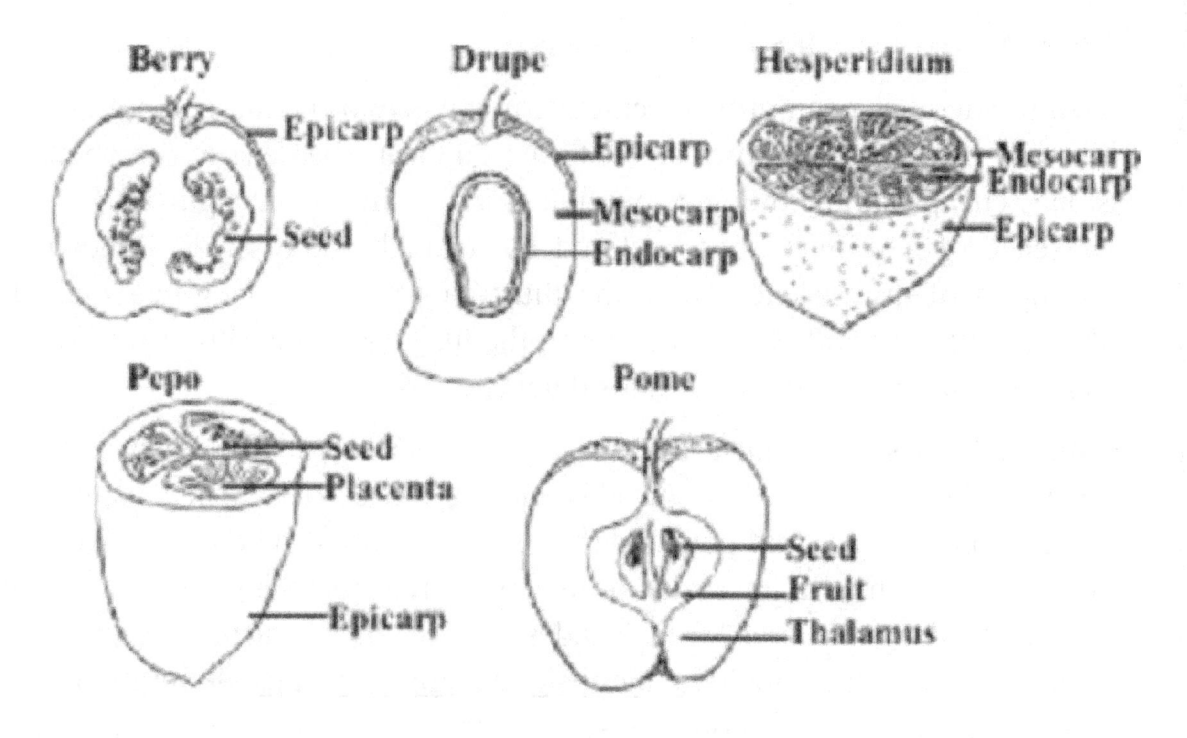

15.6.4 AGGREGATE FRUITS

Aggregate fruits represent a distinctive category within the classification of fruits. These fruits are derived from multiple ovaries within a single flower, giving rise to a unique structure known as an aggregate fruit. The aggregate fruit is composed of a cluster or aggregation of individual fruits, referred to as fruitlets, which collectively form a single cohesive unit.

Unlike other fruit types, aggregate fruits exhibit a fascinating complexity resulting from the fusion of multiple fruitlets. Each fruitlet originates from a separate ovary within the same flower, and together they contribute to the overall composition of the aggregate fruit. This distinctive feature sets aggregate fruits apart from other fruit types that develop from a single ovary.

Prominent examples of aggregate fruits include blackberries (Rubus spp.) and strawberries (Fragaria spp.). In the case of blackberries, the fruit is composed of a cluster of small drupelets, each originating from an individual ovary. Similarly, strawberries are characterized by multiple achenes, each representing a separate fruitlet that develops from an individual ovary and is arranged on the receptacle of the flower.

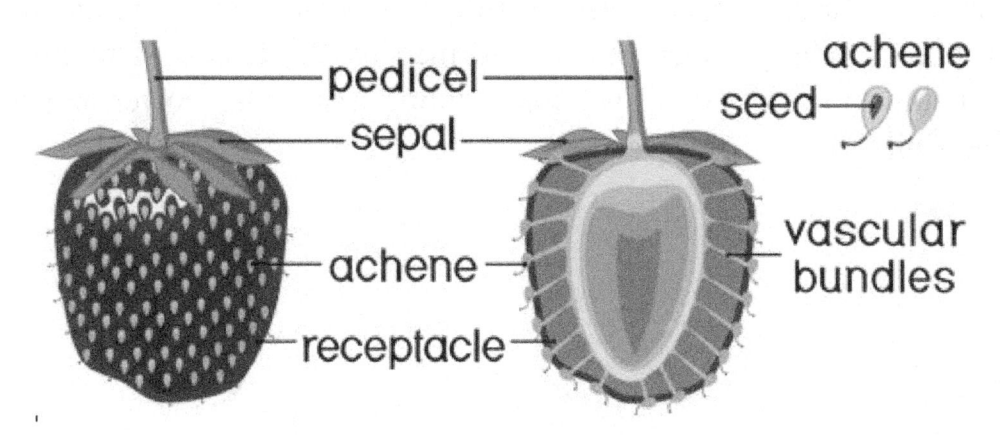

15.6.5   MULTIPLE FRUITS

0Multiple fruits, also known as false or composite fruits, are a unique category of fruits that result from the fusion of multiple individual flowers within an inflorescence. Unlike aggregate fruits, which develop from multiple ovaries within a single flower, multiple fruits are formed by the fusion of entire flowers, including their ovaries, receptacles, and other floral parts. This fusion gives rise to a single large fruit structure that incorporates the reproductive structures of multiple flowers.

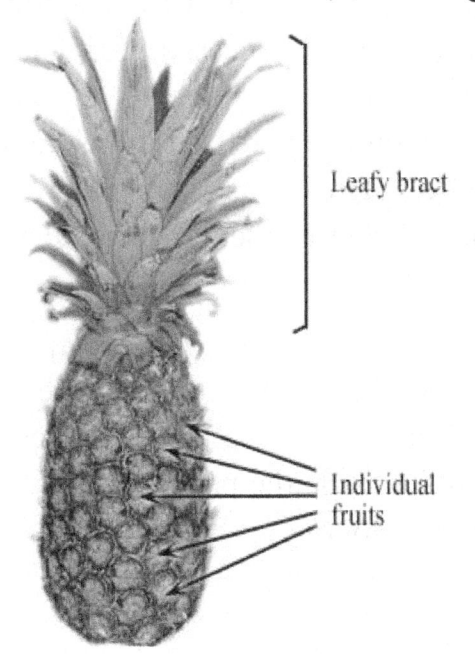

The process of multiple fruit formation begins with an inflorescence, which is a cluster of flowers grouped closely together. As the flowers undergo pollination and fertilization, their ovaries start to develop and enlarge. Over time, these individual fruits merge and coalesce, forming a single composite structure that appears as a large fruit.

Examples of multiple fruits include mulberries (Morus spp.) and pineapples (Ananas comosus). In the case of mulberries, the fleshy fruit is composed of numerous small drupes, each derived from a separate flower within the inflorescence. Similarly, pineapples are formed by the fusion of multiple individual flowers, with each "eye" representing a separate flower and its associated fruit.

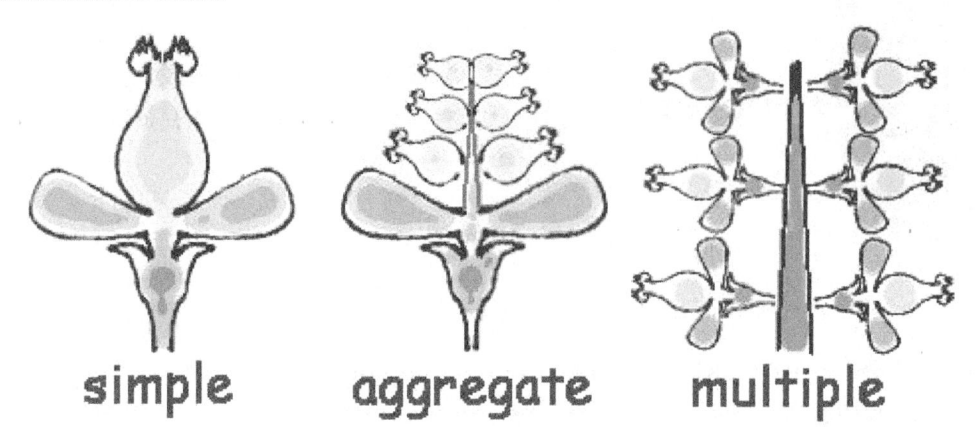

simple   aggregate   multiple

15.6.6 CLASSIFICATION OF FRUITS BASED ON FORMATION

True Fruits (Eucarps): True fruits, also known as eucarps, are formed from the mature or ripened ovaries of flowers after successful fertilization. The ovary undergoes various transformations and ripening processes, resulting in the development of a fruit. Examples of true fruits include mango, maize (corn), grape, and many others. True fruits play a vital role in seed dispersal and are commonly consumed due to their nutritional value and appealing flavors.

False Fruits (Pseudocarps): False fruits, also called pseudocarps, are derived from floral parts other than the ovary. The edible part of these fruits is not the ripened ovary itself but other floral structures that have become fleshy or enlarged. Examples of false fruits include the peduncle in cashew nut, thalamus (receptacle) in apple, pear, gourd, and cucumber, fused perianth (multiple flowers) in mulberry, and calyx in Dillenia. These structures mimic the appearance and function of true fruits but differ in their origin.

Parthenocarpic Fruits: Parthenocarpic fruits are seedless fruits that develop without fertilization. They are formed from unfertilized ovaries or other floral parts and do not require pollination or the presence of viable seeds. Banana is a well-known example of a parthenocarpic fruit. In modern horticulture, efforts have been made to develop seedless varieties of fruits such as grapes, oranges, and watermelons through parthenocarpy. Parthenocarpic fruits are valued for their convenience and consistent quality, as they do not contain seeds that can affect texture or taste. These fruits are often preferred for culinary purposes, as they eliminate the need for seed removal.

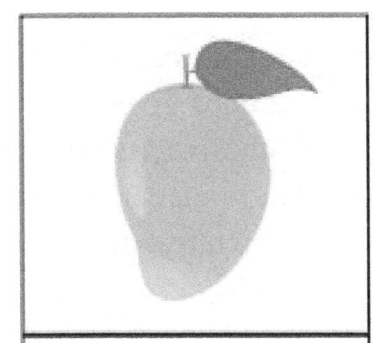

True fruit: Originates from ripened ovary	False fruit: Originates from flower's thalamus	Parthenocarpic fruit: Originates from ovary, but lacks seed

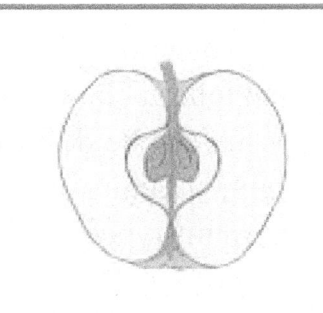

CHAPTER 16

PLANTS ANATOMY

16.1 WHAT IS PLANT ANATOMY?

Plant anatomy is a branch of botany that encompasses the study of the internal structure of various plant organs. It delves into the intricate details of plant cells, which serve as the fundamental building blocks of all living organisms, including plants. To investigate these structures, scientists typically employ high-powered light microscopes or electron microscopes. The examination of cells and their organization within tissues is known as histology.

Within a plant, cells of similar shape tend to aggregate and merge to form groups that fulfill a shared function. Each of these cell groups gives rise to a distinct tissue. Therefore, a tissue can be defined as a cohesive assembly of cells that share a common origin, structure, and function. In plants, the body is typically composed of two main types of tissues: vegetative tissue and reproductive tissue.

Vegetative tissue refers to the cells and tissues responsible for the growth, support, and maintenance of the plant. It includes various types of tissues, such as dermal tissue, ground tissue, and vascular tissue. Dermal tissue, also known as the epidermis, forms the protective outer covering of the plant. Ground tissue comprises the bulk of the plant body and performs functions such as photosynthesis, storage, and support. Vascular tissue, including xylem and phloem, facilitates the transport of water, minerals, and nutrients throughout the plant.

Reproductive tissue, as the name suggests, is involved in the production of plant offspring. It encompasses the cells and tissues responsible for sexual reproduction, such as flower structures, pollen grains, and ovules.

Reproductive tissue is essential for the formation of seeds and fruits, which are vital for plant propagation.

Tissues can be categorized into three groups:
- Meristematic
- Permanent
- Secretory

16.2 MERISTEMS OR MERISTEMATIC TISSUE

Meristematic tissue, also known as meristems, is a specialized group of cells that remain in a continuous state of division or retain their ability to divide. These meristematic cells are responsible for the growth and development of plants throughout their life cycle. Unlike most cells in mature plant tissues, which have specific functions, meristematic cells are undifferentiated and maintain the potential to differentiate into various specialized cell types.

The primary characteristic of meristematic tissue is its ability to undergo mitotic cell division, generating new cells. This perpetual division ensures the production of cells necessary for plant growth and the formation of new organs, such as leaves, stems, and roots. As meristematic cells divide, they produce two daughter cells: one retains its meristematic properties, while the other undergoes differentiation, becoming part of the plant's permanent tissues.

Meristematic tissue is typically found in regions called meristems, which serve as centers of cell division and growth. There are two main types of meristems in plants:

Apical Meristems: Located at the tips of stems and roots, apical meristems are responsible for primary growth. They enable the elongation of shoots and roots, allowing the plant to increase in length. Apical meristems give rise to primary meristems, which then develop into the various tissues and organs of the plant.

Lateral Meristems: Found in the lateral or peripheral regions of stems and roots, lateral meristems are responsible for secondary growth. They contribute to the increase in girth or thickness of the plant. The two main types of lateral meristems are vascular cambium, which produces secondary xylem and phloem for increased transport capacity, and cork cambium, which produces the protective cork tissue in stems and roots.

16.2.1 CHARACTERISTICS OF MERISTEMATIC TISSUES

Meristematic tissues possess several distinct characteristics that set them apart from other plant tissues. These characteristics contribute to their role in continuous growth and development.

Immature and Dividing Cells: Meristematic tissues consist of immature cells that are in a state of continuous division and growth. These cells have not yet specialized into specific cell types and retain their ability to divide actively. Their division provides a constant supply of new cells for plant growth and the formation of new tissues.

Absence of Intercellular Spaces: Unlike mature plant tissues, meristematic tissues lack intercellular spaces. The cells are densely packed together, facilitating close communication and coordination between neighboring cells. This compact arrangement ensures efficient exchange of signals and nutrients during growth and development.

Living and Thin-Walled Cells: Cells in meristematic tissues are generally living and have thin cell walls. This characteristic allows for flexibility and ease of cell division. The cells exhibit varying shapes, including round, oval or polygonal, depending on the specific meristematic region.

Abundant Cytoplasm and Nuclei: Each cell within a meristematic tissue possesses abundant cytoplasm, containing essential cellular components and organelles. Furthermore, they typically contain one or more nuclei, which are responsible for genetic information and cellular processes.

Small or Absent Vacuoles: The vacuoles within meristematic cells may be small in size or completely absent. Vacuoles, which are fluid-filled organelles, play a role in cellular storage, maintaining turgor pressure, and various physiological functions. The reduced presence or absence of vacuoles allows meristematic cells to maximize their space for cytoplasm and cell division.

Differentiation from Animal Cells: The presence of meristematic tissue in plants distinguishes them from animals. Meristematic tissues enable plants to undergo continuous growth throughout their lifespan, a characteristic not observed in animals. The perpetual division and growth potential of meristematic cells contribute to the unique growth patterns and regenerative abilities exhibited by plants.

Localization of Cell Division: During the early stages of plant embryo development, cell division occurs throughout the young organism. However, as the embryo matures and develops into an independent plant, the addition of new cells becomes gradually restricted to specific regions of the plant body. Other parts of the plant are engaged in activities other than growth, such as photosynthesis or reproduction.

Occurrence at Apices: Meristems are predominantly found at the apices (tips) of both main and lateral shoots and roots. This localization results in a large number of meristematic regions within a single plant. The collective activities of these meristems contribute to the expansion of the plant body, as well as the production of reproductive structures.

16.2.2 CLASSIFICATION OF MERISTEMS

Meristems, the regions of active cell division in plants, can be classified based on various factors such as their stage of development, position in the plant body, origin, function, and topography.

1. MERISTEMS BASED ON STAGE OF DEVELOPMENT

This classification categorizes meristems based on their developmental stage and their role in initiating the formation of new organs or parts of organs.

Promeristem or Primordial Meristem: Also known as urmeristem or embryonic meristem, this is the initial region in the plant body where the foundation of new organs or their parts is initiated. The cells within this region are called initials and their immediate derivatives. At this stage, the cells of the promeristem are actively dividing and maintaining their undifferentiated state. However, as these cells begin to change in size, shape and characteristics of the cell wall and cytoplasm, initiating tissue differentiation, they are no longer considered part of a typical meristem. They have transitioned to a different developmental stage.

2. MERISTEMS BASED ON ORIGIN OF INITIATING CELLS

Meristems can also be classified based on the origin of the initiating cells. This classification distinguishes between primary and secondary meristems, which differ in their developmental timing and location within the plant body.

a. **Primary Meristems:** Primary meristems are responsible for building up the primary parts of the plant and include the promeristem mentioned earlier They are located in the apical regions of roots, stems, leaves, and similar appendages. The primary meristems initiate growth and development during the early stages of plant formation. They contribute to the structural and functional completion of the plant body. These meristems give rise to the primary tissues that make up the essential organs of the plant such as the epidermis, cortex, pith, and vascular tissues.

b. **Secondary Meristems:** Secondary meristems, on the other hand, appear at a later stage of organ development within the plant body. They originate from mature or permanent tissues and are found in lateral positions, adjacent to the stems or roots. Unlike primary meristems, secondary meristems arise within existing tissues and often play a role in secondary growth and tissue modification. Some of the tissues that give rise to secondary meristems are primary permanent tissues that acquire the ability to divide and become meristematic. One example of a secondary meristem is the phellogen or cork cambium. It is formed from matured cortical, epidermal, or phloem cells.

The primary meristems contribute to the early structural and functional establishment of the plant body, while the secondary meristems are added later and contribute to the formation of supplementary tissues. These supplementary tissues may replace or modify the functions of early-formed tissues or serve in the protection and repair of wounded regions. The distinction between primary and secondary meristems highlights the dynamic nature of plant development and the continuous adaptation of plants to their environment.

3. MERISTEMS BASED ON POSITION IN PLANT BODY

Meristems can also be classified based on their position within the plant body. This classification distinguishes between three groups: apical meristem, intercalary meristem, and lateral meristem.

a. Apical Meristem: Apical meristems are located at the apices (tips) of the stems, roots, and leaves of vascular plants. These meristems play a crucial role in the primary growth of the plant. Their activities result in an increase in length of plant organs. The cells within apical meristems, known as apical cells or apical initials, maintain their individuality and specific position within the meristem. The apical meristems are responsible for the elongation of the plant body and the formation of new tissues during primary growth.

b. Intercalary Meristem: Intercalary meristems are portions of the apical meristems that become separated from the apex during development, as layers of more mature or permanent tissues form and progress in growth.

These meristems are typically found in the internodal regions of stems and leaves, particularly in grasses and other monocotyledonous plants. Intercalary meristems contribute to the elongation and growth of these plant parts, allowing for continuous and indeterminate growth even after the apical meristems have moved on.

c. Lateral Meristem: Lateral meristems are composed of cells that contribute to the increase in diameter, or thickness, of an organ. They add to the bulk of existing tissues or give rise to new tissues responsible for growth in girth of the plant body. The primary lateral meristem is the cambium, which is responsible for secondary growth in woody plants. The cambium produces secondary tissues, such as secondary xylem (wood) and secondary phloem. Another example of a lateral meristem is the cork cambium, which gives rise to cork cells for the formation of the protective outer bark.

4. MERISTEMS BASED ON FUNCTION

Meristems can also be classified based on their function within the plant body. This classification highlights the specific roles of meristems in giving rise to different types of tissues and organs.

a. Protoderm: The primary meristem located at the apex of the stem and root, known as the protoderm, is responsible for the formation of the epidermis. The protoderm gives rise to the outermost tissue layer of the plant, which serves as a protective barrier against environmental stresses and regulates water loss through specialized structures such as stomata.

b. Procambium: Another primary meristem at the apex of the stem and root is the procambium. The procambium is responsible for the development of primary vascular tissues, including the primary xylem and primary phloem. These tissues play essential roles in water and nutrient transport within the plant.

c. Ground Meristem: The ground meristem, also known as the fundamental meristem, is the primary meristem that develops into the ground tissue of the plant. Ground tissue refers to the parenchyma, collenchyma, and sclerenchyma cells that make up the bulk of the non-specialized tissue in plants. Additionally, the ground meristem contributes to the formation of the pith, which is the central region of the stem or root.

By classifying meristems based on their function, we can understand how different meristematic tissues contribute to the formation of specific types of tissues and organs. The protoderm gives rise to the protective epidermis, the procambium forms the vascular tissues, and the ground meristem develops into the ground tissue and pith.

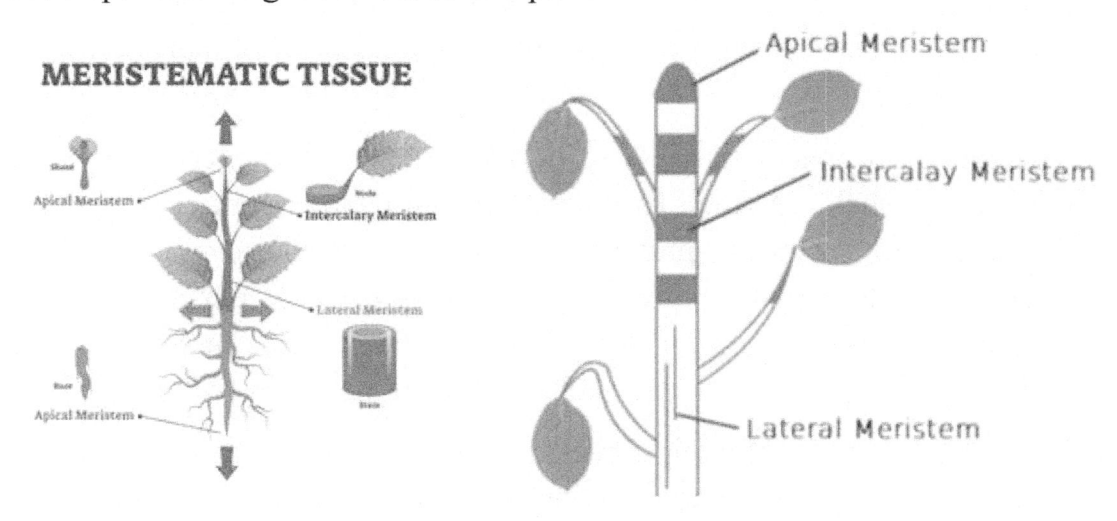

16.3 PERMANENT TISSUES

Permanent tissues in plants are composed of cells that have completed their division and have attained their final form and size. These tissues no longer possess the ability to divide further. Permanent tissues can exhibit various characteristics such as being living or dead, thin-walled or thick-walled, and can be categorized as either primary or secondary.

The primary permanent tissues are derived from the apical meristems located at the tips of stems and roots. These tissues are initially formed from the divisions of cells in the apical meristem. As these cells undergo differentiation and specialization, they gradually assume distinct forms and give rise to different types of permanent tissues. Examples of primary permanent tissues include the epidermis, cortex, pith, primary xylem, and primary phloem.

Secondary permanent tissues are derived from lateral meristems. Lateral meristems, such as the vascular cambium and cork cambium, are responsible for secondary growth in plants. The vascular cambium produces secondary xylem (wood) towards the inside of the stem and secondary phloem towards the outside, contributing to the increase in stem girth. The cork cambium gives rise to cork cells that form the protective outer bark.

In the early stages of their development, cells in permanent tissues are relatively similar in structure. However, as they undergo differentiation and specialization, they assume various forms and functions, contributing to the diversity of permanent tissues in plants. The differentiation of permanent tissues allows for the division of labor within the plant, enabling different tissues to perform specific functions such as support, conduction, storage, and protection.

Apical meristems, located at the tips of shoots and roots in all vascular plants, play a crucial role in plant growth and development. They give rise to three types of primary meristems, which in turn produce the mature primary tissues of the plant.

The mature primary tissues derived from apical meristems are classified into three categories: dermal, vascular, and ground tissues.

Dermal Tissues (Epidermis): The primary dermal tissue, known as the epidermis, forms the outer layer of all plant organs, including stems, roots, leaves, and flowers. The epidermis serves as a protective barrier against excessive water loss, physical damage, and invasion by insects and microorganisms. It consists of a single layer of tightly packed cells that may contain specialized structures like trichomes (hair-like structures) and stomata (pores for gas exchange).

Vascular Tissues: The vascular tissues are responsible for the transport of water, nutrients, and organic substances throughout the plant. They include two types: xylem and phloem.

Xylem: The primary xylem is involved in water and mineral transport from the roots to the aerial parts of the plant. It is composed of specialized cells called tracheary elements, which include vessel elements and tracheids. These cells form long tubes or interconnected networks for efficient water movement.

Phloem: The primary phloem is responsible for the transport of organic compounds, such as sugars and nutrients, from photosynthetic tissues (source) to the growing and storage tissues (sink) of the plant. It consists of sieve tube elements and companion cells, which form a system of sieve tubes for efficient nutrient transport.

Ground Tissues: The ground tissues constitute the bulk of the plant's internal tissue, excluding the dermal and vascular tissues. They serve various functions such as support, storage, and photosynthesis. The three types of ground tissues are parenchyma, collenchyma, and sclerenchyma.

Permanent tissues are classified as simple and complex tissues according to the number of cell types that they comprise.

Simple Tissues: Simple tissues are composed of a single type of cells that form a homogenous or uniform mass. They work collectively to perform specific functions in the plant. The common simple tissues are parenchyma collenchyma and sclerenchyma. They are also known as the ground tissues.

Complex Tissues: Complex tissues are composed of more than one type of cell working together as a functional unit. The two main types of complex tissues are the xylem and the phloem, also known as the vascular tissues. The common examples of complex tissues are the xylem and the phloem also known as the vascular tissues. These tissues consist of parenchymatous and sclerenchymatous cells but not collenchymatous cells.

SIMPLE TISSUES (GROUND TISSUES)

1. PARENCHYMA

Parenchyma is a type of simple permanent tissue composed of living cells that exhibit a wide range of morphological and physiological characteristics The cells of parenchyma have thin cell walls and a polyhedral shape, often resembling a cube or irregularly shaped polygon. They are responsible for various vegetative activities within the plant. Parenchyma tissue is widely distributed throughout the plant, occurring in the soft parts such as the leaves, stems, roots, and fruits. It serves several functions essential to plant growth and development. These functions include photosynthesis, storage of nutrients, secretion, and gas exchange. Parenchyma cells contain chloroplasts, enabling them to carry out photosynthesis and contribute to the production of carbohydrates.

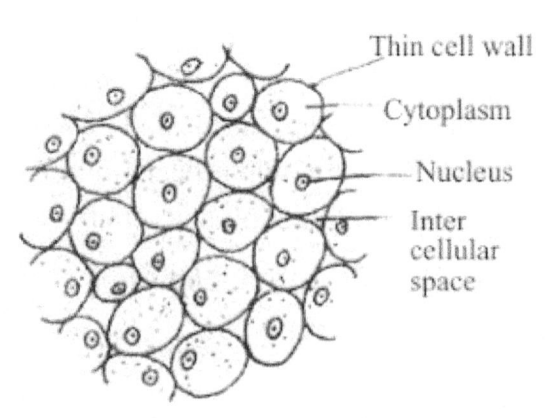

The cell walls of parenchyma cells are relatively thin, allowing for efficient exchange of materials between adjacent cells. They also have a large central vacuole that aids in storage of water, minerals, and other substances. Additionally, parenchyma cells can undergo cell division and dedifferentiation, enabling them to regenerate and differentiate into other cell types when necessary.

FUNCTIONS OF PARENCHYMA

a. Storage of food material: Parenchyma cells can store various types of organic substances, such as starch, proteins, oils, and sugars. These stored materials serve as a source of energy and nutrients for the plant during times of need, such as during periods of dormancy or when resources are limited.

b. Manufacturing of food material: Parenchyma cells containing chloroplasts, known as chlorenchyma, are responsible for photosynthesis. They utilize sunlight, carbon dioxide, and water to produce glucose and other organic compounds, which serve as a source of energy for the plant. Chlorenchyma is primarily found in green tissues, such as the mesophyll cells of leaves.

c. Buoyancy and gaseous exchange: In submerged water plants, parenchyma cells can form a specialized type of tissue called aerenchyma. Aerenchyma consists of parenchyma cells with large air spaces between them. This tissue provides buoyancy to the plant, allowing it to float in water. The air spaces within the aerenchyma tissue also facilitate gaseous exchange between the plant and its surroundings, ensuring an adequate supply of oxygen for respiration.

d. Wound healing and regeneration: Parenchyma cells have the remarkable ability to undergo cell division and dedifferentiation, allowing them to contribute to wound healing and regeneration in plants. When a plant is injured or wounded, parenchyma cells can undergo mitotic division and differentiate into specific cell types to repair and regenerate the damaged tissues. In this way, parenchyma plays a crucial role in the recovery and survival of the plant.

2.   COLLENCHYMA

Colenchyma is a living tissue composed of elongated cells with thick primary non-lignified walls. These cells are typically thickened at the corners adjacent to the intercellular spaces. Collenchyma tissue can be found in various parts of plants such as stems, leaves, floral parts, and fruits. However, in many monocotyledonous plants, collenchyma is absent in stems and leaves and is instead replaced by sclerenchyma, which develops at an early stage. In roots, collenchyma cells may develop when they are exposed to light.

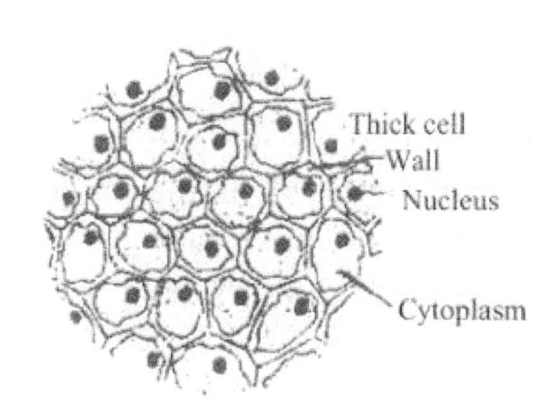

There are three main types of collenchyma:

Angular collenchyma: In this type, the cells have thickened corners, giving them an angular appearance. The cell walls are unevenly thickened, providing mechanical support and flexibility to the plant.

Lamellar collenchyma: Lamellar collenchyma cells have thickened walls that are arranged in concentric layers or lamellae. These cells often occur in cylindrical strands, providing support to the plant while allowing some flexibility.

Lacunar collenchyma: Lacunar collenchyma consists of elongated cells with irregularly thickened walls, creating intercellular spaces or lacunae. These spaces are filled with air and provide additional buoyancy to the plant particularly in aquatic or floating plants.

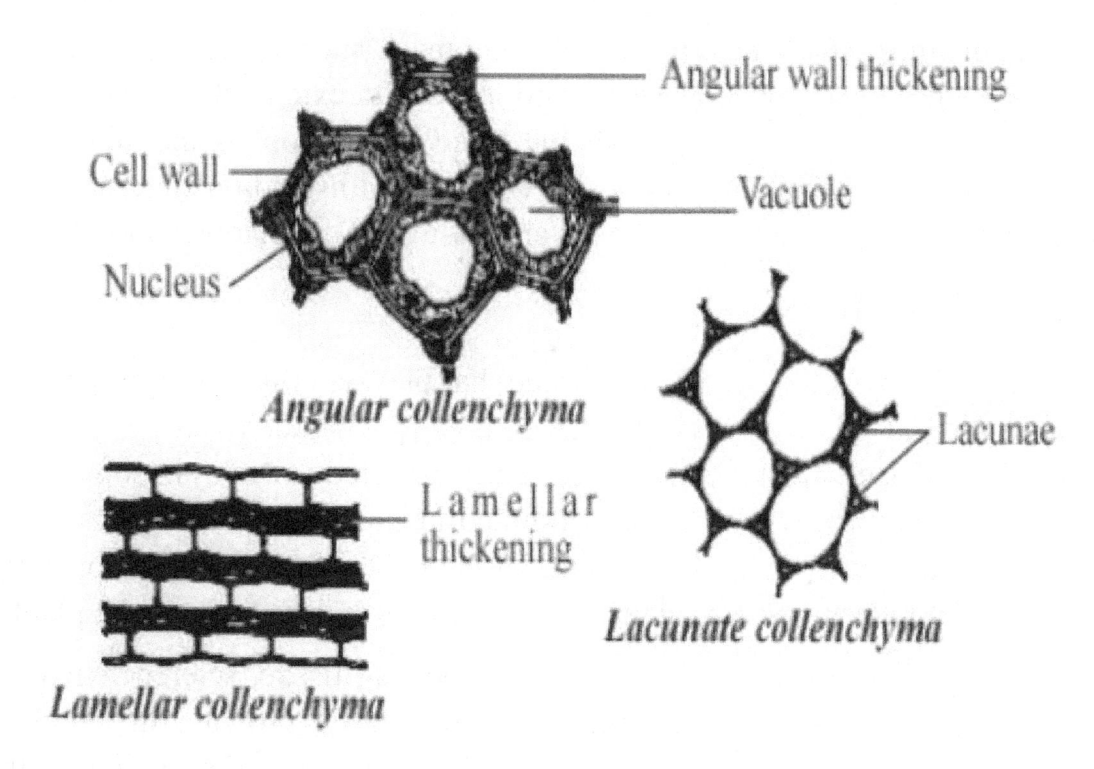

Angular collenchyma — Cell wall, Nucleus, Angular wall thickening, Vacuole

Lamellar collenchyma — Lamellar thickening

Lacunate collenchyma — Lacunae

THE FUNCTIONS OF COLLENCHYMA

a. Providing mechanical support: Collenchyma plays a crucial role in providing mechanical support to growing and expanding stems and leaves. Its elongated cells with thickened walls help resist bending and the pulling action of wind, thereby preventing the plant from collapsing or breaking under external forces. This support is particularly important in young, actively growing plant tissues.

b. Participation in photosynthesis: In certain cases, collenchyma cells may contain chloroplasts, the organelles responsible for photosynthesis. When chloroplasts are present in collenchyma cells, they contribute to the process of photosynthesis, converting light energy into chemical energy and producing sugars and oxygen. This allows the collenchyma tissue to actively participate in the manufacturing of food for the plant.

c. Providing tensile strength: Collenchyma tissue imparts tensile strength to growing organs, such as stems and leaves. Its extensible nature allows it to adapt and stretch along with the rapid elongation of the stem during growth. This tensile strength is vital in withstanding internal forces generated by the expansion of cells and tissues as the plant develops.

3. SCLERENCHYMA

Sclerenchyma is a type of plant tissue that plays a crucial role in providing strength and rigidity to the plant. Its cells are characterized by their long, narrow shape, thick cell walls, and lignified (woody) composition. Sclerenchyma cells are typically pointed at both ends and have a fibrous appearance, earning them the name "sclerenchymatous fibres" or simply "fibres." These fibres contribute to the hardness and stiffness of the plant.

Unlike collenchyma, another type of supporting tissue, sclerenchyma cells undergo differentiation after the plant has ceased growing in length. As sclerenchyma cells mature, their protoplasts (living contents) disintegrate, resulting in the cells becoming dead. This characteristic distinguishes sclerenchyma from collenchyma, which retains its living status even at maturity.

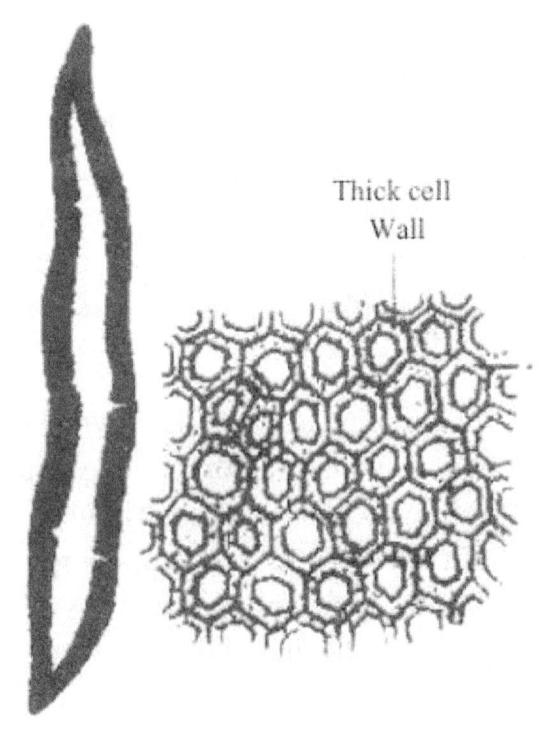

Sclerenchyma cells exhibit variations in shape, structure, origin, and development. The shape of these cells may differ depending on their specific function within the plant. For example, some sclerenchyma cells may be elongated and spindle-shaped, while others can be shorter and more irregularly shaped. Additionally, the structure of sclerenchyma cells is notable for their thick, secondary cell walls that contain lignin, a complex polymer that confers strength and rigidity.

The origin and development of sclerenchyma cells can occur through various processes. Some sclerenchyma cells originate from the meristematic tissues, which are regions of actively dividing cells in plants. These cells undergo elongation and lignification as they differentiate into mature sclerenchyma cells. Other sclerenchyma cells may arise from the transformation of other cell types, such as parenchyma or collenchyma cells, through a process known as metasclereids. This diversity in origin and development contributes to the versatility of sclerenchyma in providing structural support to different plant parts.

Sclerenchyma can be further classified into two main types: fibres and sclereids.

FIBRES: are characterized by their elongated shape, with long and narrow cells that possess sharply tapering end walls. These fibres are distributed throughout various parts of the plant body. While they are commonly observed among vascular tissues, they can also be well-developed within the ground tissues. Fibres often occur in groups or bundles, contributing to their role in providing structural support. They can be further categorized into two basic types based on their location within the plant body: xylary fibres and extraxylary fibres.

Xylary fibres are predominantly found in the xylem, the vascular tissue responsible for conducting water and minerals from the roots to the rest of the plant. Xylary fibres play a vital role in strengthening the xylem and providing mechanical support to the plant's water transport system.

Extraxylary fibres, on the other hand, are located outside the xylem and can be present in various plant tissues. They are commonly found in the ground tissues, which include parenchyma, collenchyma, and sclerenchyma cells. Extraxylary fibres contribute to the overall rigidity and strength of the plant, aiding in its ability to withstand mechanical stress and maintain an upright posture.

SCLEREIDS:, the other type of sclerenchyma cells, exhibit a diverse range of shapes and can be categorized as isodiametric, irregular, or branched. They are present in various positions within the plant body, serving different functions depending on their location. Sclereids are frequently observed as individual cells or in groups in the cortical (outer) and pith (central) regions of many dicotyledonous plants. Additionally, they can be found in the phloem and xylem tissues, as well as the hard shells of fruits and the tough coats of seeds. The classification of sclereids is based on the form and structure of the cells. There are four main types of sclereids:

a) **Brachysclereids:** These sclereids have a relatively short and thick shape, often with rounded ends. They are commonly found in the seed coats and fruit pulp, providing mechanical strength and protection.

b) **Macrosclereids:** Macrosclereids are elongated and slender sclereids. They can occur in various plant parts such as the cortex, pith, and vascular bundles. Macrosclereids contribute to the structural integrity of the tissues they are present in.

c) **Osteosclereids:** Osteosclereids are sclereids with a star-like or branched shape. They are typically found in the cortex and pith of stems, leaves, and fruits. Osteosclereids reinforce these plant organs and provide support against mechanical stresses.

d) **Astrosclereids:** Astrosclereids are characterized by their irregular and star-shaped appearance, resembling astrocyte cells in the nervous system. They are commonly found in the epidermis and cortex of leaves and stems, where they enhance tissue strength and help protect against herbivory and physical damage.

e) **Trichosclereids:** Trichosclereids are a specific type of sclereid, which are sclerenchyma cells with a specialized shape and function.

Trichosclereids are characterized by their elongated and hair-like structure, often resembling tiny hairs or bristles. They are found in various plant organs, such as the epidermis, leaves, stems, and fruits.

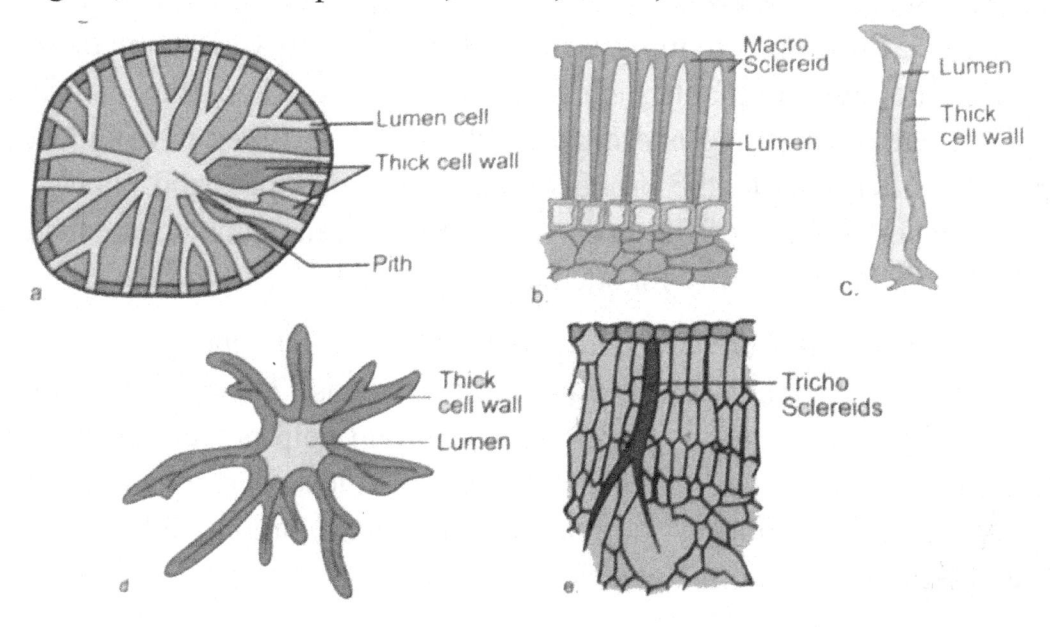

FUNCTIONS OF SCLERENCHYMA

a. Mechanical Support and Protection: Sclerenchyma cells, including both fibres and sclereids, contribute to the overall mechanical support and rigidity of mature plants. Their long, fibrous nature and thick cell walls help provide structural strength, enabling plants to withstand mechanical stresses such as wind, gravity, and external pressures. Sclerenchyma cells also offer protection to delicate plant tissues by forming strong coats around seeds, as well as hard shells or husks in fruits.

b. Strengthening of Cell Walls: The secondary walls of sclerenchyma cells are notably thickened and often impregnated with lignin, a complex polymer. Lignin deposition strengthens the cell walls, making them more rigid and resistant to deformation. Additionally, lignin imparts waterproofing properties, allowing sclerenchyma cells to withstand water-related stresses and protect underlying tissues from excessive water loss or absorption.

c. Fibre Formation: Certain types of sclerenchyma cells, known as fibres, play a crucial role in providing structural integrity to various plant parts. Fibres can be found in plants such as hemp and flax, where they contribute to the formation of strong and durable fibers. These fibres have industrial applications, such as in the production of textiles, ropes, and paper. The long and tapered shape of fibres derived from sclerenchyma cells enhances their strength and suitability for such uses.

d. Seed Coat Formation: Sclereids, another type of sclerenchyma cells, are responsible for forming strong coats around seeds. These coats provide mechanical protection to the developing seed and help prevent desiccation or damage. The tough and durable nature of sclereids ensures that the seeds are well-protected during dormancy and germination, allowing for successful propagation of plant species.

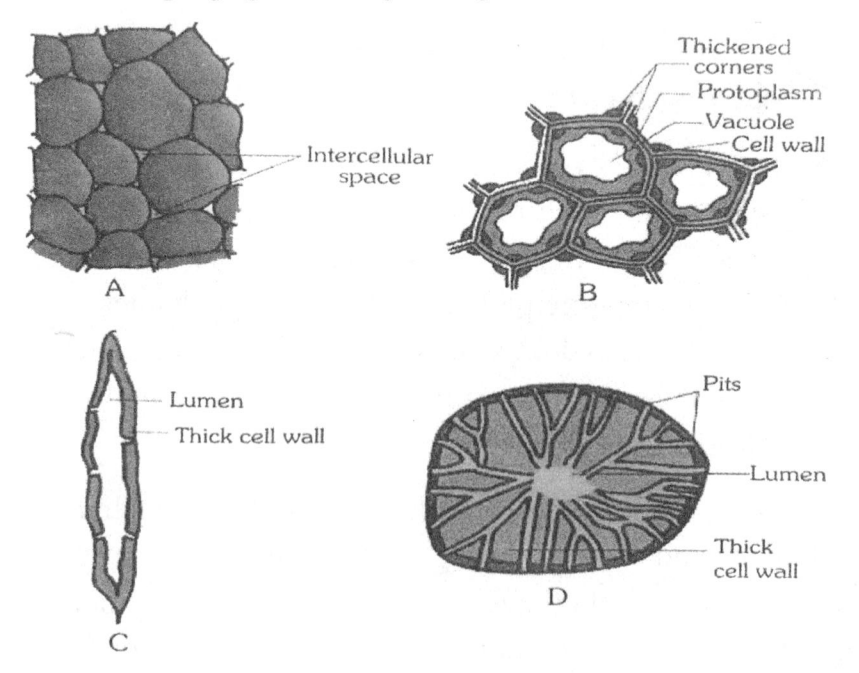

A-Collenchyma, B-Parenchyma, C-Fibre (Sclerenchyma), D-Sclereid (Sclerenchyma)

COMPLEX TISSUES

XYLEM

Xylem is a complex vascular tissue responsible for the upward conduction of water and mineral nutrients in plants. It plays a crucial role in the transport of these substances from the roots to the leaves, where they are utilized for various physiological processes.

The primary function of xylem is the long-distance transport of water, facilitated by a process called transpiration. Transpiration occurs when water evaporates from the leaf surfaces through tiny pores called stomata. This creates a negative pressure gradient, or tension, which pulls water upward through the xylem from the roots. This continuous flow of water is known as the transpiration stream. It's also transports mineral nutrients absorbed by the roots from the soil. These minerals, dissolved in the water, are carried along with the transpiration stream and delivered to the different parts of the plant where they are required for growth, metabolism, and other essential functions.

Xylem also provides mechanical support to the plant body. The xylem vessels and tracheids, the two main types of cells in xylem, have thick, lignified secondary cell walls that confer strength and rigidity. The arrangement of these cells in long, interconnected tubes or vessels adds structural support to the plant, preventing collapse under its own weight and resisting mechanical stresses caused by wind or other external forces.

The xylem is composed of five different kinds of elements. They are: Tracheids, Fibres and fibre-tracheids, Vessels or tracheae, wood fibres and wood parenchyma.

1. TRACHEIDS:

Tracheids are a fundamental cell type found in xylem tissue. They are elongated, tube-like cells that have tapering, rounded, or oval ends.

Compared to other xylem cells, the walls of tracheids are not heavily thickened or lignified. At maturity, tracheids lack a protoplast, meaning they are non-living cells. The cell cavity or lumen of a tracheid is relatively large and empty, allowing for the efficient transport of water and dissolved nutrients.

Tracheids possess various types of thickenings within their cell walls, which can be classified as annular (ring-like), spiral, scalariform (ladder-like), reticulate (net-like), or pitted tracheids. These different patterns of thickenings provide additional strength and support to the cell walls. In certain plant groups, such as ferns and gymnosperms, tracheids alone make up the xylem tissue. However, in angiosperms (flowering plants), tracheids are found in conjunction with other xylem elements, such as vessels.

Tracheids are important in the conduction of water and dissolved nutrients through the xylem tissue, contributing to the upward movement of water from the roots to the aerial parts of the plant. They also provide support to the plant body.

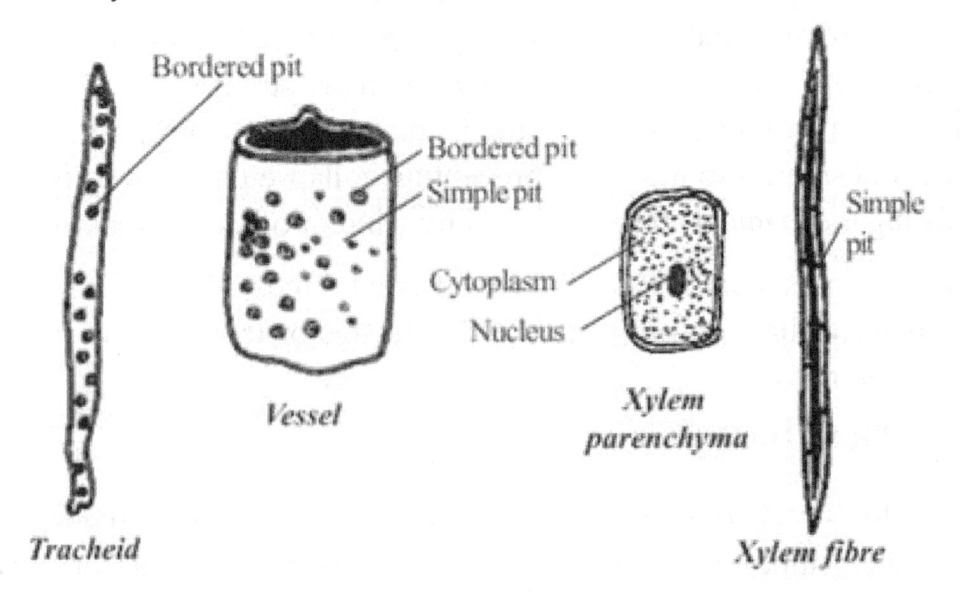

2. FIBRES AND FIBRE-TRACHEIDS:

The development of fibres in plants follows a distinct phylogenetic pattern. As the fibre develops, the thickness of its cell wall increases, while the diameter of the cell lumen decreases. This results in a narrowing of the cell and a reduction in the overall length of the fibre. Furthermore, the number and size of pits found on the cell walls also decrease as the fibre matures.

In some cases, the lumen of the fibre can become so narrow and the pits so small that the conduction of water through these cells is greatly reduced or completely blocked. This specialization towards reduced water conduction leads to the formation of fibres.

Within the progression from tracheids to fibres, there exist transitional forms known as fibre-tracheids. These cell types exhibit characteristics that are intermediate between typical tracheids and fibres. They share some features of tracheids, such as larger lumens and more abundant and larger pits, while also displaying some characteristics of fibres, such as thicker cell walls. The evolutionary trend from tracheids to fibre-tracheids and ultimately to fibres highlights the gradual shift towards cells with thicker walls and reduced water-conducting capabilities. This transition is associated with an increasing emphasis on mechanical support rather than active water transport.

3. VESSELS:

Vessels are a specialized type of conducting element found in the xylem tissue of plants. They are formed from procambium cells or derivatives of cambium during the final stages of development. The key feature of vessels is their significantly larger diameter compared to tracheids, another type of xylem cell.

Vessels function as long tubes that facilitate the efficient transport of water from the roots to the leaves. Their large diameter allows for a higher volume of water to be transported through the xylem, increasing the overall efficiency of water conduction in plants.

While vessels are a characteristic feature of angiosperms (flowering plants), it's important to note that there are exceptions. Certain angiospermic families, such as Winteraceae, Trochodendraceae, and Tetracentraceae, lack vessels in their xylem. In many monocotyledonous plants (monocots), vessels are absent from the stems and leaves. Instead, monocots rely on other xylem elements, such as tracheids, to transport water and nutrients. This absence of vessels in monocots is a distinctive feature that distinguishes them from dicotyledonous plants (dicots) where vessels are commonly present.

4. WOOD FIBRES:

Wood fibres are specialized cells found in the xylem tissue of woody plants. They are elongated, spindle-shaped cells with tapered ends and are primarily responsible for providing mechanical strength and support to the plant. Wood fibres, also known as libriform fibres, are derived from the differentiation and maturation of cambial cells in the secondary xylem.

The walls of wood fibres are thick and reinforced with lignin, a complex polymer that enhances their structural integrity. The lignin deposition in the cell walls increases their rigidity and resistance to compression and bending forces, making wood fibres an essential component of the plant's structural framework.

Wood fibres are interconnected through pits, which are small openings in the cell walls that allow for the lateral movement of water and nutrients. This interconnected network facilitates the efficient transport of water and dissolved substances within the wood. They also play a role in water conduction and storage. The interconnectedness of wood fibres allows for the movement of water between cells, aiding in the transport of water from the roots to the leaves. The cell lumens of wood fibres can store water, contributing to the plant's ability to withstand periods of drought or water scarcity.

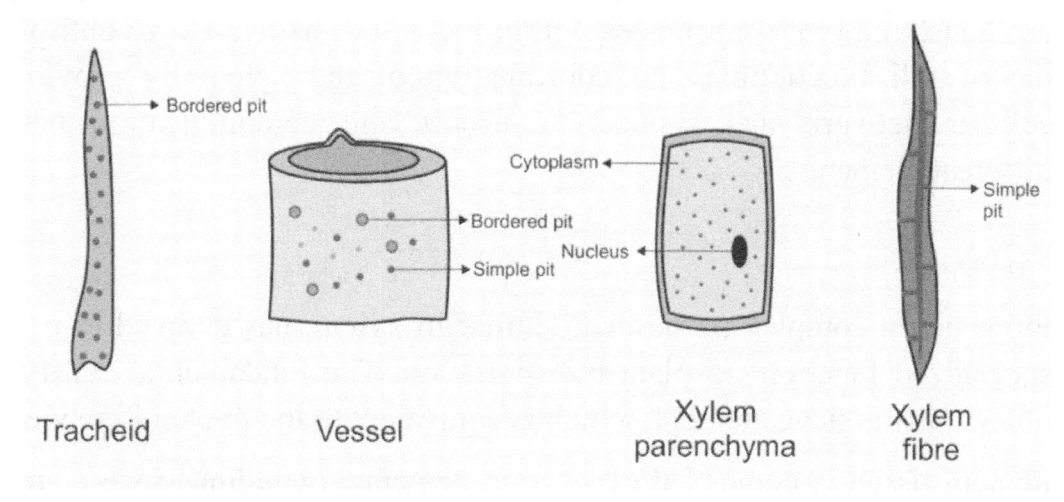

Tracheid — Vessel — Xylem parenchyma — Xylem fibre

5. WOOD PARENCHYMA:

Wood parenchyma, also referred to as xylem parenchyma, is a type of plant tissue that is found in the xylem of woody plants. It serves multiple functions within the plant.

One notable function of wood parenchyma is the storage of food reserves. These cells contain abundant cytoplasm and can accumulate starch or fats as energy reserves. The stored starch and fats can be utilized by the plant during periods of growth, development, or energy demand. The cells can also contain other substances such as tannins and crystals. Tannins are organic compounds that contribute to the defense mechanisms of plants, protecting them against herbivores and pathogens. Crystals, on the other hand, may serve various functions including calcium storage or mechanical support.

Wood parenchyma cells also play a role in the conduction of water within the xylem. While vessels and tracheids are the primary water-conducting elements, wood parenchyma cells assist in this process directly or indirectly. They can facilitate the movement of water between adjacent vessels or tracheids by creating pathways or promoting water transfer through their interconnected cell walls. They can also participate in the exchange of gases nutrients, and other substances within the xylem tissue. They contribute to the overall functioning and coordination of the xylem by providing a cellular network that supports transport and communication between different elements.

PHLOEM

Phloem is a complex tissue that, similar to xylem, has evolved to perform specialized functions in plant transportation. The fundamental cell type in phloem is the sieve element, which is equivalent to the tracheid in xylem.

Phloem tissue is composed of several elements, including **sieve elements, companion cells, phloem fibres, and phloem parenchyma**. These elements work together to facilitate the transport of organic nutrients, such as sugars, from the site of production (source) to the sites of utilization or storage (sinks) in the plant.

In pteridophytes (ferns) and gymnosperms (cone-bearing plants), the phloem tissue consists of sieve cells and phloem parenchyma. Sieve cells are the main conducting cells in these groups, responsible for the translocation of nutrients. Phloem parenchyma provides support and performs metabolic functions within the tissue.

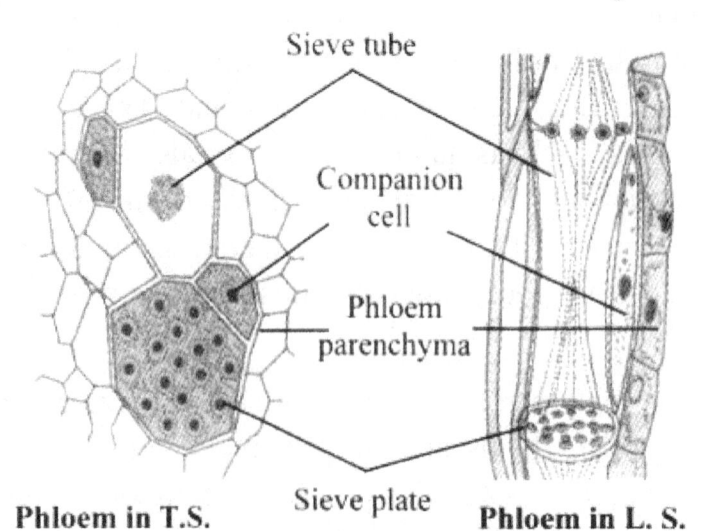

In some gymnosperms, such as conifers, additional elements are

Phloem tissue

present in the phloem tissue. Along with sieve cells and phloem parenchyma, phloem fibres are also found. Phloem fibres are elongated cells that provide mechanical support to the phloem tissue, assisting in maintaining its structural integrity.

In angiosperms (flowering plants), the phloem tissue is more complex and consists of multiple cell types. In addition to sieve elements and phloem parenchyma, angiosperms possess companion cells, phloem fibres, sclereids, and secretory cells.

1. SIEVE ELEMENTS:

Sieve elements are the specialized cells responsible for conducting organic nutrients, such as sugars, through the phloem tissue. There are two main types of sieve elements: sieve cells and sieve tube elements.

Sieve cells are less specialized and are typically found in lower vascular plants and gymnosperms. They have a long and slender shape with tapering ends. In tissues, sieve cells overlap each other, forming a continuous network. Sieve areas, which are specialized regions of the cell wall involved in nutrient transport, are numerous and distributed throughout the cell wall.

In angiosperms (flowering plants), sieve tube elements are the predominant type of sieve element. They are more highly specialized compared to sieve cells. Sieve tube elements are arranged end to end in long series, forming structures known as sieve tubes. The sieve areas in sieve tube elements are highly differentiated and localized in the form of sieve plates. Sieve plates are porous regions composed of modified cell wall material that allow for the movement of nutrients between adjacent sieve tube elements.

The arrangement of sieve tube elements in sieve tubes facilitates the efficient long-distance transport of nutrients in angiosperms. The series of sieve tube elements create a continuous pathway for the flow of sugars and other organic molecules from source regions, where they are produced (such as leaves), to sink regions, where they are utilized or stored (such as roots, fruits, or developing tissues).

2. COMPANION CELLS:

Companion cells are specialized parenchyma cells that are closely associated with sieve tube elements in the phloem tissue. They share a common origin, position, and function with sieve tube elements. In transverse section, companion cells are typically smaller and have distinctive shapes such as triangular, rounded, or rectangular, located adjacent to a sieve tube element. One key characteristic of companion cells is that they are living cells. Unlike some other phloem cells, companion cells do not contain starch granules. Instead, they have a highly active cytoplasm that is rich in organelles and metabolic enzymes. This metabolic activity supports the functioning of the sieve tube elements by providing energy and maintaining the physiological processes required for efficient nutrient transport.

The close association between companion cells and sieve tube elements is crucial for the proper functioning of the phloem tissue. Companion cells are connected to sieve tube elements by numerous plasmodesmata, which are microscopic channels that allow for communication and transport of substances between cells. This close proximity enables companion cells to actively regulate the loading and unloading of sugars and other organic molecules into and out of the sieve tubes.

Companion cells play a vital role in supporting sieve tube elements in their nutrient transport function.

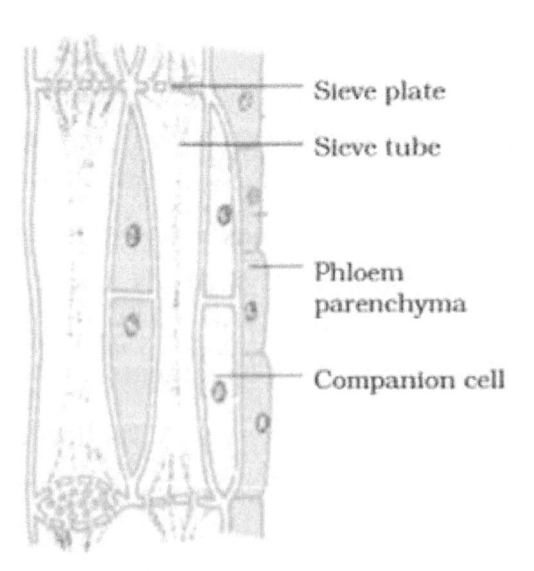

Section of phloem

They provide energy and metabolic resources to sieve tube elements, which lack many of the necessary organelles for cellular processes. Companion cells help to maintain the osmotic balance within the sieve tubes, ensuring the efficient flow of nutrients through the phloem.

3. PHLOEM FIBRES:

Fibres are an important component of both primary and secondary phloem tissues. However, their presence varies among different plant groups. In living pteridophytes (ferns), fibres are rarely found or may be absent altogether in the phloem tissue. Similarly, in some gymnosperms and angiosperms, fibres may be lacking in the phloem.

Phloem fibres, also known as bast fibres, are characterized by their strength and durability. They have thick, lignified cell walls that provide mechanical support and rigidity to the phloem tissue. The fibres are elongated cells that are often arranged in bundles or strands within the phloem.

The strength and toughness of phloem fibres make them valuable in various applications. They are commonly used in the manufacturing of cords, ropes, mats, and cloth. The fibres can be extracted from certain plant sources and processed into fibers that have desirable properties for these purposes. The high tensile strength and flexibility of bast fibres make them suitable for applications where strength and durability are required.

The term "bast fibres" is often used to specifically refer to the fibres derived from the phloem tissue of certain plants. These fibres are typically obtained from the inner bark or bast region of the plant stem. Examples of plants with commercially important bast fibres include flax, hemp, jute, and ramie.

4. PHLOEM PARENCHYMA:

Phloem parenchyma cells are a type of living parenchyma cells that are found in the phloem tissue. These cells are responsible for performing various functions associated with living parenchyma, such as the storage of starch, fats, and other organic substances.

Phloem parenchyma cells have a relatively thin cell wall and a large central vacuole that allows for storage. Within their cytoplasm, they contain various organelles involved in metabolic activities. These cells play a role in storing reserves of energy-rich substances, such as starch and fats, which can be utilized by the plant as needed. Additionally, they can store other organic substances, such as proteins or secondary metabolites.

It's important to note that phloem parenchyma cells may not be present in all plant species. In many monocotyledonous plants (plants with a single seed leaf), phloem parenchyma cells are absent or occur in limited amounts. This absence of phloem parenchyma is a characteristic feature of some monocotyledons.

The storage function of phloem parenchyma cells is significant for the overall metabolic activities of the plant. They act as reservoirs of stored energy and organic compounds that can be mobilized and transported to various plant parts as required. This enables the plant to have a flexible and efficient allocation of resources for growth, development, and reproduction.

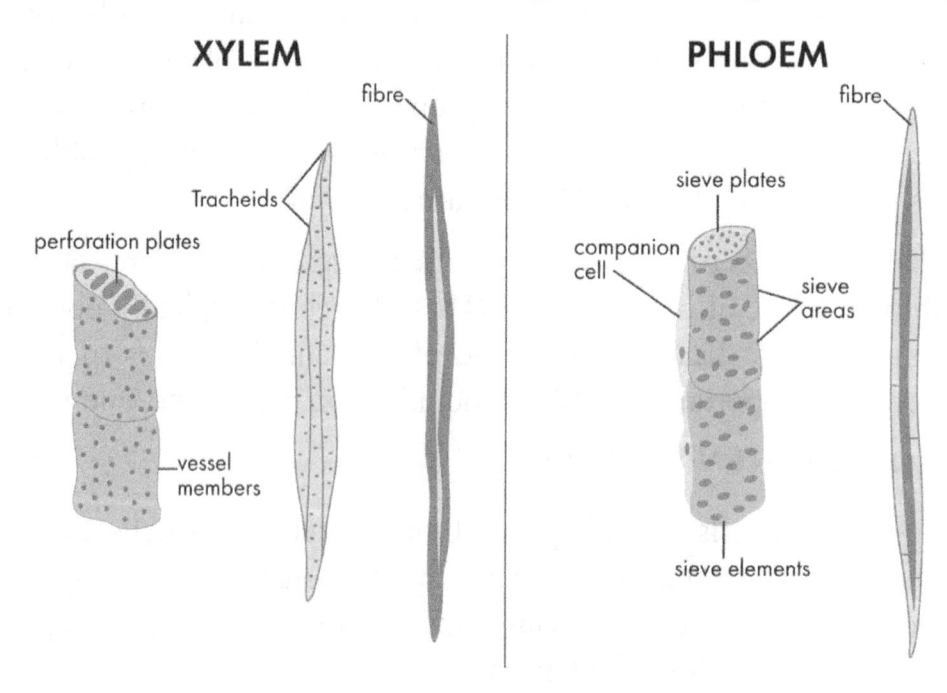

16.4 SECRETORY TISSUE

Secretory tissues are specialized tissues in plants that are responsible for the production and secretion of various substances. These tissues play a crucial role in plant defense, communication, and interaction with the environment. Some common examples of secretory substances include gums, resins, volatile oils, nectar, and latex.

Gums are complex carbohydrates produced by secretory cells and are often involved in wound healing and protection. They form a sticky substance that can help seal and protect injured plant tissues. Resins, on the other hand, are a mixture of organic compounds produced by specialized secretory cells. They often have protective properties and can act as a defense mechanism against herbivores or pathogens.

Volatile oils, also known as essential oils, are aromatic compounds produced by secretory tissues. These oils are typically found in specialized structures such as glandular trichomes or secretory glands. They contribute to the characteristic fragrance and flavor of many plants and often play a role in attracting pollinators or repelling herbivores.

Nectar is a sweet liquid produced by secretory tissues in flowers. It serves as a reward for pollinators, attracting them to the flowers for pollination. Nectar production is an important adaptation that promotes successful pollination and ensures the reproductive success of many flowering plants.

Latex is a milky fluid that contains various substances such as proteins, carbohydrates, and secondary metabolites. It is produced by specialized cells called laticifers, which are found in specific plant tissues. Latex can have diverse functions, including defense against herbivores, wound sealing, and antimicrobial properties.

The location and distribution of secretory tissues vary among different plant species. They can be found in various plant organs, such as leaves, stems, flowers, and roots, depending on the specific type of secretion and its function. Secretory tissues often have specialized structures, such as glandular trichomes, secretory glands, or laticifers, which facilitate the production and release of secreted substances.

Secretory tissue are further divided into two groups – Laticiferous tissue and Glandular tissue.

LATICIFEROUS TISSUE:

Laticiferous tissue is a specialized tissue found in many flowering plants that is involved in the production and secretion of latex. Latex is a viscous fluid that can vary in color, such as being white, yellow, or pinkish. It is characterized as a colloidal substance, meaning it consists of small particles dispersed in a liquid.

The presence of latex in certain plants holds great significance, particularly as a source of rubber. Latex from rubber trees, for example, is essential in the production of natural rubber, which has numerous industrial applications.

Laticiferous tissue is composed of laticiferous ducts, which can be classified into two main types: latex cells or non-articulate latex ducts, and latex vessels or articulate latex ducts. Both types serve the same functions and contain similar contents, but they differ in their structure and morphology.

Latex cells, also known as non-articulate latex ducts, consist of individual cells or chains of cells that contain latex. These cells are typically elongated and interconnected, forming a network throughout the plant tissue. Latex cells are commonly found in plants belonging to the Apocynaceae, Asclepiadaceae, and Moraceae families, among others.

Latex vessels, also called articulate latex ducts, are larger ducts that are formed by the fusion of cells end-to-end. They form continuous tubular structures within the plant tissue. Latex vessels are found in plants of the Euphorbiaceae family, which includes rubber trees.

Both types of laticiferous ducts function in the production, storage, and transportation of latex. Latex serves various purposes, including defense against herbivores, protection against pathogens, and wound healing. The contents of latex can vary but often include a mixture of proteins, carbohydrates, lipids, alkaloids, and other secondary metabolites.

GLANDULAR TISSUE:

Glandular tissue is a specialized tissue found in plants that contains structures called glands, which produce and secrete various substances. These substances can serve different functions, such as digestion, protection, attraction, or excretion. Glandular tissues can be categorized into internal glands and external glands.

Internal glands are located within the interior tissues of the plant. They are often embedded within parenchyma cells or other specialized tissues. Examples of internal glands include oil glands, mucilage-secreting glands, glands that secrete gums, resins, and tannins, as well as digestive glands that produce enzymes. Oil glands, for instance, produce and store essential oils, while mucilage-secreting glands release a gelatinous substance that can provide hydration or assist in seed dispersal. Digestive glands play a role in breaking down complex substances into simpler forms for absorption by the plant.

External glands, on the other hand, occur on the surface of the plant, specifically on the epidermis. Glandular epidermal hairs are one example of external glands. These specialized hairs can be found on various plant parts, such as leaves, stems, or reproductive structures. They secrete substances that serve specific functions, such as protection against herbivores or pathogens, attraction of pollinators, or absorption of moisture from the environment. Nectaries are another type of external gland that secrete nectar a sugary liquid that attracts pollinators.

Glandular tissues and their secretory products can vary greatly among different plant species and even within different parts of the same plant. The presence and nature of glandular tissues are often influenced by environmental factors, genetic factors, and plant adaptations to specific ecological niches.

16.5 DERMAL TISSUE

Dermal tissue, also known as the epidermis, is the outermost layer of cells in plants. It covers the entire surface of the plant body, including the leaves, stems, roots, and reproductive structures. The primary function of the dermal tissue is to provide protection to the underlying tissues and regulate various physiological processes. The dermal tissue is composed of a single layer or multiple layers of tightly packed cells. These cells are specialized for different functions and exhibit specific adaptations depending on their location and role within the plant. The main cell types found in the dermal tissue include:

EPIDERMIS

The epidermis is a specialized tissue that is typically located in the outermost layer of the plant body, including leaves, flowers, stems, and roots. It consists of a single layer of cells known as epidermal cells. The epidermis is covered by a protective layer called the cuticle, which is composed of a waxy substance called cutin. The cuticle serves as a waterproof barrier that helps prevent excessive water loss from the plant and protects it from desiccation.

Primary functions of the epidermis is to provide protection to the underlying tissues of the plant. It acts as a physical barrier against pathogens, pests, and environmental stresses such as drought, excessive sunlight, and mechanical damage. The cuticle, along with other structural adaptations of the epidermal cells such as trichomes and specialized cell wall thickenings, further enhances the protective role of the epidermis.

CORK

Cork is a specialized tissue that is formed in the outermost layer of roots and stems as they mature and increase in girth. It is produced by a secondary meristem called the cork cambium, which generates cork cells toward the periphery of the plant.

Cork cells are dead at maturity and lack intercellular spaces. Their walls are heavily thickened due to the deposition of a waxy substance called suberin. This suberin layer makes the cork cells impermeable to water and gases, providing excellent insulation and preventing desiccation. The thickened cell walls also provide mechanical protection against physical injuries and external pathogens.

The primary function of cork is to protect the underlying tissues of the plant from water loss, infections, and mechanical damage. It forms a durable and impermeable barrier, acting as a protective shield. The suberized cork cells effectively seal off the outer surface of the plant, reducing water loss through evaporation and providing a protective shield against pathogens and harsh environmental conditions.

Cork has several commercial uses due to its unique properties. It is harvested from the cork oak tree (Quercus suber) in a sustainable manner. Cork is commonly used for making bottle stoppers, as it provides an excellent seal for wine bottles and other containers. It is also utilized in the production of insulation boards, sports equipment, flooring, and other products that benefit from its insulation and protective properties.

STOMATA

Stomata are specialized structures found on the epidermis of leaves, stems, and other aerial parts of plants. They play a crucial role in regulating gas exchange, including the uptake of carbon dioxide (CO_2) for photosynthesis and the release of oxygen (O_2) and water vapor.

Each stoma is formed by two specialized epidermal cells called guard cells, which are uniquely shaped to create an opening or pore. The guard cells surround the stomatal pore and control its opening and closing. When the guard cells are turgid or filled with water, they expand and cause the stomatal pore to open, allowing the exchange of gases between the plant and the external environment. When the guard cells lose water and become flaccid, they shrink, leading to the closure of the stomatal pore and reducing the loss of water vapor through transpiration.

The opening and closing of stomata are regulated by various environmental and internal factors. Factors such as light intensity, humidity, temperature, and the concentration of CO_2 influence the movement of ions and the resulting water movement into or out of the guard cells. Hormones such as abscisic acid (ABA) play a role in signaling the closure of stomata under conditions of water stress or drought.

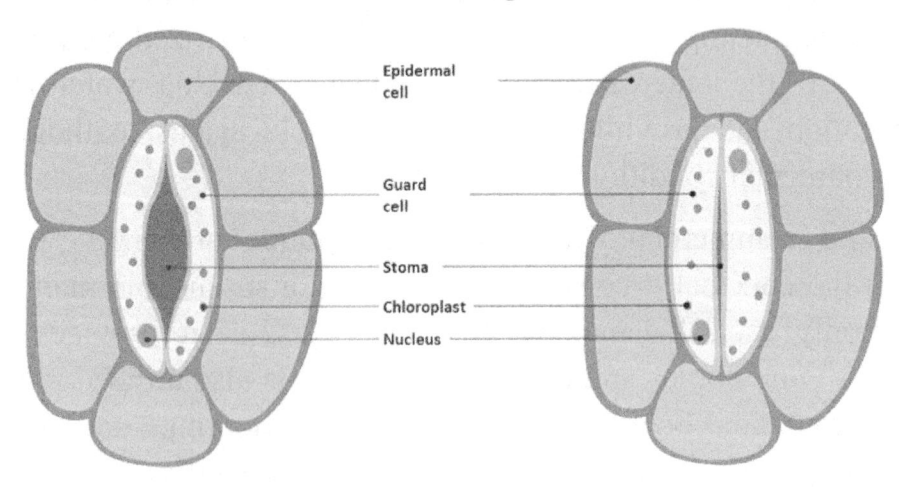

The presence of stomata allows for the exchange of CO_2, which is needed for photosynthesis, and the release of O_2, which is a byproduct of photosynthesis. This exchange enables the plant to take in the necessary carbon dioxide for energy production while releasing oxygen into the atmosphere. Stomata also play a role in regulating water vapor loss through transpiration, as they can close to reduce excessive water loss during periods of water scarcity or high temperatures.

OPENING & CLOSING OF STOMATA :

The opening and closing of stomata are regulated by a complex interplay of various factors. While light, temperature, and CO_2 concentration can influence stomatal behavior, the primary driving force behind the opening and closing of stomata is the osmotic movement of water into or out of the guard cells.

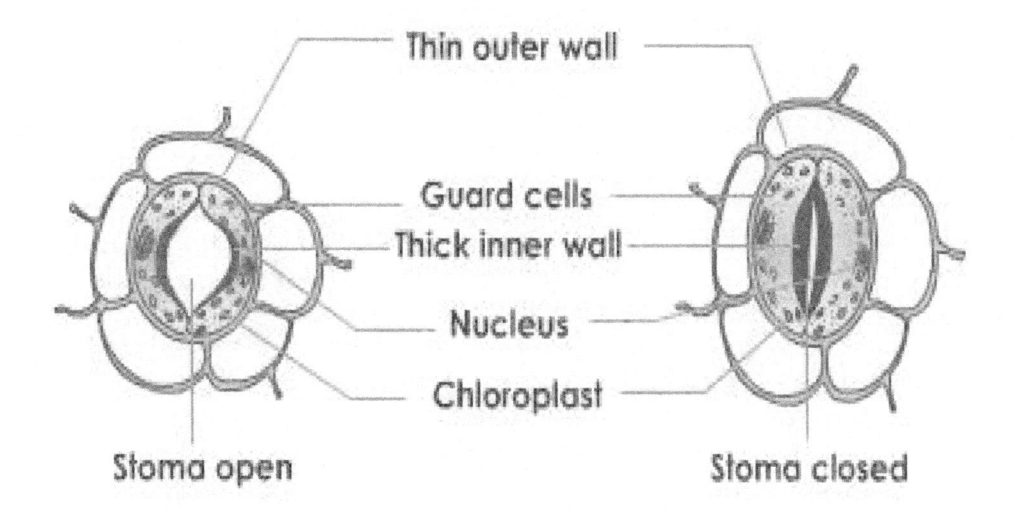

In the presence of light, especially in the blue light spectrum, photosynthesis is active, leading to the accumulation of sugars in the guard cells. This accumulation of sugars increases the osmotic pressure within the guard cells, causing water to enter the cells through osmosis. As a result, the guard cells become turgid, leading to the opening of the stomatal pore.

High temperatures can also trigger stomatal opening. When temperatures rise, the guard cells activate various ion channels, allowing the influx of potassium ions (K+) into the guard cells. This influx of ions leads to the accumulation of solutes and an increase in osmotic pressure, resulting in water uptake and turgidity of the guard cells, thereby opening the stomatal pore.

When CO_2 levels decrease, the concentration of sugars in the guard cells decreases as well. This decrease in osmotically active solutes leads to a decrease in osmotic pressure, causing water to move out of the guard cells through osmosis. As a result, the guard cells become flaccid, leading to the closure of the stomatal pore.

SUMMARY OF PLANT TISSUES

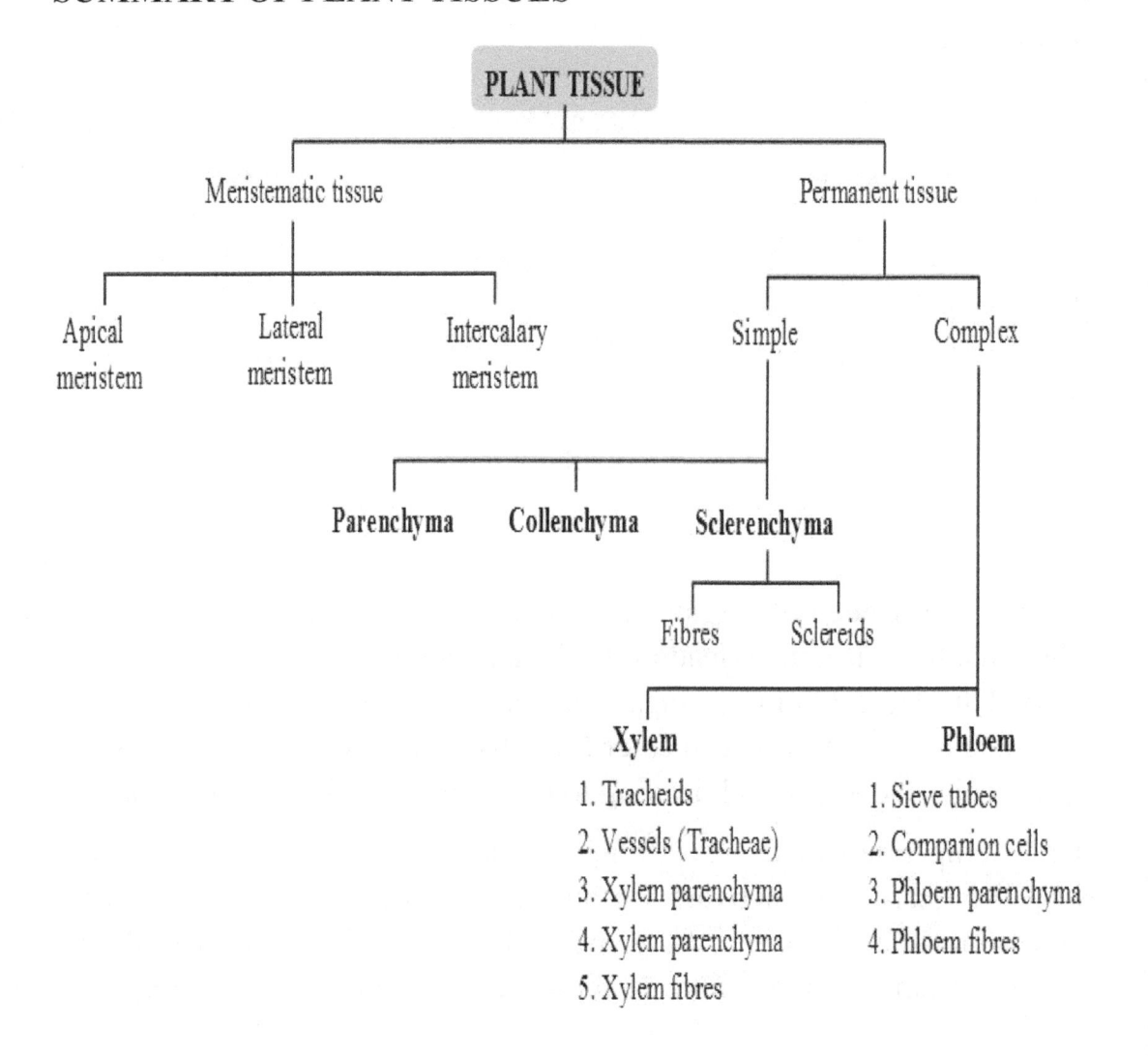

```
                          PLANT TISSUE
                    ┌──────────┴──────────┐
            Meristematic tissue      Permanent tissue
          ┌─────────┼─────────┐       ┌──────┴──────┐
      Apical    Lateral   Intercalary  Simple      Complex
     meristem  meristem   meristem
                          ┌──────────┼──────────┐
                      Parenchyma  Collenchyma  Sclerenchyma
                                             ┌──────┴──────┐
                                           Fibres        Sclereids

                                     Xylem                 Phloem
                              1. Tracheids          1. Sieve tubes
                              2. Vessels (Tracheae) 2. Companion cells
                              3. Xylem parenchyma   3. Phloem parenchyma
                              4. Xylem parenchyma   4. Phloem fibres
                              5. Xylem fibres
```

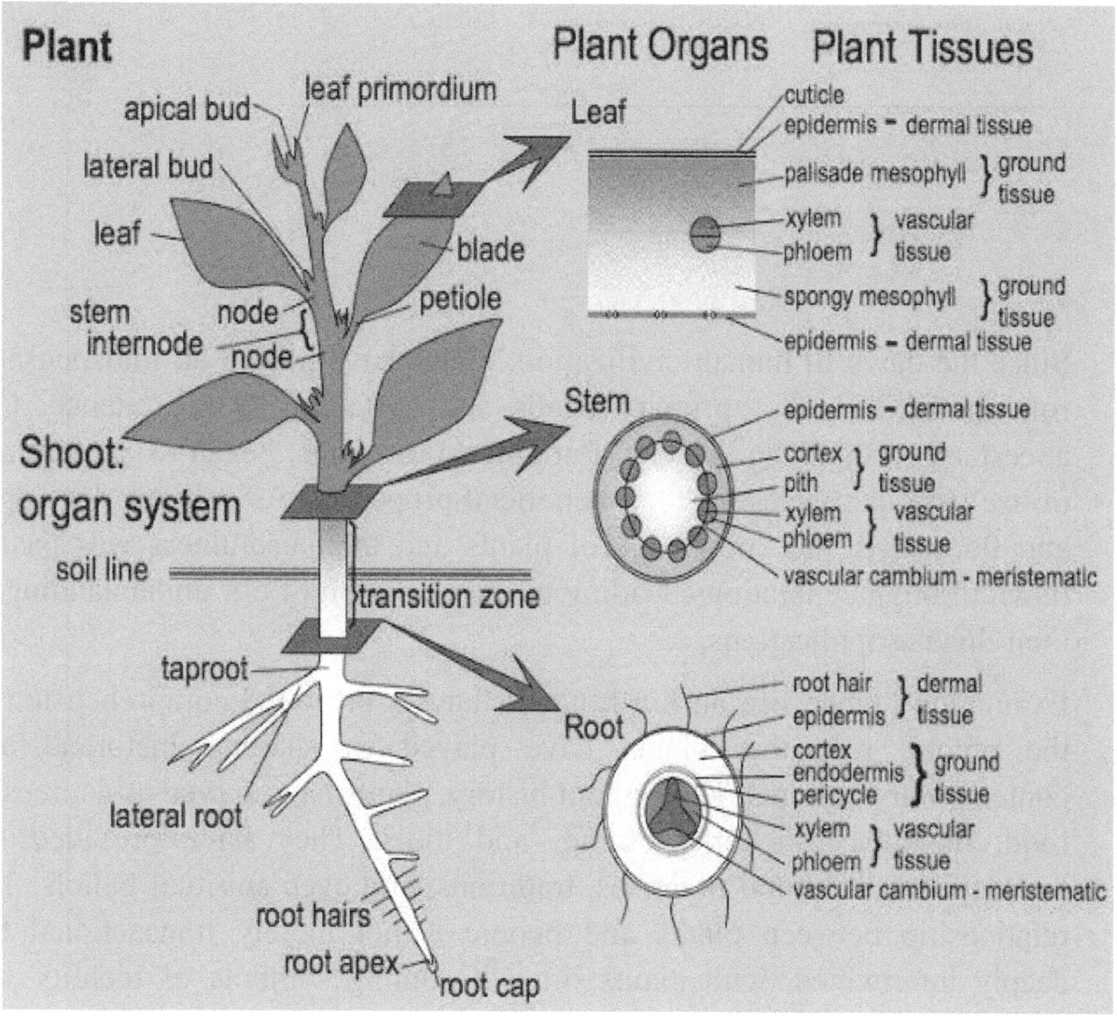
Plant
Plant Organs
Plant Tissues
apical bud
leaf primordium
lateral bud
leaf
blade
petiole
stem
node
internode
node
Shoot:
organ system
soil line
transition zone
taproot
lateral root
root hairs
root apex
root cap
Leaf
cuticle
epidermis - dermal tissue
palisade mesophyll
ground tissue
xylem
phloem
vascular tissue
spongy mesophyll
ground tissue
epidermis - dermal tissue
Stem
epidermis - dermal tissue
cortex
pith
ground tissue
xylem
phloem
vascular tissue
vascular cambium - meristematic
Root
root hair
epidermis
dermal tissue
cortex
endodermis
pericycle
ground tissue
xylem
phloem
vascular tissue
vascular cambium - meristematic

CHAPTER 17

INTRODUCTORY TOPICS IN ETHNOBOTANY

17.1 PLANTS AND PEOPLE

Since the dawn of human civilization, plants have played an indispensable role in fulfilling our primary needs and sustaining our existence. Our ancestors, recognizing the vital importance of plants, began to explore and utilize various species for their beneficial properties. As cultures developed and flourished, the knowledge of plants and their usefulness was passed down through generations, leading to the evolution of our understanding of their diverse applications.

Examining human life on Earth necessitates a profound comprehension of the pivotal role that plants have played in shaping historical and contemporary cultures. Throughout history, plants have served as sources of food, medicine, shelter, clothing, and tools. They have provided the foundation for cultural practices, traditions, and even spiritual beliefs. The relationship between plants and people is not merely transactional but deeply intertwined, with plants often becoming symbols of identity and cultural heritage. Even in the present day, our dependence on plants remains paramount for our survival. They continue to supply us with sustenance, nourishing our bodies with essential nutrients. Plants contribute to the regulation of the Earth's climate and the purification of our air and water. They serve as the primary source of energy for many ecosystems, supporting the intricate web of life that sustains biodiversity.

The significance of plants extends beyond their direct contributions to human well-being. They act as essential hosts and resources for numerous pollinators, such as bees, butterflies, and birds, ensuring the continuation of vital ecological processes and the propagation of many plant species.

In the field of ethnobotany, the study of the relationship between plants and people, we explore the rich tapestry of human-plant interactions. Ethnobotanists delve into traditional knowledge systems, indigenous practices, and local ecological wisdom to understand the intricate connections between cultures and the plant world. Through ethnobotanical research, we aim to conserve traditional knowledge, promote sustainable plant use, and foster a deeper appreciation for the diverse ways in which plants have shaped and continue to shape human societies.

17.2 DEFINITIONS AND CONCEPTS: ETHNOBOTANY

Ethnobotany, a discipline with roots dating back to ancient times, holds a significant place in the study of the relationship between humans and plants. Throughout history, humans have relied on plants and their vital pollinators for their survival and well-being. To comprehend human life on Earth, it is essential to understand the profound role of plants in both past and present cultures.

Ethnobotany provides valuable insights into the traditional uses of plant resources, offering information that can be utilized for integrated tribal development. The term "ethnobotany" derives from the combination of "ethno," referring to the study of people, and "botany," the study of plants. In 1895, J.W. Harshberger coined the term to encompass the study of plants used by primitive and aboriginal communities.

Over the course of the twentieth century, ethno botany has emerged as a distinct academic discipline within the natural sciences. Although various definitions have been proposed, the fundamental concept of ethno botany has remained relatively consistent, while its scope continues to expand. Ethnobotany now encompasses a broad range of research areas, including indigenous plant knowledge, cultural practices, traditional medicine, food systems, and the preservation of traditional ecological knowledge.

At its core, ethnobotany is the study of how people from specific cultures and regions utilize native plants. This interdisciplinary field integrates knowledge and methods from botany, anthropology, ecology, pharmacology and other related disciplines. By exploring the intricate relationship between humans and plants, ethnobotany sheds light on the diverse ways in which plants have shaped and continue to shape human societies worldwide. It serves as a bridge between traditional knowledge systems and modern scientific approaches, aiming to promote sustainable plant use, conservation of biodiversity, and cultural heritage preservation.

According to Schultes (1962), ethno botany is defined as the study of the relationships between people belonging to primitive societies and plants. It focuses on understanding how these societies utilize and interact with plants in their cultural, social, and ecological contexts.

Alcom (1984) defines ethno botany as the study of plant use within specific cultural and environmental contexts. This approach emphasizes the importance of considering the cultural, social, and ecological factors that shape the use and management of plants by different communities.

Jain (1987) further expands the definition, stating that ethno botany encompasses the comprehensive natural and traditional relationship between humans and the plant diversity in their surroundings. It highlights the intricate interactions and dependencies between people and the plant wealth that surrounds them.

In Wickens' (1990) perspective, ethno botany is defined as the study of useful plants before they undergo commercial exploitation or domestication This definition emphasizes the exploration of plants' potential uses and the understanding of their traditional significance before they become subject to commercial interests.

Indeed, ethno botany represents the initial knowledge of plants acquired by primitive and aboriginal communities through necessity, intuition, observation, and experimentation in forest environments. It serves as the foundation of human understanding and utilization of plant resources.

In contemporary times, ethno botany has become a significant and crucial area of research and development, particularly in the fields of resource management, sustainable utilization, biodiversity conservation, and socioeconomic development. Botanists, social scientists, anthropologists, and practitioners of indigenous medicine from around the world are actively engaged in studying the intricate interactions between humans and plants in their natural environments.

17.3 DIFFERENT ASPECTS OF ETHNOBOTANY

The field of ethnobotany encompasses a wide range of disciplines and aspects due to its interdisciplinary nature and profound socio-economic impacts. Its linkages have proliferated, leading to diverse areas of study and application. Here are some of the different aspects and connections of ethnobotany:

Sociology and Anthropology: Ethnobotany is closely linked to sociology and anthropology as it explores the cultural, social, and traditional practices associated with plant use. It investigates the relationships between plants and human societies, including the cultural significance of plants, traditional knowledge systems, and the role of plants in rituals, ceremonies, and cultural expressions.

Taxonomy and Phytochemistry: Ethnobotany relies on taxonomy, the classification and identification of plants, to accurately document and understand the species being utilized by different cultures. Phytochemistry, the study of plant chemistry, is another relevant aspect of ethnobotany as it investigates the chemical composition of plants and their potential medicinal, nutritional, or toxic properties.

Archaeology: Ethnobotanical studies often collaborate with archaeologists to analyze plant remains found at archaeological sites. By examining ancient plant materials, such as seeds, pollen, and plant fibers, ethnobotanists can gain insights into past human-plant interactions, ancient plant uses, and the domestication of plants.

Ecology: Ethnobotany has strong ties to ecology as it explores the relationships between plants and their natural environments. Understanding the ecological context of plant use is crucial for sustainable resource management and conservation efforts. Ethnobotanical studies contribute to the knowledge of how human activities impact plant populations and ecosystems.

Agriculture: Ethnobotany provides valuable insights into traditional agricultural practices, crop diversity, and local farming systems. It examines indigenous methods of plant cultivation, seed saving, and the selection of plant varieties suited to specific environments. This knowledge contributes to sustainable agriculture practices, crop improvement, and the preservation of traditional farming techniques.

Medicine: Ethnobotany has significant relevance to the field of medicine. Traditional medicinal practices and knowledge of plants' therapeutic properties are studied to discover potential new drugs or validate the effectiveness of traditional remedies. Ethnobotanical research contributes to the documentation and conservation of traditional medicinal practices and supports the integration of traditional medicine into modern healthcare systems.

Linguistics: The study of language and its connection to plants is another aspect of ethnobotany. Many indigenous cultures have rich botanical vocabularies and intricate plant naming systems, reflecting their deep knowledge and cultural significance of plants. Ethnobotanists work closely with linguists to document and preserve these languages, ensuring the transmission of traditional plant knowledge.

Economic and Conservation Perspectives: Ethnobotany has implications for economic development, poverty alleviation, and the conservation of biodiversity. By identifying and promoting sustainable uses of plant resources, ethnobotanical studies contribute to the development of cottage industries, income generation for local communities, and the preservation of traditional knowledge. Furthermore, understanding the cultural and economic value of plants fosters conservation efforts to protect biodiversity and ensure the sustainable use of plant resources.

17.4 IMPORTANCE OF ETHNOBOTANY

Traditional Knowledge and Cultural Preservation: Ethnobotany provides valuable information regarding the traditional uses of plant resources. This knowledge is deeply rooted in the cultural heritage of indigenous and local communities. By documenting and preserving this traditional knowledge, ethnobotany contributes to the conservation of cultural practices, beliefs, and customs associated with plants.

Integrated Tribal Development: Ethnobotany informs integrated tribal development by identifying the potential uses of plant wealth. It offers insights into the utilization of plants for various purposes, including food processing, fiber production, oil extraction, medicinal applications, and more. By recognizing and promoting these uses, ethnobotanical studies support the socioeconomic development and empowerment of tribal communities.

Sustainable Resource Management: Ethnobotany promotes the sustainable use and conservation of plant resources. By understanding the traditional knowledge and practices of communities, it provides insights into sustainable harvesting techniques, plant propagation methods, and ecosystem management. This knowledge is crucial for maintaining the biodiversity of plant species and ecosystems for future generations.

Medicinal Plant Research: Ethnobotany has immense significance in the field of medicine. Traditional knowledge of medicinal plants, gathered through ethnobotanical studies, contributes to the discovery of new drugs and the validation of traditional remedies. It serves as a valuable resource for modern healthcare systems, supporting the development of herbal medicines and providing alternative treatment options.

Industrial and Economic Applications: Ethnobotany uncovers novel uses of plants that can be exploited for various industrial applications. The study of plant fibers, oils, resins, dyes, and other valuable plant products contributes to the development of agro-based industries, such as food processing, textiles, cosmetics, and natural product manufacturing. This enhances economic opportunities for local communities, particularly those relying on plant-based livelihoods.

Practical Contributions to Human Welfare: Ethnobotanical knowledge has made substantial contributions to human welfare. It is estimated that around 30% of modern medicines have been developed based on ethnobotanical information and knowledge. Many conventional drugs are derived directly or indirectly from naturally occurring substances, predominantly of plant origin. Ethnobotanical studies facilitate the identification and exploration of potential medicinal plants, contributing to advancements in healthcare and pharmaceutical research.

Environmental Awareness and Conservation: Ethnobotany enhances our understanding of human dependence on plants and the impact of human activities on plant populations and ecosystems. By recognizing the interconnectedness between humans and the natural world, ethnobotany promotes environmental awareness and conservation efforts. It emphasizes the need to conserve plant diversity, protect endangered species, and maintain the delicate balance of ecosystems.

17.5 AIMS AND OBJECTIVES OF ETHNOBOTANY

Ethnobotany, as a multidisciplinary field, has several aims and objectives that contribute to the understanding, preservation, and utilization of the complex relationships between cultures and the uses of plants. Some of the specific aims and objectives of ethnobotany are:

Documentation and Description: The primary aim of ethnobotany is to document, describe, and explain the intricate relationships between cultures and the uses of plants. This involves recording and documenting traditional knowledge, practices, and beliefs regarding plants and their uses. Ethnobotanists strive to create comprehensive records that capture the diversity and richness of traditional plant knowledge systems.

Medicinal Plant Documentation: Ethnobotany aims to properly document indigenous knowledge about medicinal plants. This involves identifying and recording traditional medicinal practices, including the preparation and administration of herbal remedies. The documentation of medicinal plant knowledge contributes to the preservation and validation of traditional healthcare systems.

Preservation of Traditional Knowledge: Ethnobotany aims to preserve unwritten traditional knowledge about herbal plants. This includes recording traditional plant uses, cultivation techniques, harvesting methods, and preparation processes that have been passed down through generations orally. By documenting and preserving this traditional knowledge, ethnobotany helps safeguard cultural heritage and prevent the loss of valuable information.

Conservation of National Heritage: Ethnobotany seeks to conserve our national heritage before it faces the risk of extinction. This includes identifying and documenting rare or endangered plant species, as well as traditional practices that contribute to the preservation of plant diversity. By promoting the conservation of plant resources and their associated cultural practices, ethnobotany contributes to biodiversity conservation efforts.

Awareness and Education: Ethnobotany aims to create awareness about the role of plants in cultural, social, and health contexts. This includes educating communities, policymakers, and the general public about the value and significance of traditional plant knowledge. By raising awareness, ethnobotany advocates for the recognition and respect of traditional knowledge systems and encourages the integration of traditional practices into modern society.

Herbal Drug Development: Ethnobotany aims to promote the increased manufacturing and utilization of herbal drugs. By conducting research on traditional medicinal plants and their potential therapeutic applications, ethnobotany contributes to the development of herbal medicines. This includes exploring the safety, efficacy, and sustainability of herbal remedie as well as integrating traditional medicine into mainstream healthcare practices.

Research and Job Opportunities: Ethnobotany provides opportunities for research and employment in various fields. It offers avenues for conducting interdisciplinary research, including botany, anthropology, ecology, pharmacology, and more. Ethnobotanical research contributes to the advancement of scientific knowledge, conservation efforts, and the development of sustainable practices. Additionally, it provides job opportunities in academia, conservation organizations, pharmaceutical companies, and related industries.

17.6 SCOPE OF ETHNOBOTANY

Ethnobotany, as a field of study, has a wide-ranging scope that encompasses various aspects of human life and the relationship between people and plants. In recent times, the scope of ethnobotany has expanded to address several pressing issues and challenges.

Rural Health: Ethnobotany plays a significant role in understanding traditional medicinal practices in rural areas. It focuses on documenting and studying the use of plants for healthcare purposes, including the treatment of ailments and the promotion of overall well-being. Ethnobotanical research contributes to improving rural health by identifying effective traditional remedies and supporting their integration into primary healthcare systems.

Drugs and Abuses: Ethnobotany examines the cultural, social, and economic dimensions of plant use in relation to drugs and substance abuse. It investigates traditional practices, rituals, and beliefs associated with psychoactive plants and substances, as well as the implications for public health. Ethnobotanical studies can inform substance abuse prevention and intervention strategies.

Social Customs: Ethnobotany explores the relationship between plants and social customs, rituals, and traditions. It investigates the cultural significance of plants in various societies, including their use in ceremonies, festivals, and traditional practices. By studying the role of plants in social customs, ethnobotany helps to preserve cultural heritage and promotes a deeper understanding of cultural diversity.

Cottage Industries: Ethnobotany recognizes the importance of plants in cottage industries, which often involve small-scale, locally-based production. It examines the use of plants in crafts, textiles, natural product manufacturing, and other traditional industries. Ethnobotanical knowledge contributes to the sustainable development of cottage industries by identifying locally available plant resources and supporting their responsible utilization.

Conservation of Ecosystems: Ethnobotany has a strong focus on the conservation of ecosystems and biodiversity. It investigates the traditional knowledge and practices related to sustainable use and management of plant resources. By studying the interactions between people and plants, ethnobotany promotes the conservation of valuable plant species, habitats, and ecosystems.

Nutrition: Ethnobotany explores the traditional uses of plants for nutrition and food security. It investigates the diversity of edible plants, traditional food preparation methods, and the nutritional value of different plant-based foods. Ethnobotanical research supports efforts to promote sustainable and culturally appropriate dietary practices, addressing nutritional challenges in various communities.

Energy: Ethnobotany investigates the traditional uses of plants as sources of energy. It examines the utilization of plant materials for cooking, heating, and other energy needs in different cultural contexts. Ethnobotanical studies contribute to the exploration of sustainable energy practices, including the use of renewable plant-based resources.

17.7 DESCRIPTION OF SOME USEFUL PLANTS.

i. Cannabis sativa L.; Common Names: Indian Hemp

Ethnomedicinal uses: Cannabis sativa is used for various medicinal purposes, including reducing general body inflammation, alleviating intoxication, and stimulating appetite. It has been traditionally employed in certain cultures for its therapeutic properties. Seed: The seeds of Cannabis sativa are primarily used as feed for caged birds, providing them with essential nutrients. Flower: The flowers of Cannabis sativa contain psychoactive and physiologically active chemical compounds called cannabinoids. These cannabinoids have been consumed for recreational, medicinal, and spiritual purposes. Preparations of the flowers, leaves, and resinous extracts derived from the plant (such as marijuana and hashish) are commonly consumed through smoking, vaporizing, or oral ingestion.

ii. Mangifera indica L.; Common Names: Mango

Ethnobotanical uses: Mangifera indica, commonly known as mango, has various ethnobotanical uses. The fruit can be eaten raw, offering a refreshing and delicious summer treat. Mangoes are also commonly used to make juices, ice cream, fruit bars, and other culinary delights. Due to their sweet and tangy flavor, mangoes are highly valued in many cuisines and are enjoyed for their tropical taste.

iii. Calotropis procera (Aiton) W.T; Common Names: Sodom's Apple

Ethnobotanical Uses: Calotropis procera, also known as Sodom's Apple, has various medicinal uses. Different parts of the plant, including the root skin, latex, flowers, and leaves, are utilized for their therapeutic properties. The dried leaves powder, when topically applied, can expedite wound healing. The application of latex helps reduce inflammation in glandular swellings. Fermentation of the leaves, slightly warmed with a thin coat of castor oil, is beneficial for relieving abdominal pain. The local application of latex is recommended for hair fall and dental aches.

iv. Aloe vera (L.) Burm.f.; Common Names: Kwargandal

Ethnobotanical Uses: Aloe vera is a versatile plant with various ethnobotanical uses. It is commonly used in cosmetic and skincare products, such as moisturizers, soaps, sunscreens, and lotions. Aloe vera is known for its moisturizing properties and is used to alleviate body weakness and treat conditions like pimples or acne. It is a popular ingredient in commercially available products like yogurt, beverages, and desserts. Aloe vera is used in tissues, incense, shaving cream, shampoos, and makeup. It is also utilized as a moisturizer and anti-irritant in face tissues, particularly for individuals suffering from hay fever or cold to reduce chafing of the nose.

The ethnic plant wealth may be divided into the following groups on the basis of their uses:

(1) Wild edible plants (food plants)

(2) Ethno medicinal plants

(3) Fibre and floss yielding plants

(4) Oil yielding plants

(5) Dye yielding plants

(6) Gum and resin yielding plants

(7) Ethno toxic and fish poison plants

(8) Timber and wood work

(9) Tannin yielding plants

(10) Plants used in basketry and brooms.

CHAPTER 18

BIODIVERSITY

18.1 DEFINITION OF BIODIVERSITY

Biodiversity, derived from the contraction of "biological diversity," refers to the incredible variety of life forms, including plants, animals, and microorganisms, found on Earth or within a specific location. It encompasses the full range of biological entities and their interactions, including their ecological roles and genetic variability. Biodiversity is not uniform and varies significantly across different ecosystems and climates.

The regions with humid tropical climates typically exhibit the highest levels of biodiversity, while biodiversity tends to decrease towards the poles as temperatures decrease. Biodiversity holds immense importance and value across various dimensions, including evolutionary, ecological, economic, social, cultural, and intrinsic aspects. It serves as an indicator of the health and resilience of biological systems.

When exploring biodiversity, it is often analyzed and studied at three interconnected levels: genetic diversity, species diversity, and ecosystem diversity. These levels of biodiversity work in conjunction, contributing to the intricate complexity of life on Earth. Genetic diversity refers to the variety of genes within a population or species, which enables adaptation and evolutionary processes. Species diversity reflects the number and abundance of different species present in a particular area. Ecosystem diversity encompasses the range of distinct ecosystems, such as forests, grasslands, wetlands, and marine environments, each characterized by unique biological communities and ecological processes.

18.2 IMPORTANCE OF BIODIVERSITY

Biodiversity holds immense significance for human beings and the planet as a whole, providing numerous benefits and services that are vital for our well-being and the health of ecosystems. The importance of biodiversity can be understood through the following aspects:

Ecological Life Support: Biodiversity plays a critical role in supporting and maintaining functioning ecosystems. It contributes to the production of oxygen, ensures the availability of clean air and water through natural purification processes, facilitates pollination of plants by insects and other organisms, aids in pest control by maintaining natural predator-prey relationships, and assists in wastewater treatment by microbial communities.

Ecosystem Services: Biodiversity provides essential ecosystem services that directly benefit humans. It serves as a source of food, supplying us with a diverse range of crops, livestock, and seafood. Biodiversity also offers natural materials for shelter, such as timber and fibers for construction and clothing.

Regulation: Biodiversity is intricately involved in regulating ecological processes. It plays a crucial role in nutrient cycling, ensuring the recycling and availability of vital elements like carbon, nitrogen, and phosphorus in ecosystems. Biodiversity also contributes to the maintenance of fertile soils, which are crucial for agricultural productivity and food security.

Provision of Leisure Activities: Many people derive immense value and enjoyment from engaging with biodiversity in various leisure activities. Exploring natural environments, hiking through diverse ecosystems, birdwatching, or studying natural history are examples of how people connect with and appreciate biodiversity.

Provision of Inspiration: Biodiversity serves as a source of inspiration for artistic and creative endeavors. Musicians, painters, sculptors, writers, and other artists often draw inspiration from the beauty and diversity of the natural world, reflecting it in their works.

Stability of Ecosystems: Biodiversity contributes to the stability and resilience of ecosystems over time. Diverse ecosystems are better able to withstand and recover from disturbances, such as human exploitation or natural disasters. Higher biodiversity levels enhance ecosystem stability, reducing the risk of ecosystem services being disrupted.

18.3 HUMAN IMPACT ON BIODIVERSITY

Human activities have had significant and often detrimental impacts on biodiversity, contributing to the decline and loss of numerous species and ecosystems. It is crucial for individuals to recognize the consequences of their actions on biodiversity and understand the importance of preserving the remaining biodiversity on Earth. Some key factors highlighting human impact on biodiversity include:

Population Pressure: The sheer size and growth of the human population have exerted immense pressure on natural resources and ecosystems. Increased human population puts greater demands on land, water, energy, and other natural resources, leading to habitat destruction and fragmentation.

Land Use Changes: Human activities, particularly in the form of urbanization, deforestation, agriculture, and infrastructure development, have resulted in the conversion of natural habitats into human-dominated landscapes. These land use changes destroy and degrade habitats, displacing many species and disrupting ecological processes.

Exploitation and Overconsumption: Human demand for resources, including timber, minerals, water, and wildlife, has led to overexploitation and unsustainable harvesting practices. Overfishing, illegal hunting, and logging contribute to the decline of species and the disruption of ecosystems.

Pollution and Climate Change: Pollution from industrial activities, agriculture, improper waste disposal, and the release of chemicals into the environment has detrimental effects on biodiversity. It contaminates habitats affects water quality, and harms species directly or indirectly. Additionally, human-induced climate change, primarily driven by greenhouse gas emissions, poses significant threats to biodiversity by altering ecosystems, increasing the frequency of extreme weather events, and disrupting species' distribution patterns.

Introduction of Invasive Species: Human activities, such as the intentional or accidental introduction of non-native species into new environments, have led to the establishment of invasive species. These invasives can outcompete native species, disrupt ecological relationships, and cause significant ecological and economic damage.

Fragmentation and Habitat Loss: Fragmentation of habitats due to human infrastructure development, such as roads, dams, and urban areas, disrupts natural connectivity between ecosystems. This fragmentation reduces the size and quality of habitats, making them less suitable for many species and impeding their movement and gene flow.

18.4 THREAT TO BIODIVERSITY

Biodiversity is facing significant threats worldwide, and its decline has been observed over the past century. Several factors contribute to the erosion of biodiversity. The major threat to Biodiversity is extinction. The factors that threaten bio diversity have been categorized. Biodiversity is declining because of (i) Habitat loss/Destruction of Habitat (ii) Invasive Species (iii) Pollution (iv) Population growth (v) Over-Consumption (Unstainable use) (vi) Climate change.

(i) Destruction/Loss/Fragmentation of Habitat: Habitat loss is one of the biggest threats to Biodiversity. Most of the species extinctions from

1000 Ad to 2000 AD are due to human activities, in particular destruction of plants and animal habitats – tropical forest.

(ii) Exotic/Invasive species – Introduced species- Invasive species: Invasive species are the second largest threat to Biodiversity loss after habitat loss. An invasive species is a species that is not native to a particular area, but arrives, establishes a population, and spread on its own. Invasive species have much larger impacts on an ecosystem than other species. Loss of indigenous species allows introduced species to flourish. E.g. *Tithonia diversifolia, Chromolaena odorata.* The widespread introduction of exotic species by human is a potent threat to biodiversity.

(iii) Genetic Pollution: Pollution introduces contaminants into the environment that can maim or even kill plant and animal species. Pollution contaminates natural ecosystems and again poses a threat to Biodiversity. Purebred naturally evolved region-specific wild species can be threatened with extinction through the process of genetic pollution i.e. uncontrolled hybridization

(iv) Effect of Climate Change on Plant Biodiversity: The most pronounced environmental problem in the tropics today is climate change and global warming. The recent phenomenon of global warming is also considered to be a major threat to global biodiversity. Climate change is causing huge changes to Biodiversity. Global warming disrupts normal weather pattern creating hotter and drier weather, exposure of soil to intense tropical sun and torrential rains, desertification, flooding. Climate change degrades biodiversity. Temperature increase makes certain environment uninhabitable to previously indigenous species.

(v) Over-Consumption and Unsustainable Use: Unsustainable practices, such as over-harvesting, overfishing, overhunting, and illegal wildlife trade, result in the overexploitation of species. These activities exceed the natural reproduction rates of species, leading to population declines and even extinction. Unsustainable agricultural practices, excessive water consumption, and over-extraction of resources also contribute to the loss of biodiversity.

18.5 SOCIETY'S ROLE IN SUPPORTING BIODIVERSITY

Conserving biodiversity is a collective responsibility that requires the participation and involvement of society as a whole. Here are some ways in which society can contribute to supporting biodiversity:

Awareness and Education: Raising awareness about the importance of biodiversity and its conservation is crucial. Educational campaigns, public outreach programs, and incorporating biodiversity topics into school curricula can help foster a sense of responsibility and understanding among individuals.

Sustainable Practices: Adopting sustainable practices in various sectors is vital for reducing the negative impact on biodiversity. This includes sustainable agriculture, responsible fishing, ethical wildlife tourism, and responsible consumption and production. Choosing sustainable and eco-friendly products can help reduce demand for resources that contribute to biodiversity loss.

Conservation Efforts: Society can actively participate in conservation efforts by supporting and volunteering with organizations and initiatives dedicated to biodiversity conservation. This can involve activities such as habitat restoration, reforestation, and wildlife monitoring.

Advocacy and Policy Support: Engaging in advocacy and supporting policies that prioritize biodiversity conservation is important. Encouraging governments and policymakers to implement and enforce legislation and regulations that protect habitats, control invasive species, and promote sustainable practices can have a significant impact.

Sustainable Land Use Planning: Society can promote responsible land use planning, advocating for the protection of natural areas, creation of wildlife corridors, and the integration of biodiversity considerations into urban planning and infrastructure development.

Sustainable Tourism: Promoting and practicing sustainable tourism can help minimize the negative impact on biodiversity. Responsible travelers can choose eco-friendly accommodations, engage in wildlife-friendly activities, and respect local ecosystems and cultures.

Collaboration and Partnerships: Collaboration among governments, NGOs, academic institutions, local communities, and individuals is essential for effective biodiversity conservation. Sharing knowledge, resources, and expertise can lead to more comprehensive and successful conservation efforts.

18.6 BIODIVERSITY CONSERVATION METHODS

Biodiversity conservation is a critical endeavor aimed at protecting the variety of life forms on Earth and ensuring their sustainable existence. It is increasingly evident that biodiversity serves as the foundation of our existence and is worthy of conservation for reasons of curiosity and aesthetic appreciation. Conservation encompasses the sustainable management of resources, including both protection and responsible utilization.

Preservation is an important aspect that entails maintaining something in its original state without alteration or change. To conserve biodiversity, two primary approaches are employed: ex-situ conservation, which involves protecting elements of biodiversity outside their natural habitats, and in-situ conservation, which focuses on the preservation of habitats, species, and ecosystems within their natural environment.

In-situ conservation, also known as "on-site" conservation, aims to safeguard habitats, species, and ecosystems in their natural settings. By practicing in-situ conservation, not only are the natural processes and interactions within these systems preserved, but the inherent biodiversity is also safeguarded. This approach offers the benefit of maintaining recovering populations in their surroundings, allowing them to continue developing their distinctive characteristics. Furthermore, in-situ conservation supports ongoing evolutionary processes and adaptation within their native environments. Consequently, conserving species in their natural habitats through in-situ conservation is widely considered the most appropriate method for biodiversity conservation.

Ex-situ conservation, on the other hand, involves protecting elements of biodiversity outside their natural habitats. This "off-site conservation" approach encompasses various methods, including the establishment of zoos, botanical gardens, and seed banks. These institutions serve as repositories for endangered species and play a crucial role in conserving biodiversity. Zoos and botanical gardens have long been recognized as effective ex-situ conservation measures. These facilities provide valuable research opportunities and contribute to the protection of biological diversity.

While ex-situ conservation methods have their merits, they can be more expensive to maintain compared to in-situ conservation. Therefore, they should be seen as complementary rather than standalone approaches. However, a significant challenge with ex-situ collections lies in the incomplete representation of important species, particularly those of significant value in tropical regions. Addressing this gap is crucial to ensure comprehensive biodiversity conservation efforts worldwide.

18.7 CONSERVATION AT THE NATIONAL LEVEL

Conservation of biodiversity at the national level plays a crucial role in ensuring the protection and preservation of species and habitats. National governments have the responsibility to enact laws and regulations that require the safeguarding of biodiversity. These laws serve as the legal framework for conservation efforts and provide the necessary foundation for preserving species and their habitats.

The establishment of laws alone, however, is not sufficient. It is imperative that governments demonstrate the will and allocate sufficient resources to enforce these laws effectively. Without the commitment to enforce conservation measures, the effectiveness of the legal framework is compromised. Adequate funding, personnel, and infrastructure are essential to ensure the implementation and enforcement of conservation regulations.

Public opinion also plays a significant role in shaping national conservation efforts. The support and awareness of the general public regarding the importance of biodiversity conservation can influence government policies and decision-making. Public pressure and advocacy for conservation can lead to the development and strengthening of laws and regulations aimed at protecting species and habitats.

Also, international conventions and agreements have been established to address the preservation of biodiversity. These conventions provide a platform for countries to collaborate and coordinate efforts to conserve biodiversity on a global scale. Examples of such conventions include the Convention on Biological Diversity (CBD) and the Convention on International Trade in Endangered Species of Wild Fauna and Flora (CITES). Through these international agreements, countries commit to implementing measures for the conservation and sustainable use of biodiversity and to cooperate in addressing common conservation challenges.

GLOSSARY

A

Abscission: The natural process by which plants shed leaves, flowers, fruits, or other plant parts.

Adventitious: Referring to plant structures, such as roots or buds, that arise from an unusual or atypical location.

Allele: One of the alternative forms of a gene that occupies a specific position on a chromosome.

Allopatric: Describing a population of organisms that are geographically separated and thus unable to interbreed.

Angiosperm: A type of flowering plant that produces seeds enclosed within a protective structure called a fruit.

Annual: A plant that completes its life cycle within a single year or growing season.

Anther: The part of the flower's stamen that produces and releases pollen grains.

Apical meristem: A region of actively dividing cells located at the tips of plant roots and shoots, responsible for primary growth.

Apomixis: The asexual reproduction in plants through seeds without fertilization, resulting in offspring genetically identical to the parent.

Aquatic: Pertaining to plants or organisms that live in or are adapted to an aquatic or water-based environment.

Arboreal: Referring to plants that grow or live in trees or elevated habitats.

Autotroph: An organism, typically a plant, capable of synthesizing its own organic molecules using inorganic substances and energy from the environment, usually through photosynthesis.

Auxin: A class of plant hormones that regulate various aspects of growth and development, including cell elongation, root formation, and tropisms.

Axil: The upper angle formed between a leaf and the stem, often giving rise to branches, flowers, or buds.

Axillary bud: A bud found in the axil of a leaf, capable of developing into a new shoot or flower.

Adaptation: A characteristic or trait that helps an organism survive and reproduce in its specific environment.

Aerobic: Requiring or relating to the presence of oxygen, as in aerobic respiration.

Aggregates: Small, loosely clustered groups of individual particles or structures.

Allelopathy: The production and release of biochemical compounds by plants to inhibit or suppress the growth of neighboring plants.

Amylase: An enzyme that breaks down starch into simpler sugars, such as glucose.

Anatomy: The study of the internal structure and organization of living organisms.

Angiospermy: The process of seed production within the enclosed ovary of a flowering plant.

Anthocyanin: A group of pigments responsible for the red, purple, or blue colors in flowers, fruits, and leaves.

Aperture: An opening or pore, typically found in pollen grains or spores.

Aquaporin: A protein channel that allows the passage of water across cell membranes.

Asexual reproduction: The production of offspring by a single parent, resulting in genetically identical or nearly identical offspring.

B

Biennial: A plant that completes its life cycle in two years, typically germinating and growing vegetatively in the first year and flowering and producing seeds in the second year.

Bark: The protective outer covering of woody stems and roots, composed of dead cells.

Basal: Referring to the base or lowest part of a structure, such as basal leaves or basal rosette.

Biome: A large ecological region characterized by distinct climate, vegetation, and organisms.

Bulb: A modified underground stem with fleshy leaves that store nutrients, typically giving rise to new shoots and roots.

Biodiversity: The variety of life forms within a given ecosystem, including species diversity, genetic diversity, and ecosystem diversity.

Bryophyte: A group of non-vascular plants, including mosses, liverworts, and hornworts.

Bud: An undeveloped or embryonic shoot, often covered by protective bud scales.

Bulblet: A small bulb-like structure produced by certain plants, capable of developing into a new plant.

Butterfly pollination: Pollination carried out by butterflies, typically attracted to brightly colored flowers and feeding on nectar.

Bark: The protective outer covering of woody stems and roots, composed of dead cells.

Bract: A modified leaf, often appearing near or surrounding a flower or inflorescence.

Biodegradable: Capable of being decomposed by natural processes, typically by microorganisms.

Biomass: The total mass of living matter within a given area or ecosystem.

Basidiomycetes: A group of fungi that produce spores on club-shaped structures called basidia, including mushrooms and toadstools.

Biotic: Relating to living organisms and their interactions within an ecosystem.

Bryophyte: A group of non-vascular plants, including mosses, liverworts, and hornworts.

Bulbil: A small bulb-like structure produced in leaf axils or flower clusters, capable of developing into a new plant.

Bioluminescence: The production of light by living organisms, often used for communication or attracting prey.

Biological control: The use of natural enemies, such as predators, parasites, or pathogens, to control populations of pests or invasive species.

Basipetal: Describing the direction of growth or movement from the tip toward the base of a plant or structure.

Buttress root: Large, above-ground roots that provide stability and support to tall trees in tropical rainforests.

Basal rosette: A cluster of leaves arising from the base of a plant, often seen in plants with a rosette growth habit.

Bark beetle: A group of small beetles that bore into the bark of trees, often causing damage or disease.

C

Calyx: The outermost whorl of a flower, consisting of sepals that protect the developing bud.

Cambium: A layer of meristematic cells in the vascular tissue of plants, responsible for secondary growth in stems and roots.

Carpel: The female reproductive organ in a flower, consisting of the stigma, style, and ovary.
Cell wall: A rigid layer surrounding plant cells, providing structural support and protection.
Chlorophyll: A green pigment found in chloroplasts that absorbs light energy for photosynthesis.
Chromosome: A thread-like structure in the nucleus of a cell that carries genetic information.
Companion cell: A specialized plant cell located adjacent to sieve elements in phloem, responsible for supporting and nourishing these cells.
Cotyledon: The embryonic leaf of a seedling, which provides nutrients to the developing plant.
Cross-pollination: The transfer of pollen from the anther of one flower to the stigma of another flower, often facilitated by wind, insects, or other agents.
Cuticle: A waxy, waterproof layer covering the outer surface of leaves and stems, reducing water loss.
Cytoplasm: The gel-like substance within a cell, excluding the nucleus, where various cellular processes occur.
Cytokinin: A class of plant hormones involved in cell division, shoot initiation, and other growth processes.
Cytoskeleton: A network of protein filaments within the cytoplasm of a cell, providing structural support and facilitating cellular movement.
Chloroplast: An organelle found in plant cells that is responsible for photosynthesis, converting light energy into chemical energy.
Coleoptile: A protective sheath covering the emerging shoot of a seedling, commonly found in monocot plants.
Collenchyma: A type of plant tissue composed of elongated cells with unevenly thickened cell walls, providing flexible support.
Compound leaf: A leaf composed of multiple leaflets attached to a common stalk or petiole.
Cotyledonary node: The part of a stem where the cotyledons are attached, typically found above the point of attachment of the seed coat.
Cross-section: A cut or slice made through an object, such as a stem or leaf, to reveal its internal structure.

D

Deciduous: Referring to plants that shed their leaves seasonally, typically in response to changing environmental conditions.

Dicotyledon: A type of flowering plant that possesses two cotyledons (seed leaves) in its embryo.

Diffusion: The passive movement of molecules or ions from an area of higher concentration to an area of lower concentration.

Drought-tolerant: Describing plants that are adapted to survive and thrive in arid or water-limited environments.

Dormancy: A period in a plant's life cycle where growth and development are temporarily suspended, often to withstand unfavorable environmental conditions.

Double fertilization: A unique process in flowering plants where two sperm cells fertilize two different nuclei within the embryo sac, leading to the formation of the embryo and endosperm.

Drupe: A type of fruit with a fleshy outer layer (exocarp), a stony or hard inner layer (endocarp), and a seed in the center.

Dicot: Informal term for a dicotyledonous plant, which is a flowering plant belonging to the class Magnoliopsida.

Dormant bud: A bud that is in a state of dormancy, typically protected by scales or other structures, and capable of producing new growth under favorable conditions.

Dicotyledonous stem: The stem of a dicot plant, characterized by the arrangement of vascular bundles in a ring and the presence of pith in the center.

Deoxyribonucleic acid (DNA): A molecule that carries the genetic information in all living organisms.

Dioecious: Referring to plant species in which male and female reproductive organs are borne on separate individuals.

Drought stress: The physiological and metabolic changes that occur in plants when subjected to prolonged periods of water scarcity or drought conditions.

Dominant: In genetics, a trait or allele that is expressed or observed when present in the genotype.

Drip irrigation: A method of irrigation that delivers water directly to the roots of plants through a network of tubes or emitters.

Dispersal: The movement or spreading of seeds, fruits, or other reproductive structures away from the parent plant to new locations.

Determinate growth: A type of growth in which the size or structure of an organ or organism is predetermined and ceases once a certain stage or size is reached.

Dry fruit: A type of fruit that does not have a fleshy or juicy pericarp, typically characterized by a dry and hard texture.

Decentralized stomata: Stomata that are evenly distributed on the surfaces of leaves, allowing for efficient gas exchange and reduced water loss.

E

Ecosystem: A community of living organisms (plants, animals, and microorganisms) interacting with each other and their physical environment.

Endosperm: The nutrient-rich tissue in the seeds of flowering plants, providing nourishment to the developing embryo.

Epidermis: The outermost layer of cells covering the surfaces of leaves, stems, roots, and other plant organs.

Evaporation: The process by which liquid water is converted into a gaseous state (water vapor) and released into the atmosphere.

Embryo: The early stage of development in a multicellular organism, following fertilization and preceding the formation of the complete organism.

Ethylene: A gaseous plant hormone involved in various physiological processes, including fruit ripening, leaf abscission, and senescence.

Ecological succession: The process of change in the structure and composition of a community over time, typically following a disturbance or on a barren substrate.

Evergreen: Referring to plants that retain their leaves throughout the year, continuously replacing old leaves with new ones.

Etiolation: The process of plant growth in low light conditions, characterized by elongated, pale stems and reduced chlorophyll production.

Endangered species: Species that are at high risk of extinction in the near future due to declining population numbers and threats to their habitat.

Exocarp: The outermost layer of a fruit, often referred to as the skin or peel.

Epiphyte: A plant that grows on the surface of other plants, usually trees, without deriving nutrients from the host plant.

Ecosystem services: The various benefits and resources that ecosystems provide to humans, such as clean air and water, pollination, and nutrient cycling.

Evergreen forest: A forest ecosystem dominated by trees that retain their leaves year-round, typically found in temperate and tropical regions.

Evapotranspiration: The combined process of water evaporation from soil, water bodies, and plant surfaces, including transpiration from plant tissues.

Elongation: The process of cell elongation or expansion, leading to growth in plant organs such as stems, roots, and leaves.

Endodermis: The innermost layer of cells in the cortex of plant roots, responsible for controlling the movement of water and solutes into the vascular tissue.

Estivation: A state of dormancy or inactivity during hot and dry periods, typically observed in certain plants and animals.

Excretion: The elimination of waste products or metabolic byproducts from the cells or body of an organism.

Epigeous germination: The type of seed germination in which the cotyledons emerge above the soil surface.

F

Fertilization: The process of fusion between the male gamete (sperm) and the female gamete (egg) to form a zygote.

Fibrous root: A type of root system consisting of numerous thin, branching roots that spread out in all directions, often found in monocot plants.

Flower: The reproductive structure of flowering plants, typically consisting of sepals, petals, stamens, and carpels.

Fruit: The mature ovary of a flowering plant, containing seeds and typically developed from the fertilized ovule.

Filament: The slender stalk of a stamen that supports the anther, where pollen grains are produced.

Fermentation: An anaerobic metabolic process in which sugars are converted into simpler compounds, such as alcohol or lactic acid, to produce energy.

Fungi: A group of eukaryotic organisms that includes mushrooms, molds, and yeasts, characterized by their ability to obtain nutrients through absorption.

Fragrance: The pleasant or distinctive smell emitted by certain flowers, often attracting pollinators.

Frond: The leaf of a fern or palm, typically divided into smaller leaflets or pinnae.

Fertilizer: A substance added to soil or plants to provide essential nutrients for growth and improve fertility.

Forest: A complex ecosystem dominated by trees and other woody vegetation, with high biodiversity and a dense canopy cover.

Food web: The interconnected feeding relationships among different organisms within an ecosystem, illustrating the flow of energy and nutrients.

Fragmentation: The breaking up of a habitat into smaller, isolated patches, often resulting in the loss of biodiversity and disruption of ecological processes.

Flagellum: A whip-like appendage that some microorganisms, such as certain algae and bacteria, use for movement.

Flowering plant: Also known as angiosperms, these are plants that produce flowers and bear seeds within a protective fruit.

Fossil fuel: A natural fuel formed from the remains of ancient plants and animals, such as coal, oil, and natural gas.

Flavonoid: A class of plant secondary metabolites that are often responsible for pigmentation in flowers and play various roles in plant defense and signaling.

Freezing tolerance: The ability of a plant to survive exposure to sub-freezing temperatures without significant damage.

Foliar feeding: A method of fertilizing plants by applying nutrient solutions directly to the leaves, allowing for rapid nutrient uptake.

Flowering time: The period when a plant produces flowers, often influenced by environmental cues such as day length, temperature, and hormone signaling.

Fruit set: The process by which a flower's ovary develops into a fruit after successful fertilization.

G

Gamete: A reproductive cell, such as a sperm or an egg, that fuses with another gamete during fertilization.

Germination: The process by which a seed begins to develop into a new plant, usually involving the absorption of water and the emergence of the embryonic shoot (plumule) and root (radicle).

Germination inhibitor: A substance that inhibits or delays seed germination by preventing the activation of metabolic processes required for growth.

Gibberellins: A class of plant hormones that regulate various aspects of plant growth and development, including stem elongation, seed germination, and flowering.

Glucose: A simple sugar and the primary source of energy for many organisms, produced through photosynthesis in plants.

Guard cells: Specialized cells found in the epidermis of leaves that control the opening and closing of stomata, thereby regulating gas exchange and water loss.

Gymnosperm: A group of seed-producing plants that do not produce flowers or fruits, including conifers, cycads, and ginkgos.

Gene: A segment of DNA that carries the genetic instructions for the synthesis of a particular protein or trait.

Genotype: The genetic makeup or combination of alleles that an organism possesses.

Germination rate: The percentage of seeds that successfully undergo germination under specific conditions.

Genome: The complete set of genetic material (DNA) present in an organism or cell.

Gravitropism: The growth response of plants in response to gravity, resulting in the upward growth of shoots (negative gravitropism) and downward growth of roots (positive gravitropism).

Ground tissue: The primary tissue system in plants that includes parenchyma, collenchyma, and sclerenchyma cells, performing various functions such as storage, photosynthesis, and support.

Growth hormone: A type of plant hormone that regulates overall plant growth, including cell division, elongation, and differentiation.

Geriatric: Referring to the study of the aging and senescence processes in plants, including the physiological and biochemical changes associated with aging.

Grana: Stacks of flattened, membranous structures (thylakoids) inside chloroplasts where the light-dependent reactions of photosynthesis take place.

Genetically modified organism (GMO): An organism that has had its genetic material altered using genetic engineering techniques.

Greenhouse effect: The process by which certain gases in the Earth's atmosphere trap heat, leading to a rise in global temperatures.

Groundwater: Water present beneath the Earth's surface in saturated soil or rock layers.

H

Habitat: The specific environment or locality in which a plant or animal species naturally occurs and thrives.

Herbaceous: Referring to plants that have soft, non-woody stems and usually die back to the ground at the end of each growing season.

Hormone: A chemical substance produced in one part of an organism and transported to another part, regulating various physiological processes, such as growth, development, and response to stimuli.

Hybrid: The offspring resulting from the crossbreeding of two different plant species or varieties, often possessing a combination of traits from both parents.

Hydroponics: A method of growing plants without soil, where the roots are immersed in a nutrient-rich water solution.

Hypertonic: Referring to a solution or environment with a higher solute concentration compared to the cytoplasm of a cell, resulting in water loss from the cell.

Hypotonic: Referring to a solution or environment with a lower solute concentration compared to the cytoplasm of a cell, resulting in water uptake by the cell.

Hyphae: The thread-like structures that make up the body of a fungus, responsible for nutrient absorption and mycelium formation.

Heterotroph: An organism that obtains its energy and nutrients by consuming other organisms or organic matter.

Homozygous: Having identical alleles for a particular gene, either dominant or recessive.

Herbivore: An animal that primarily feeds on plants or plant material.

Hormone signaling: The process by which plant hormones transmit information and regulate physiological responses in various parts of the plant.

Halophyte: A plant that is adapted to and can tolerate high salt concentrations in its environment.

Host plant: A plant species on which a particular organism, such as an insect or parasite, depends for its survival, reproduction, or feeding.

Hydrophilic: Having an affinity for water; capable of absorbing or interacting with water molecules.

Hydrophobic: Having a lack of affinity for water; repelling or not readily mixing with water.

Homeostasis: The ability of an organism or cell to maintain internal stability and balance, even in the face of external changes or fluctuations.

Hormone transport: The movement of plant hormones throughout the plant body, facilitated by vascular tissues and other transport mechanisms.

Haploid: Referring to a cell or organism that contains only one set of chromosomes, typically denoted as "n" in the context of sexual reproduction.

I

Invasive species: Non-native species that spread aggressively and have negative impacts on the native ecosystems they invade.

Indeterminate growth: A type of growth exhibited by certain plants in which they continue to grow throughout their lifespan, without a distinct endpoint or limit.

Insect pollination: The process of transferring pollen from the anther to the stigma of a flower by insects, such as bees, butterflies, and beetles.

Inflorescence: A cluster or arrangement of flowers on a plant, often displaying a specific pattern or structure.

Internode: The portion of a stem between two adjacent nodes.

Inbreeding: The mating or reproduction between individuals that are closely related genetically, resulting in a higher frequency of homozygous traits.

Imbibition: The process of absorbing water by dry seeds, causing them to swell and become activated for germination.

Inorganic nutrients: Essential elements required for plant growth that are derived from non-living sources, such as minerals and gases.

Insecticide: A substance or chemical compound used to control or eliminate insects that pose a threat to plants or human health.

Insectivorous plant: A plant that obtains nutrients by capturing and digesting small insects and other arthropods.

In vitro culture: The cultivation of plant cells, tissues, or organs in a controlled laboratory environment, typically involving the use of a nutrient medium.

Isotope: Different forms of an element that have the same number of protons but different numbers of neutrons, resulting in different atomic masses.

Induced mutation: A mutation that is deliberately induced or triggered by external agents, such as radiation or chemicals, for the purpose of creating genetic variation.

Inoculation: The introduction of beneficial microorganisms, such as nitrogen-fixing bacteria, to enhance plant growth and nutrient availability.

Intercropping: The practice of growing two or more different crops in close proximity to each other, resulting in mutual benefits, increased productivity, and efficient use of resources.

Irrigation: The artificial application of water to plants, typically through methods such as sprinklers, drip systems, or flood irrigation.

Inbreeding depression: The reduced fitness or vigor observed in offspring resulting from the mating of closely related individuals, often due to the expression of deleterious recessive traits.

Ion exchange: The process by which ions are exchanged between soil particles and plant roots, allowing plants to take up essential nutrients.

Immunity: The ability of a plant to resist or defend against pathogens, pests, or diseases through various mechanisms, including physical barriers, chemical defenses, and immune responses.

J

Jackfruit: A tropical fruit that belongs to the mulberry family (Moraceae), known for its large size and sweet, fleshy arils.

Juncaceae: A family of flowering plants commonly known as the rush family, comprising grass-like plants with cylindrical stems and often found in wetland habitats.

Juglandaceae: A family of flowering plants commonly known as the walnut family, including trees such as walnuts, hickories, and pecans.

Juvenile: Referring to the early or immature stage of a plant's growth or development, characterized by specific morphological and physiological features distinct from the adult stage.

Juvenile hormone: A hormone found in insects that plays a crucial role in regulating growth and development, particularly during the larval stage.

Justicia: A genus of flowering plants in the family Acanthaceae, comprising various species of herbs and shrubs known for their colorful flowers.

K

Karyotype: The number, size, and shape of chromosomes in a cell or organism, typically displayed in a visual arrangement.

Kingdom: The highest taxonomic rank in biological classification, categorizing organisms into five major groups: Animalia, Plantae, Fungi, Protista, and Monera.

Kinetin: A type of plant growth regulator, or cytokinin, that promotes cell division and delays senescence in plants.

Krebs cycle: Also known as the citric acid cycle or tricarboxylic acid cycle, it is a series of chemical reactions in aerobic respiration that generate ATP and carbon dioxide.

Keel: A specialized petal or sepal structure found in some flowers, forming a boat-like shape, often associated with plants in the pea family (Fabaceae).

Kinetochore: A protein structure located on the centromere of a chromosome, essential for spindle fiber attachment during cell division.

Karyokinesis: The division of the cell nucleus during mitosis or meiosis, involving the separation and distribution of chromosomes.

L

Lateral meristem: A meristem located at the sides of a plant, responsible for secondary growth in thickness, leading to the formation of secondary tissues like vascular cambium and cork cambium.

Lenticel: Small, raised areas on the surface of stems and roots of woody plants that allow for gas exchange between the internal tissues and the external environment.

Leaching: The process by which water-soluble substances, such as minerals or nutrients, are washed out or dissolved from the soil, potentially leading to nutrient depletion.

Legume: A plant belonging to the family Fabaceae (Leguminosae), characterized by its ability to form nitrogen-fixing symbiotic associations with nitrogen-fixing bacteria (rhizobia) in root nodules.

Lignin: A complex, rigid polymer found in the cell walls of plants, providing strength and rigidity to the tissues.

Lipid: A class of organic compounds that includes fats, oils, and waxes, serving as a major energy source and playing structural and signaling roles in cells.

Locus: The specific position or location of a gene or DNA sequence on a chromosome.

Lysosome: A membrane-bound organelle within plant cells that contains digestive enzymes responsible for breaking down waste materials and cellular components.

M

Macronutrients: Essential nutrients required by plants in relatively large quantities for growth and development, including elements such as nitrogen (N), phosphorus (P), and potassium (K).

Meiosis: A type of cell division that produces gametes (sex cells) with half the number of chromosomes as the parent cell, leading to genetic variation and the formation of haploid cells.

Meristem: A region of undifferentiated cells in plants that is responsible for growth and development, giving rise to various plant tissues.

Micropropagation: The process of multiplying plants through tissue culture techniques, such as the use of small tissue explants and growth hormones.

Mitochondria: Organelles found in the cells of plants and other organisms that are responsible for energy production through cellular respiration.

Monocotyledon: A type of flowering plant characterized by having a single cotyledon (seed leaf) in the embryo, parallel-veined leaves, scattered vascular bundles in the stem, and floral parts usually in multiples of three.

Monoculture: The practice of cultivating a single crop species over a large area, often associated with industrial agricultural systems.

Monohybrid cross: A genetic cross between two individuals that differ in a single trait or gene.

Mycorrhiza: A mutualistic association between the roots of plants and specialized fungi, benefiting both organisms through enhanced nutrient uptake and improved soil structure.

Mutation: A change in the DNA sequence of a gene or chromosome, resulting in genetic variation and the potential for the expression of new traits.

Mycelium: The vegetative, thread-like part of a fungus, composed of a network of hyphae that serves as the main body of the fungus.

N

Nastic movement: A non-directional movement in plants in response to external stimuli, such as changes in temperature, light, or touch.

Nitrogen fixation: The conversion of atmospheric nitrogen gas into a form that plants can use, facilitated by nitrogen-fixing bacteria or through industrial processes.

Node: The part of a stem where leaves, buds, and branches emerge.

Nucleus: The membrane-bound organelle within eukaryotic cells that contains the genetic material (DNA) and controls cellular activities.

Nectar: A sugary solution produced by flowers to attract pollinators, serving as a reward for their visitation.

Nitrate: A form of nitrogen that is readily taken up and utilized by plants as a nutrient.

Nutrient uptake: The process by which plants absorb and take in nutrients from the soil or other sources.

Nymphaea: A genus of aquatic plants commonly known as water lilies, characterized by their floating leaves and showy flowers.

Nymph: The immature stage of some insects, such as grasshoppers and dragonflies, which undergoes incomplete metamorphosis.

O

Ovary: The female reproductive organ in flowering plants that contains ovules and develops into a fruit after fertilization.

Ovule: The structure within the ovary of a flowering plant that contains the female reproductive cells (eggs) and develops into a seed after fertilization.

Operculum: A protective covering or lid that opens and closes in certain plants, such as mosses and some fruits, to release spores or seeds.

Organic matter: The decomposed remains of plants, animals, and other organic materials in the soil, which provide nutrients and contribute to soil fertility.

Osmosis: The movement of water molecules across a semipermeable membrane from an area of lower solute concentration to an area of higher solute concentration.

Ovule fertilization: The process in which a pollen grain fuses with the egg cell within the ovule, leading to the formation of a zygote and eventual seed development.

Ozone layer: A layer of ozone (O3) molecules located in the Earth's stratosphere that helps protect living organisms from harmful ultraviolet (UV) radiation.

Ovary wall: The outer layer of tissue surrounding the ovary in a flower, which develops into the fruit wall after fertilization.

Open pollination: The natural process of pollination in which pollen is transferred between flowers of the same species by wind, insects, or other natural agents.

Organism: Any individual living entity, such as a plant, animal, or microorganism, capable of carrying out life processes and reproduction.

P

Palisade mesophyll: The layer of elongated, chloroplast-rich cells found in the upper part of a plant leaf, responsible for photosynthesis.

Parenchyma: A type of simple plant tissue composed of thin-walled cells that perform various functions, such as photosynthesis, storage, and secretion.

Parasite: An organism that lives and feeds on another organism (host) to obtain nutrients and other resources, often causing harm or damage to the host.

Perennial: A plant that lives for more than two years, usually flowering and producing seeds multiple times during its lifetime.

Petiole: The stalk that connects a leaf to the stem of a plant.

Phloem: The vascular tissue responsible for transporting sugars, organic nutrients, and other substances throughout the plant.

Photosynthesis: The process by which plants and some other organisms convert sunlight, carbon dioxide, and water into glucose (energy-rich molecules) and oxygen, utilizing chlorophyll and other pigments.

Phototropism: The growth or movement of a plant in response to light, such as bending towards a light source.

Phylogenetics: The study of evolutionary relationships and the construction of phylogenetic trees or diagrams to represent the evolutionary history of organisms.

Phyllotaxis: The arrangement of leaves on a stem or branch, including patterns such as alternate, opposite, and whorled.

Pollination: The transfer of pollen from the male reproductive organs (anthers) to the female reproductive organs (stigma) of a flower, leading to fertilization.

Pollinator: An organism, such as a bee, butterfly, bird, or wind, that facilitates the transfer of pollen between flowers.

Primary growth: The growth of a plant in length, occurring primarily at the apical meristem and resulting in the elongation of roots and shoots.

Prokaryote: A single-celled organism lacking a nucleus and membrane-bound organelles, including bacteria and archaea.

Propagation: The process of reproducing or multiplying plants through various methods, such as seeds, cuttings, grafting, or tissue culture.

Q

Quiescent center: A region of meristematic cells in the root apical meristem that remains inactive and serves as a source of cells for root growth and repair.

Quorum sensing: A communication system used by bacteria to coordinate gene expression and behavior based on the density of the bacterial population.

Quinone: A class of organic compounds that play essential roles in electron transfer processes during photosynthesis and cellular respiration.

R

Radial symmetry: A type of symmetry found in some flowers and plant structures where multiple planes can divide the structure into roughly identical halves.

Radicle: The embryonic root of a plant that emerges from the seed during germination and develops into the primary root.

Respiration: The metabolic process by which plants and other organisms release energy from organic compounds, typically using oxygen and producing carbon dioxide as a byproduct.

Rhizome: An underground stem that grows horizontally, producing roots and shoots at intervals and serving as a storage organ in some plants.

Root hair: Small, elongated outgrowths of root epidermal cells that increase the surface area for absorption of water and nutrients from the soil.

Rosette: A cluster of leaves arranged in a circular or rosette-like pattern at the base of a plant, often seen in certain plant families like the Rosaceae.

S

Stamen: The male reproductive organ of a flower, composed of anther and filament, responsible for producing and releasing pollen.

Sepal: One of the outermost, usually green, floral parts that protect the developing flower bud.

Stigma: The receptive part of the female reproductive organ (carpel) in a flower, where pollen grains land during pollination.

Style: The elongated part of the female reproductive organ (carpel) in a flower, connecting the stigma to the ovary.

Self-pollination: The transfer of pollen from the anther to the stigma of the same flower or a different flower on the same plant.

Sexual reproduction: The process of reproduction involving the fusion of gametes (sex cells), typically from different individuals, resulting in offspring with genetic variation.

Secondary growth: The growth in girth or thickness of stems and roots, primarily occurring in woody plants through the activity of lateral meristems.

Stomata: Microscopic openings, typically found on the surface of leaves, stems, and other plant organs, responsible for gas exchange (such as carbon dioxide uptake and oxygen release) and water vapor loss.

Spore: A reproductive structure produced by certain plants, fungi, and some microorganisms that can develop into a new individual without the need for fertilization.

Stem: The main structural part of a plant that provides support, transportation of water, nutrients, and sugars, and serves as a site for leaf and flower attachment.

T

Tissue: A group of cells with a similar structure and function that work together to perform specific tasks in an organism.

Transpiration: The process by which water is lost from the aerial parts of plants, primarily through stomata, in the form of water vapor.

Turgor pressure: The pressure exerted by the fluid inside the vacuoles of plant cells against the cell walls, providing structural support and maintaining cell shape.

Taproot: A main, central root that grows vertically and gives rise to smaller lateral roots, commonly found in dicotyledonous plants.

Tropism: A growth or movement response of a plant in response to a directional stimulus, such as light (phototropism), gravity (gravitropism), or touch (thigmotropism).

Trichome: Small, hair-like structures that can be found on the surface of plant leaves, stems, and other plant parts, often serving various functions such as protection, secretion, or reducing water loss.

Thigmotropism: A plant's response to touch or physical contact, resulting in growth or movement in response to mechanical stimulation.

Translocation: The movement of sugars and other organic compounds through the phloem from sources (such as leaves) to sinks (such as growing regions or storage organs) in a plant.

Thylakoid: A membrane-bound compartment within the chloroplast where the light-dependent reactions of photosynthesis take place.

U

Umbel: A type of inflorescence in which multiple flower stalks arise from a common point, resembling the ribs of an umbrella.

Unicellular: Referring to organisms or structures composed of a single cell.

Uptake: The process of absorbing or taking in substances, such as water, minerals, or nutrients, by plant roots from the surrounding environment.

Utricle: A small, bladder-like structure found in certain plants, such as some aquatic plants, that aids in buoyancy and dispersal.

V

Vacuole: A membrane-bound organelle found in plant cells that stores water, ions, sugars, pigments, and other substances.

Vascular bundle: A group of specialized tissues in plants that transport water, nutrients, and sugars throughout the plant.

Vegetative propagation: A form of asexual reproduction in plants where new individuals are produced from non-reproductive plant parts, such as stems, leaves, or roots.

Vernalization: The process of exposing certain plants to a prolonged period of cold temperature to induce flowering or promote other physiological changes.

Vernation: The arrangement of leaves in a bud before they unfurl, characterized by various patterns, such as folded, rolled, or twisted.

Vessel element: A type of water-conducting cell found in the xylem of vascular plants, characterized by their long, cylindrical shape and perforated end walls.

Vine: A climbing or trailing plant with long, flexible stems that typically requires support to grow upward.

Virus: A small infectious agent that consists of genetic material (DNA or RNA) enclosed in a protein coat, capable of infecting and replicating inside host cells.

Volatile organic compounds (VOCs): Organic chemicals that are released into the air by plants, often as a means of communication, defense, or attraction of pollinators.

W

Wilt: A condition in plants characterized by drooping or limp leaves and stems due to a loss of turgor pressure, often caused by water deficiency or excessive transpiration.

Woody plant: A plant characterized by having a strong, rigid stem composed of wood, such as trees and shrubs.

X

Xerophyte: A plant species that is adapted to arid or dry environments, typically possessing structural and physiological adaptations to conserve water.

Xylem: The vascular tissue in plants that conducts water and dissolved minerals from the roots to the rest of the plant, providing support and contributing to the transport of nutrients.

Y

Yeast: Single-celled fungi that are commonly used in baking and brewing and play a role in fermentation processes.

Yellowing: The condition in plants where leaves or other plant parts exhibit a yellow color, often indicating nutrient deficiencies, disease, or other stress factors.

Yield: The amount of crop or plant product produced per unit of land area, often measured in weight or volume.

Z

Zygote: The diploid cell formed by the fusion of gametes (sperm and egg) during fertilization, which develops into an embryo.

Zymology: The branch of science that deals with the study of fermentation processes, including those carried out by microorganisms in the production of food and beverages.

Zone of inhibition: The area surrounding an antimicrobial agent (such as an antibiotic) on a culture plate where the growth of microorganisms is inhibited.

Zonation: The spatial arrangement of different plant communities or species along an environmental gradient, such as from a water source to dry land.

Made in the USA
Las Vegas, NV
13 December 2023